*Also by Variorum:*

**Ed. VALENTÍN VÁZQUEZ DE PRADA and IGNACIO OLÁBARRI**
Understanding Social Change in the Nineties: Theoretical Approaches and Historical Perspectives

**Ed. PENELOPE GOUK**
Wellsprings of Achievement: Cultural and Economic Dynamics in Early Modern England and Japan

*In the Collected Studies Series:*

**ROY MacLEOD**
Public Science and Public Policy in Victorian England

**WILLIAM H. BROCK**
Science for All: Studies in the History of Victorian Science and Education

**BRIAN PULLAN**
Poverty and Charity: Europe, Italy and Venice, 1400–1700

**MAURICE CROSLAND**
Studies in the Culture of Science in France and Britain since the Enlightenment

**ROBERT FOX**
Science, Industry and Social Order in Post-Revolutionary France

**ANTHONY PAGDEN**
The Uncertainties of Empire: Essays in Iberian and Ibero-American Intellectual History

**A.W. SKEMPTON**
Civil Engineers and Engineering in Britain before 1840

**JOHN HARRIS**
Essays in Industry and Technology in the 18th Century: England and France

**JACOB M. PRICE**
Tobacco in Atlantic Trade: The Chesapeake, London and Glasgow, 1675–1775

**E. WILLIAM MONTER**
Enforcing Morality in Early Modern Europe

COLLECTED STUDIES SERIES

# The Biological Standard of Living in Europe and America, 1700–1900

John Komlos

# The Biological Standard of Living in Europe and America, 1700–1900

## Studies in anthropometric history

Routledge
Taylor & Francis Group
LONDON AND NEW YORK

1995

First published 1995 by Variorum, Ashgate Publishing

Published 2017 by Routledge
2 Park Square, Milton Park, Abingdon, Oxon, OX14 4RN
711 Third Avenue, New York, NY 10017, USA

*Routledge is an imprint of the Taylor & Francis Group, an informa business*



**British Library CIP Data**
Biological Standard of Living in Europe and America, 1700–1900: Studies in Anthropometric History
(Variorum Collected Studies Series: 508)
I. Komlos, John II. Series
573.6

**US Library of Congress CIP Data**
Komlos, John, 1944–
Biological Standard of Living in Europe and America, 1700–1900: Studies in Anthropometric History / John Komlos.
p. cm. — (Variorum Collected Studies Series: CS508)
Includes bibliographical references (p. ) and index.
ISBN 0-86078-457-6 (cloth : alk. paper)
1. Anthropometry—Cross-cultural studies. 2. Human growth—Economic aspects—Cross-cultural studies. 3. Human growth—Environmental aspects—Cross-cultural studies. 4. Quality of life—Cross-cultural studies. I. Title. II. Series: Collected Studies: CS508.
GN51.K65 1995 95–36744
016.7822'2—dc20 CIP

ISBN 13: 978-0-86078-457-9 (hbk)

COLLECTED STUDIES SERIES CS508

# CONTENTS

This volume contains xvi + 282 pages

# PUBLISHER'S NOTE

The articles in this volume, as in all others in the Collected Studies Series, have not been given a new, continuous pagination. In order to avoid confusion, and to facilitate their use where these same studies have been referred to elsewhere, the original pagination has been maintained wherever possible.

Each article has been given a Roman number in order of appearance, as listed in the Contents. This number is repeated on each page and quoted in the index entries.

# PREFACE

Though the efforts of anthropometric historians to unearth the broad patterns of biological well-being of the populations of the last two and a half centuries are probably quite well known to many social scientists, a retrospective edition of articles that appeared in widely scattered journals should introduce the topic to a still wider audience. The notion of using historical evidence on the physical stature of a population to understand socio-economic processes goes back, of course, to the nineteenth century. Medical doctors, political reformers, politicians, academics, and other contemporary observers of economic growth and the concomitant momentous changes in social organization were struck by changes in physical stature, and used that evidence to analyse the effects of income distribution, factory work, or of industrialization on the well-being of the workers and their families. Hence, the essence of anthropometric analysis is by no means new, even if great advances have been made in the various aspects of the discipline, including in the theory of the measurement of human proportions, a separate field known as auxiology.[1]

Within recent memory, French historians of the *Annales* tradition were the first to resurrect the humanistic relevance of the history of human height.[2] The true renaissance of the field, however, might be dated with the inception of an independent research program on American soil some seventeen years ago by Richard Steckel. The fact that the research agenda had to be rediscovered in America shows the obvious disadvantages of having academic communities separated from one another by language. Because of its far reaching consequences, Steckel's publications of 1978 and 1979 on the physical stature of slaves in antebellum America, arguably can be considered the manifesto of this research program. His interest in the demographic experience of the slaves, and their well-being, was the initial inspiration of his analysis.[3] Such a genesis was hardly accidental. After all, the treatment of those unfortunate millions on the quaintessential soil of free-

[1] James Tanner, *A History of the Study of Human Growth* (Cambridge: Cambridge University Press, 1981).

[2] E. Le Roy Ladurie, N. Bernageau, and Y. Pasquet, 'Le conscrit et l'ordinateur. Perspectives de recherches sur les archives militaires du XIXe siècle français', *Studi Storici* 10 (1969): 260–308.

[3] Richard Steckel, 'Slave Height Profiles from Coastwise Manifests', *Explorations in Economic History* 16 (1979): 453–81.

dom went, and still goes, to the very heart and soul of the American historical profession's concerns.

With the publication of a separate issue of *Social Science History* devoted entirely to anthropometric history in 1982, which included several classic articles on the empirical and theoretical underpinnings of the topic, the interest in the research program grew substantially. With the encouragement and support of Robert Fogel many students and collaborators, including Roderick Floud then of the University of London, devoted increased attention to the complex issues at hand. My interest in the topic was also awakened by Fogel's lectures at the University of Chicago. I realized the need to explore the possibilities of anthropometric history within the regional setting of East-Central Europe which had been the focus of my attention until then. I was helped all along not only by Robert Fogel's encouragement and support, but perhaps just as importantly also by the generosity of James Tanner, then of the University of London and coeditor of the *Annals of Human Biology*, whose resolve to help newcomers to the field understand the intricacies of human growth was exemplary.

The outcome of immensely complex socio-economic and biological processes, the understanding of the history of human height is complicated even more by difficult statistical problems associated with most of the extant samples (see chapter 3). In fact, even many of the biological aspects of physical growth are not completely known. No wonder then, that seeking socio-economic explanations for the history of human height has been by no means straightforward (see chapters 1 and 2). I am consequently particularly grateful to Professor Tanner, not only for devoting considerable effort to educating all of us about the details of human growth without regard to time constraints affecting ordinary mortals, but also for opening up the pages of his journal to our writings. Without his and Fogel's support this niche of the discipline would certainly not have had the reception it ultimately received.

The results of my first researches were published three years later (1985) (chapter 12). I outline in that article the main features of the secular trend in the physical stature of the populations living in the Habsburg Monarchy, and explore its relationship with nutritional status, demographic change and economic activity. In that study I first explored what has become known as the 'early industrial growth puzzle', namely the divergence between conventional and biological indicators of the standard of living.[4]

After lengthy searches in the Viennese archives I was fortunate to find some data on the height of children in Austria. Using that evidence I was able to confirm in 1986 my main findings of the previous year based on adults,

[4] The contemporaneous downturn was subsequently found in Swedish samples as well. See Lars Sandberg and Richard Steckel, 'Heights and Economic History: the Swedish Case', *Annals of Human Biology* 14 (1987): 101–10.

namely, that nutritional status deteriorated in Central Europe at the end of the eighteenth century (chapter 11). These corraborating results were extremely important at the time, because most military height samples are difficult to analyse on account of the fact that the military did not accept short men, and as a consequence one frequently has to work with truncated samples. In other words, I was able to show that the 'puzzle' was not an artifact of the biases of the military data, but that the downward trend is evident in a civilian sample as well.

My discovery of cycles in physical stature in East-Central Europe naturally led me to become interested in cycles in human height in other contexts as well, including those in North America.[5] Discovered in 1983, the correlates of the American cycles also needed more attention: what were their causes; what determined their turning points? My study of the height and weight of West Point cadets (1987) was devoted to this very issue. I argued in that paper that not only in the Habsburg Monarchy, but also in North America the downward phase of the long cycle in physical stature coincided with the acceleration in the rate of population growth, of urbanization, as well as with the rapid expansion of the industrial sector both absolutely and relatively to agriculture. In the late eighteenth century, just as the 'classical phase' of the Industrial Revolution was underway, the average height of the (male) population began to decline in the Habsburg Monarchy. Although the decline in the United States first became apparent some two generations later, during the second half of the 1830s, it, too, coincided with the similar demographic and economic trends as the one on the European continent: an acceleration in industrial, demographic and urban growth. These ideas on the 'antebellum puzzle' are published here in chapter 13.

Thereafter I showed that the manumitted slaves of Maryland experienced a decline in nutritional status similar to that of the white population (chapter 15). In other words, the 'antebellum puzzle' affected the free population of the United States, both white and black.

I was not very happy about the statistical procedures available at the time to analyse truncated height samples, which seemed to me in practice to be much too inaccurate. Therefore, together with J.H. Kim of the University of Soul, I turned my attention to the statistical issues and developed a new method which turned out to yield more accurate time trends in truncated height samples than other methods (chapter 3).

I returned to the analysis of the cycles in human physical stature several years later in the context of the United Kingdom. Roderick Floud and his collaborators had published a monograph (1990) in which they argued that the physical stature of the population of the United Kingdom increased

[5] Robert Margo and Richard Steckel, 'Heights of Native Born Northern Whites during the Antebellum Period', *Journal of Economic History* 43 (1983): 167–174.

between the mid-eighteenth and the early nineteenth centuries. This was quite unexpected, given my results in both the Habsburg and American populations. If the downturns in physical stature were associated with urbanization, industrialization, and an acceleration in demographic growth, then one would have expected heights to decline in Britain as well, rather than increase, inasmuch as these processes were even more pronounced there in many respects than in most other places. In short, either my thesis was wrong, or Floud's statistical procedures were wrong. It was worth finding out which was the case. As it turns out, using the Komlos-Kim method for analyzing the trend in truncated height distributions I demonstrated that Floud's statistical procedures were actually what caused the divergent results (chapters 6 and 7). In other words, Floud's contentions notwithstanding, the history of human height in Britain was not exceptional, but rather bore substantial similarities to the Habsburg, Swedish, as well as to the American experience. The trends in the heights of Irish, Scottish, and English recruits as well as those of the destitute children of London were all precisely the ones found in the Habsburg and Swedish contexts: a decline at the end of the eighteenth century, followed by a small and short-lived recovery at the end of the Napoleonic Wars.

In order to further substantiate my contention, I collected a sample of height of runaway indentured servants, mostly of English and Irish origin, and found that the evidence corroborated my, rather than Floud's, position (chapter 5). In addition, a number of other scholars were able to show independently some inaccuracies in Floud's procedures. For example, records of convicts and indentured servants transported to Australia implied the existence of a negative trend in nutritional status in the United Kingdom during the course of the second half of the eighteenth century.[6] Moreover, a number of more recent studies have also tended to contradict at least some aspects of Floud's estimates as well. For example, evidence based on the records of the East India Company's army, studied by Mokyr and O'Grada, suggests that Floud overestimated the height of the birth cohorts of the 1790s.[7] Similar differences were reported in Riggs's study of Scottish criminals.[8] The latter study confirms that the post-Napoleonic increase in stature was overestimated by Floud. In other words, on the basis of every data set so far analysed

[6] Stephan Nicholas and Richard Steckel, 'Heights and Living Standards of English Workers during the Early Years of Industrialization, 1770–1815', *Journal of Economic History* 51 (1991): 937–57.

[7] Joel Mokyr and Cormac O'Gráda, 'The Heights of the British and the Irish c. 1800–1815: Evidence from Recruits to the East India Company's Army', in John Komlos, (ed.), *Stature, Living Standards, and Economic Development: Essays in Anthropometric History* (Chicago: The University of Chicago Press, 1994): 39–59.

[8] Paul Riggs, 'The Standard of Living in Scotland, 1800–1850', in Komlos, (ed.), *Stature, Living Standards, and Economic Development*: 60–75.

one can reject the method of analysis of truncated samples advocated by Floud and his collaborators.[9]

I continued the analysis of American heights on the basis of students attending The Citadel military academy in Charleston, SC (chapter 14). The article was coauthored by Peter Coclanis of the University of North Carolina, an expert of the economic history of the American South. We found that physical stature of southerners did not increase for a long time after the Civil War. In fact, not until World War I did students at the Citadel begin to become taller than their pre-Civil War counterparts. This is a vivid reminder of the difficulties of transforming this region of the United States after emancipation. This is in contrast to the improvements in physical stature that were being experienced in the Northern states as well as in most European countries (chapter 9).

Because heights were collected only within certain institutional settings, they generally pertain to men, and usually to the lower classes, such as soldiers, sailors, orphans, convicts, servants, and slaves. Records on females as well as on upper and middle class persons are much rarer in a historical context than they are today. Therefore, two data sets on the height of middle or upper class youth were promised at least some information in this regard. The height of students attending a school in Stuttgart indicate that they were not affected by the nutritional crisis of the late eighteenth century (chapter 8). The paper was coauthored by James Tanner of the University of London and some of his assistants. The results led to the inference that the deterioration in heights was less likely to have been caused by a deterioration of the disease environment, because that would have affected all classes of a society. Instead, the cause of the diminution in nutritional status is much more likely to have been caused by economic forces, from which the upper and middle classes were immune on account of their higher income. I did another study of middle and upper class Europeans, based on the students of the Parisian École Polytechniqe, founded by Napoleon (chapter 4).

There is but a sliver of evidence on females in all of these studies. Only the study of the free blacks of Maryland have a substantial number of observations on females (chapter 15). The runaway indentured servant and runaway slave sample have only a negligible number of observations on females. So far this research program has not born extensive results on females. Because of the paucity of data on females, the analysis of the age at menarche, another biological indicator of nutritional status, is a useful addition to the literature

[9] The major advantage of most of these studies is that the samples on which they are based are drawn from the civilian population, i.e. they are not based on truncated height distributions as is Floud's military sample, and are therefore easier to evaluate statistically. As a consequence, our confidence in their results is enhanced considerably in comparison to those based on military samples.

(chapter 10). Obviously more creative thinking is needed to be able to explore the history of biological well being of females.

Already at the very beginning of this research program Steckel had shown, on the basis of shipping documents, that in the antebellum period American-born slaves were quite tall relative to their African brethren, or even compared to the common people of Europe. The difference between them and their masters was also substantially less than the difference between the gentry and the lower classes of Old World societies. In short, the high productivity of slave labor provided incentives to the slave owners to provide adequate alimentation to their human chattel. Combined with the propitious environmental conditions, and low food prices, this meant that the oppressive nature of the system notwithstanding, the biological welfare of even the very lowest segments of American society was better in international comparison than one would have anticipated a priori.[10] I was able to demonstrate that these generalizations essentially held in the eighteenth century as well (chapter 16).

Needless to say, work is continuing, and new findings are forthcoming continuously.[11] Two dissertations, by Timothy Cuff at the University of Pittsburgh (on Pennsylvania) and Jörg Baten at the University of Munich (on Bavaria) are nearing completion. Others are in the beginning stages.[12] Translations are appearing.[13] Two anthologies have just been published, and another, edited by Richard Steckel is forthcoming.[14] Just as importantly perhaps, the anthropometric research program has enabled us to view historical processes in a new light. It led to a new understanding of the nature of Malthusian crisis. It enabled us to form a new paradigm to conceptualize the Industrial Revolution. And it led to a better understanding of the nature of the biological standard of living and its relationship to conventional economic measures. These avenues of research, offshoots of the anthropometric research program, are also worth pursuing. In short, the history of human height has born fruit and it will most certainly continue to fascinate and

[10] Only later did Steckel's researches show that slave children were excepted from this generalization, and in contrast to their parents, were quite stunted.

[11] John Komlos, Paul Katzenberger and Peter Coclanis, 'On Cycles in the Biological Standard of Living: The "Antebellum Puzzle" Revisited', unpublished manuscript; John Komlos and Bjorn Alecke, 'Slave Heights in Antebellum America', *Journal of Interdisciplinary History* (forthcoming).

[12] By Paul Katzenberger on the American middle and upper classes and by Schubert on France in the eighteenth century both at the University of Munich.

[13] John Komlos, *Ernährung und Wirtschaftliche Entwicklung unter Maria Theresia und Joseph I: Eine anthropometrische Geschichte* (St Katherine: Scripta Mercaturae Verlag, 1994).

[14] John Komlos (ed.), *Stature Living Standards and Economic Development: Essays in Anthropometric History* (Chicago: The University of Chicago Press, 1994); John Komlos, (ed.), *The Bilogical Standard of Living on Three Continents: Further Explorations in Anthropometric History* (Boulder, CO: Westview Press, 1995).

reward the scholarly community. This collection of essays is a contribution in that direction.

JOHN KOMLOS

*University of Munich*
*March 1995*

# ACKNOWLEDGEMENTS

The editor and publisher would like to thank the following institutions and individuals, who have given permission to reproduce the articles in this volume: Herbert M. Johnson, Publisher, *Journal of Social and Biological Sciences* and JAI Press, Greenwich, CT and London (I); Gianni Toniolo, Giulio Einaudi Editore, Turin (II); Helen Dwight Reid Foundation and Heldref Publications, Washington, DC and Joo Han Kim (III); *Journal of Interdisciplinary History*, Massachusets Institute of Technology Press, Cambridge, MA (IV); *Economic History Review*, Basil Blackwell, Oxford (V, VI & VII); James M. Tanner, *Annals of Human Biology*, Taylor and Francis Ltd, Basingstoke (VIII & XI); Michael Silber, Hebrew University, Jerusalem (IX); Helen Dwight Reid Foundation and Heldref Publications, Washington, DC (X); American Historical Association (XII); Cambridge University Press (XIII); Duke University and Peter A. Coclanis (XIV); University of Chicago Press (XV & XVI).

# I

# Anthropometric History: What Is It?

One can think of the average height reached at a particular age or of the average adult height attained by a population as a historical record of the nutritional experience of the individuals composing that population. This line of reasoning is based on medical research, which has established beyond doubt that the cumulative nutritional intake of a population has a major influence on its average height, with the epidemiological environment also playing a part. Therefore, physical stature can be used as an indicator of how well the human organism thrives in its socio-economic environment. Specifically, nutritional status—and thus height—is related to food consumption and therefore to real family income, and therefore to wages and to prices and therefore to the standard of living broadly conceived. Therefore, we can use height as a proxy measure for these economic variables (Komlos, 1989).

Anthropometric history was born in the mid-1970s, in conjunction with efforts by economists to quantify changes in the standard of living during the course of the last two hundred years in economies for which more conventional measures of welfare such as GNP per capita were either not available, or were controversial. In contrast to real wages, anthropometric evidence is plentiful, and has the additional advantage of being available for groups such as children, housewives, subsistence peasants, aristocrats, and slaves, for whom conventional economic concepts do not always apply (Fogel, 1986).

The first results showed that American soldiers at the time of the revolutionary war were two inches taller than British soldiers of the same age. That is to say, the nutritional advantages of the New World were already evident in the eighteenth century. More startling still was the result pertaining to the most deprived of Americans, that is, those of African origin. The stature of slaves indicates that they were well nourished as young adults, although not as children. Adult slave men reached a height of 67 inches, within one inch of northern-born whites, and above contemporary European norms. In fact, their physical stature was closer to that of the European aristocrats than to European peasants. In spite of their social degradation, Blacks born in the New World were consistently taller than their African-born counterparts. This was true in the eighteenth century as well as in the nineteenth, and it indicates vividly the abundant availability of food supplies in the New World (Steckel, 1986; Komlos, 1990).

The European results were equally noteworthy. Telling undergraduates that a rise in

*Journal of Social and Biological Structures 14(3):353–356.*

bread prices caused food riots in the late eighteenth century is one thing, but being able to assert that the 'Oliver Twists' of London were so malnourished that there is hardly a civilized population in the world today of comparable stature is quite another. Speaking of a skewing of the income distribution at the turn of the nineteenth century in the midst of the Industrial Revolution is one thing, and being able to show that at age thirteen the gentry boys in the elite Sandhurst Royal Military Academy were 19 cms taller—almost 8 inches taller—than their compatriots in the slums of London is yet another (Floud, Wachter, & Gregory, 1990).

We knew how unhealthy pre-industrial and early industrial towns were, but one can now show that whether one considers the region around the city of Baltimore in Maryland, or around Vienna, in Lower Austria, or London, England, or Tokyo in Japan, the further one lived from these urban centers, the higher were the chances of reaching one's genetic potential height. Put another way, in the early-industrial period, being close to the source of nutrients, or being self-sufficient in food production, had distinct biological advantages. Whereas in modern market societies generally there is a positive correlation between height and per capita income, this was not always the case. Until the twentieth century, rural populations were invariably taller than urban ones, and underdeveloped regions tended to have better nourished populations than developed areas. The Irish, for instance, enjoyed a higher level of nutritional status than the industrial English, and the Hungarians were taller than the Czechs. In America Southerners were better nourished than Northerners. In the 1830s, farmers in America were taller than urban professionals. In other words, propinquity to food supplies was one of the most important factors in determining one's nutritional well being. At the aggregate level, higher per capita output of food was more important in determining nutritional status than was real money income, because individuals who purchased their food had to pay for transportation costs and for the efforts of middlemen, whereas subsistence farmers did not. Hence in the early industrial era one often finds a negative relationship between food consumption and the level of industrialization. This also meant that market integration, and the expansion of urban markets, posed a potential threat to nutritional status by altering the terms of trade of the agricultural sector (Mokyr & O Gráda, 1988).

The importance of anthropometric history is accentuated by the debate over the course of the standard of living during the early phases of the Industrial Revolution, when food consumption still accounted for as much as three-fourths of total income among the lower classes. The anthropometric approach has the advantage of sidestepping the dispute over the accuracy of real wage indices and a host of related problems such as how to convert hourly wages adequately into annual family incomes.

In this regard the most interesting findings are related to cycles in human height. The first recorded downturn took place in Europe during the second half of the eighteenth century and another occurred in America during the Antebellum decades. Both were accompanied by rapid population growth, by urbanization, and by the expansion of the industrial sector. In both instances the relatively slow growth of the agricultural labor force and the absence of technological breakthroughs in food production, coupled with the increased urban demand for food, meant that the price of nutrients increased both absolutely and relatively to industrial products.

In antebellum America, for example, the urban population practically doubled every decade, while the agricultural labor force grew by less than 10 percent per decade. No wonder food prices rose in real terms and no wonder both calorie and protein consumption declined. The average 18-year-old student entering the West Point military Academy weighed just under 128 pounds; 25 percent of the cadets were under 110 pounds in their late teenage years. This is less than the weight of girls today. Thus, in spite of the very rapid economic growth of the antebellum decades, and in spite of the fact that per capita national income increased by 40 percent between 1840 and 1870, the physical stature of the American population declined during the same period. These hidden costs of the early industrialized age in even resource abundant America are now becoming evident. But was the Industrial Revolution to blame?

In the European context the relationship between the Industrial Revolution and the well-being of the societies experiencing it has long been a controversial issue. It is known as the standard of living debate. Anthropometric historians have now been able to transform the debate in its essentials—showing that the diminution in human stature actually preceded the onset of industrialization of the second half of the 18th century, and therefore that, contrary to the implication of the factors described above, the decline in well being could not have been caused by the Industrial Revolution after all. That the standard of living continued to deteriorate even after the Industrial Revolution began is true. But this pattern has misled many scholars into thinking that there was a causal relationship between the Industrial Revolution and the decline in the living standards of the lower classes. Instead, we can now see clearly that the cause was the demographic expansion that began in Europe around the 1730s, i.e., prior to the Industrial Revolution, and we can plausibly suppose that the expansion of the industrial sector arrested the further deterioration in the well-being of the society, by generating additional income, which could be used to import nutrients. Without the Industrial Revolution, the nutritional status of the population would probably have deteriorated further, and growth impulses would have been dissipated once again by Malthusian population pressures as they had in the fourteenth century, and as they had more recently in the 17th century. Thus, conventional measures of living standards such as real wages or the level of gross domestic product per capita have misled scholars concerned with the impact of industrialization and economic development on the standard of living. The ultimate culprit turns out to be not the Industrial Revolution, but the demographic revolution. That the Industrial Revolution was able to counterbalance the Malthusian pressures not in the short run, only in the long run, is true. But one can demonstrate the incredible improvements in living standards over the last two centuries by showing, for example, that the height of lower-class thirteen-year-old English boys increased by nearly twenty centimeters between 1790 and 1990. Most of the increase, to be sure, was confined to the last hundred years, partly in response to improvements in public health in urban areas.

Has this line of research generated new knowledge? I think so. It has explored the gender-specific nature of the trends in living standards, an aspect of the debate that has been all but neglected until now. For example, several studies have shown that females were often affected earlier than males by a nutritional downturn. The approach has also shown that illiterate men were shorter than literate ones. It has demonstrated the existence of anthropometric cycles which were not known to exist previously. We now know about the nutritional deprivation of slave children, and of their catch-up growth in the late teens. We know more about the great gulf that divided the Old World from the New in terms of nutritional standards. We know that African-Americans were taller than Africans in Africa. We now know that German aristocrats were taller than middle-class Germans, and that the English gentry enjoyed a higher standard of living than their German counterparts who, in

I



turn, were better off than Austrian nobles. We know that the biological standard of living can diverge from more conventional indicators of well being during the early stages of industrialization, and per-capita income can be an ambiguous measure of welfare during some phases of growth. These are noteworthy contributions to social as well as economic history.

## References

Floud, R., Wachter, K. & Gregory, A. (1990) *Height, Health and History*. Cambridge, UK: Cambridge University Press.

Fogel, R. W. (1986) "Nutrition and the Decline in Mortality since 1700; Some Preliminary Findings," in Engerman, S. & Gallman, R., eds. *Long Term Factors in American Economic Growth*. Chicago, IL: University of Chicago Press, pp. 439–555.

Komlos, J. (1989) *Nutrition and Economic Development in the Eighteenth-Century Habsburg Monarchy: An Anthropometric History*. Princeton, NJ: Princeton University Press.

Komlos, J. (1990) "Height and Social Status in Eighteenth-Century Germany." *Journal of Interdisciplinary History, 20*, 607–621.

Mokyr, J. & O Gráda, C. (1988) "Poor and Getting Poorer? Living Standards in Ireland before the Famine." *Economic History Review, 2d ser., 41*, 209–235.

Steckel, R. (1986) "A Peculiar Population: The Nutrition, Health, and Mortality of American Slaves from Childhood to Maturity." *Journal of Economic History, 46*, 721–741.

# II

## On the Significance of Anthropometric History

Within the last decade interdisciplinary historians have paid increasing attention to the interaction between economic and biological processes in order to increase our understanding of development in a broad sense.[1] Such indicators as height, weight, body-mass, age at menarche, and morbidity, have all attracted our attention, not only because they are all related to nutritional status, and consequently to such demographic variables as life expectancy, but also because of their feedback effect on the economy through their impact on labor productivity.[2] Within this context, the importance of anthropometric history is accentuated by the debate over the course of the material standard of living during the early phases of the Industrial Revolution, when food consumption still accounted for as much as three fourths of total income among the laboring classes even in advanced European societies.

Although scientists have been interested in the study of human height from a medical, biological or anthropological point of view for centuries,[3] economic historians became aware of the implications of quantifying nutritional status only within our own generation.[4] Transplanted to American shores beginning with the mid 1970s,[5] anthropometric history was first applied to explore problems related to the debate over the slave experience, using principles that have become standard procedure in assessing nutritional adequacy in third-world countries.[6] Since then these methods have been used to estimate temporal changes and cross sectional differences in the nutritional status of historical populations on five continents.[7]

Anthropometric measures are useful in understanding many aspects of the historical record, because they enable one to quantify nutritional status. This line of reasoning is based on medical research, which has established beyond doubt that the net cumulative nutritional intake of a population has a major influence on its average height, with maternal nutrition also playing a significant role.[8] Thus height at a particular age as well as the terminal height attained by a population are measures of cumulative net nutrition: the food consumed during the growing years, minus the claims on the nutrients of basal metabolism, of energy expenditure, and of disease encounters.[9] The higher the nutritional status, the more calories and protein are available for physical growth, and the closer does an individual, and a population, come to its genetic potential. Because nutritional status, and therefore height, is not only an indicator of the exposure to disease, and of health in general, but also correlates positively with life expectancy and with food intake, it shows how well the human organism thrives in its economic, epidemiological, and social environment.

Of course, one ought not to overlook the fact that the terminal height reached by an individual in a given population is influenced by genetic factors as well as by nutritional status. This consideration, however, does not affect studies of the evolution of human height as long as the genetic composition of the population is not altered through large scale migration. For this reason, historical studies frequently focus on changes in terminal height, because such changes are not affected by genetic factors.

Anthropometric evidence has been used to increase our understanding of the relationship between industrialization and demographic processes, and, in particular, of the role played by food consumption in the Industrial Revolution, enabling us to reconceptualize this episode as the "escape from the Malthusian trap," a constraint which had posed formidable obstacles to economic growth in prior centuries. It has made it possible for us to learn more about the impact of industrialization, and of urbanization on the standard of living considered not only in the aggregate, but by gender, social status, occupation, and at various stages in the life cycle as well. Consequently, physical stature can be used in very different ways from economic variables such as real wages, and has the advantage in a historical context of being available for groups, such as children, housewives, the subsistence peasantry, and the aristocracy, for whom economic concepts do not always apply.

In conventional terms the standard of living has become practically synonymous with the material standard, and consequently has been most often equated to and measured by income per capita. Yet it can be interpreted much more broadly to encompass the psychological as well as the biological dimensions of human existence, i.e. the quality of life. Distinguishing among these three components of well being would not add much conceptually meaningful to our understanding of the past if these indicators correlated perfectly with one another; recent empirical evidence, however, has tended to call into question this supposition, pointing to the need not to conflate them into a single concept. Of course, in actuality the issue is even more complicated, because the paucity of relevant evidence often does not even allow one to construct a representative index of real wages, let alone of annual income. [10] The step from the one to the other is fraught with many, many uncertainties. [11] It can hardly be surprising then, that the debate over the secular change in the standard of living conventionally conceived has reached an impasse, in spite of recent noteworthy efforts to overcome the limitations of the extant data. [12]

However, by becoming explicitly concerned with the biological well being of historical populations as a component of the standard of living one can illuminate this issue from an entirely different perspective. This has been attempted also by studying directly the health of historical populations, but that approach is obviously limited by the few relevant data sets at our disposal. [13] Another avenue taken by some scholars is to consider mortality as an integral component of well being, and, to incorporate it into the conventional index of the material standard of living. [14] However, this attempt to collapse the biological and material standard of living into a single index is confronted by the inherent difficulty of gauging the monetary value of human life.

Anthropometric history has the advantage of having an abundant evidential basis beginning with the late-seventeenth century, of which only a small fraction has been analyzed until now. [15] This approach acknowledges outright the inherent multidimensionality of the concept "standard of living," and furthermore asserts that the several dimensions are orthogonal to one another, that is to say that, in principle, they ought not to be collapsed into a single indicator. The research agenda then becomes to construct indices of the biological standard of living of various populations over time, by social class to the extent possible, and to ascertain how this indicator correlates with the material standard of living conventionally conceived. [16]

The nutritional status of a population depends primarily on its food intake, which, in turn, depends on the relative price of food and on family income. Although the positive correlation between height and income has been amply documented for populations for which both variables are available, [17] the relationship has been found to be less than perfect, especially in non-commercialized societies or during the early stages of rapid economic growth. Thus some caveats are in order, because the distribution of income has also been found to affect the mean stature of a population, and because the mix of calorie and protein intake matters to the growth process. Lastly, food consumption obviously depends not only on real income, but also on the price of food relative to that of all other products.

Up to now several episodes have been found during which the trends in the biological and the material standard of living diverged from one another. They all took place during the early stages of rapid economic growth: in East-Central Europe and in Great Britain during the second half of the eighteenth century, in late-nineteenth century Montreal, and in the antebellum United States.[18] In the United States adult male stature declined, beginning with the birth cohorts of the 1830s, by more than 2 cm.[19] Men appear to have been quite underweight, with an average weight of 126 pounds (57.3 kg) in their late teen-age years,[20] although according to conventional indicators the economy was expanding rapidly during these decades: between 1840 and 1870 per capita net national product increased by more than 40 per cent.[21] In the Habsburg Monarchy the decline in stature during the second half of the eighteenth century was between 3 and 5 cm.[22] A similar pattern was found for industrializing Montreal. The birthweight of infants born there fell after the 1870s, indicating that the nutritional status of mothers was declining.[23] Such anthropometric cycles related to economic and demographic processes were not known to exist previously.

The extent to which these episodes were accompanied by a deterioration of the epidemiological environment needs further study, but it is already clear that all of them were, in fact, characterized by a decline in per capita nutrient intake. These incidents of rapid economic growth were accompanied by stress on the human organism, even though by conventional measures the standard of living was rising. This might be considered an anomaly, since an increase in per capita income normally implies that food consumption, too, ought to have been increasing; it becomes less so once one notes that while income determines the position of the demand curve for food, an individual who purchases food at higher market prices might consume less of it than a self-sufficient peasant isolated from the market by high transport and information costs, even if the income of the former is larger than that of the latter.

Hence, during episodes of market expansion and market integration the human organism was often not doing as well as one might be led to believe from conventional indicators.[24] This is particularly true because nutritional status, as measured by height, is an important determinant of life expectancy. Thus the crude death rate was increasing, and life expectancy declining, in antebellum America even as per capita income was rapidly rising, supporting thereby the notion that the biological standard of living was falling.[25] This shows vividly the pitfalls of relying exclusively on conventional indicators of well being.[26]

The divergence in the trend of the biological and conventional indicators of well being can be explained by several other factors as well. The above examples of economic growth were all accompanied by rapid population growth and by urbanization. These processes increased the demand for food at a time when the agricultural labor force grew more slowly than the industrial one and the gains in labor productivity in agriculture, too, were lagging behind. Hence, the supply response did not generally match the increased demand for food, and food prices rose relative to those of other goods.

The relative price of food also rose because rapid technological change in the industrial sector brought with it a dramatic decline in the price of manufactured goods. The rise in the relative price of food was greatest in areas which were previously self-sufficient, that is, isolated from larger markets, and the relative price of food was actually undefined. Early stages of growth were generally accompanied by market integration; this meant that in previously isolated regions the price of food could rise discontinuously after market integration, bringing with it the potential for very large shifts in the demand curve for food.[27] In addition, new products, such as coffee, were introduced, which changed tastes and at the same time provided substitutes for traditional food products. Thus while the real wage might actually rise, it often did not rise as fast as food prices, leading to a substitution away from food consumption. Even farm operators whose income rose as fast as food prices reduced their food consumption, because the price elasticity of demand for food was greater than the income elasticity.

Because the price of a calorie is much greater if purchased through meat than if purchased through grain there was a tendency to substitute away from meat consumption. This caused protein intake, an important component of nutritional status, to fall, making it more difficult for the body to fight off nutrition-sensitive diseases. Of course, during the early stages of economic growth individuals are usually not well informed about the importance of a balanced diet, and therefore they were unaware that cutting back on the consumption of certain products would impinge on their nutritional status and health.

Thus the level of utility could increase even as food consumption fell. This would not present a problem if it were not for the fact that food consumption is an important determinant of the biological well-being of the human organism. Therefore per capita income can be an ambiguous measure of welfare during the early stages of economic development, unless it moves in the same direction as the biological standard of living, because it measures only one component of well-being. Of course, in the long run the ambiguity dissolved, because the productivity of the agricultural sector improved and food prices declined sufficiently relative to income. Hence, the share of a typical budget devoted to alimentation fell in Europe from about three-quarter to about one-quarter, and then even less, of income. Yet, recent anthropometric research indicates that the conventional indicators of welfare are incomplete for the early stages of industrialization, and need to be supplemented with other measures, such as human stature, which capture the biological well-being of the population.

In the Habsburg Monarchy the fall in nutritional levels was initiated not by sectoral shifts from agriculture into industry, but by rapid population growth pressing on scarce resources. In contrast to America, in Europe the quantity of land was fixed; therefore, population growth ran again into Malthusian ceilings in the eighteenth century. The subsequent rise in food prices led to a decline in consumption, particularly of meat, because the price elasticity of demand for meat was much greater than that of grains. This caused the intake of animal proteins, an important component of nutritional status, to fall, making it more difficult for the body to fight off nutrition-sensitive diseases.

To be sure, not everyone experienced nutritional stress during the early stages of economic growth. In America, for instance, the income of the urban middle class rose sufficiently in the 1830s and 1840s to keep pace with the rise in the price of nutrients, thereby enabling that segment of the society to maintain its nutritional status at least initially. Not until the Civil War disrupted the flow of nutrients did their biological well being begin to suffer as well.[28] In a similar fashion, the Habsburg and German upper classes were not affected by the Malthusian crisis of the late-eighteenth century, and even increased their height advantage compared with the lower classes.[29] This latter pattern is consistent with Kuznets's notion that during the early stages of economic development the distribution of income is likely to become more skewed.

The issue of gender, until now completely overlooked in the standard-of-living debate, has also been introduced into the discussion. There is actually no warrant for assuming that over time changes in the standard of living of both sexes have been identical. In fact, so far all anthropometric studies have indicated substantial gender-based differences in the biological standard of living during episodes of economic change. For example, during the early nineteenth century the decline in nutritional status of women in Great Britain was greater than that of men.[30] This was precisely the pattern among the manumitted slaves of Maryland, possibly because of the impact of changes in the sex ratio.[31] In Pittsburgh during the first half of the twentieth century, the nutritional status of women improved much less than that of men, or not at all.[32] In constrast to men, Viennese women, too, seem not to have improved their nutritional status at all in the late-nineteenth century.[33] Therefore, the evidence so far indicates that females began to experience nutritional stress earlier than men during a downturn, and were less likely to show improvements in an upswing.

That height correlated positively with education and wealth has been demonstrated in the case of Frenchmen born in the late 1840s. Illiterate recruits into the army averaged 164.3 cm, while those able to read and write were 1.2 cm taller. Presumably literate men were wealthier and spent more time at education and less at work than did illiterate ones.[34]

In all studies without exception the positive relationship between social status and physical stature has been consistently documented in various societies and at different times. German aristocrats in the eighteenth century were 7 cm taller than middle class boys throughout adolescence; the latter, in turn were about 4 cm taller than lower class boys of the same age.[35] Boys attending prestigious military schools in Austria, England, France and Germany were all much taller than the population at large. English gentry adolescents were as much as 20 cm taller than the nutritionally deprived slum boys of London.[36] To be sure, these are extreme differences, but even in egalitarian America social standing did have an effect on height and it persisted throughout the nineteenth and twentieth centuries. In the wake of the New Deal, however, these effects became less pronounced, as the income distribution became less skewed.

Another important finding was that the nutritional advantages of the New World were quite pronounced by the early eighteenth century. Thereafter, the height of the American population was above European norms until the mid-twentieth century, confirming the extent to which the new continent's environment was favorable from a human biological point of view.[37]

Yet another pattern that has so far emerged is that the propinquity to nutrients is a crucial determinant of nutritional status in preindustrial societies, and in societies caught in the early stages of economic development. In such circumstances per capita income is not as important a determinant of human stature as the availability, i.e. the relative price, of food. This pattern was first noted in the case of East-Central Europe, and has since been confirmed for Japan, Great Britain and the United States as well.[38] Being isolated from markets by high transport and transaction costs had its advantages as long as population density did not exceed the carrying capacity of the land, because subsistence peasants/farmers had little choice but to consume all of their own food output. Once they became integrated into a larger market system, however, they had to compete with other segments of the population for food, which tended to impinge on their nutritional intake.

Hence the share of agricultural ouput in total income was an important determinant of stature even in the United States, where Southerners were 1.5 cm taller than men of the more industrialized North, even though per capita income was greater in the latter region than in the former. The same pattern held among England, the United States, Scotland and Ireland. Although England had a higher per capita income perhaps by as much as 25 per cent, its army was shorter than that of the United States in the eighteenth century. A century later, Irish recruits into the Union army were also taller than soldiers born in England, although English per capita income was certainly higher than the Irish one.

The remote parts of Japan also had taller populations in the late-nineteenth century than those situated near urban markets. Among the birth cohorts of the 1830s attending the West Point Military Academy sons of farmers were 1.3 per cent taller than cadets whose fathers were professionals. The importance of remoteness to nutritional status can be seen from the fact that in 1850 the tallest people in Europe were found in Dalmatia, one of the least developed areas of the continent. The relationship reappears, in an even more astounding form at the end of the nineteenth century: although Bosnia-Herzegovina was certainly among the least developped parts of the Habsburg Monarchy, its population was nonetheless the tallest.[39]

All this evidence corroborates the notion that being close to the supply of food during the early stages of industrialization had a positive effect on nutritional status, possibly because it meant that the transaction costs of obtaining food were lower. In addition, rural populations could have benefited from a lower exposure to disease than those who lived in or near an urban area. Thus, regions around London had the shortest populations in England, as did the counties around Baltimore in Maryland.[40] This finding also confirms the notion that the epidemiological environment of towns was inferior compared to rural areas prior to the improvement in sanitary conditions. Only in the last decades of the nineteenth century did this pattern reverse itself, as the effects of public health improvements began to be felt. Another probable reason for the difficulty of maintaining nutritional status during the initial phase of economic development is that the distribution of income becomes more uneven. The changes in entitlements to goods, with the rise in food prices, lead many to lose their ability "to command food."[41]

Much work has explored the nutritional status of slaves ever since Richard Steckel's path-breaking contribution of 1979.[42] From the anthropometric evidence hitherto examined, it has been concluded that slaves in nineteenth-century America were well nourished as young adults, although not as children. Slave children were below the 10th centile of modern height standards. In spite of their early nutritional deprivation, adult slaves were closer to the 25th centile. Men reached a terminal height of over 67 inches–within one inch of northern-born whites, and well above contemporary African and European norms. In fact, their physical stature was closer to that of the European aristocracy than to that of the peasantry. There can be no doubt that slaves in America, though legally deprived, had a higher nutritional status than their brethren left behind in Africa.

Before concluding one should note that a lot of work still needs to be done. We need to assemble the various pieces of evidence before we will have a complete anthropometric history of even Europe and America. Very few investigations have dealt with weight, with the exception of birthweight,[43] or with the biomass index (weight/height$^2$) which promises to be an important predictor of mortality.[44] Moreover, much corroborating evidence needs to be assembled; height samples are frequently difficult to analyze, because military data are almost always biased by the imposition of minimum height requirements for recruits. While statistical methods have been developed to deal with truncated samples,[45] such techniques do not always yield unambiguous results.[46] For instance, the sign of the trend in the English population of the late-eighteenth century is now under dispute.[47] In addition, we have not yet identified precisely when the downturn in nutritional status began in nineteenth century America, nor when the trough was actually reached, because these turning points must have been region, gender, and class specific. So the work must go on.

Yet much has already been accomplished. It is important to know that economic growth was often accompanied by a decline in nutritional status, and thus, that the biological standard of living can diverge from the material standard during the early stages of industrialization and urbanization. Therefore, per capita income can be an ambiguous measure of welfare during some phases of growth, unless it moves in the same direction as the other measures of well being. It is useful to have indices of well being at the county level, for both sexes, for children, and for social groups for which these were not available before. Thus for a decade of research quite a lot has already been accomplished, even if more issues still await exploration.[48]

1 R. W. Fogel, S. L. Engerman, J. Trussell, R. Floud, C. L. Pope, and L. T. Wimmer, "The Economics of Mortality in North America, 1650-1910: A Description of a Research Project," *Historical Methods*, 11 (1978), pp. 75-108.

2 R. W. Fogel, "Nutrition and the Decline in Mortality since 1700; Some Preliminary Findings," in S. Engerman and R. Gallman, eds., *Long Term Factors in American Economic Growth*, National Bureau of Economic Research, Studies in Income and Wealth, vol. 51 (Chicago, 1986); For a theory of anthropometric history see J. Komlos, *Nutrition and Economic Development in the Eighteenth-Century Habsburg Monarchy: An Anthropometric History* (Princeton, 1989), ch. 1; J. Komlos, "The Age at Menarche in Vienna: The Relationship between Nutrition and Fertility," *Historical Methods*, 22 (1989), pp. 158-63; R. Floud, K. Wachter, and A. Gregory, *Height, Health and History* (Cambridge, 1990); R. Steckel, "A Peculiar Population: The Nutrition, Health, and Mortality of American Slaves from Childhood to Maturity," *Journal of Economic History*, 46 (1986), pp. 721-41.

3 J. Tanner, *A History of the Study of Human Growth* (Cambridge, 1981).

4 In the English-speaking literature the issue appears to have been first raised in S. Engerman, "The Height of U.S. Slaves," *Local Population Studies*, 16 (1976), pp. 45-50.

5 E. Le Roy Ladurie, E. N. Bernageau, and Y. Pasquet, "Le conscrit et l'ordinateur. Perspectives de recherches sur les archives militaires du XIXe siècle français," *Studi Storici*, 10 (1969), pp. 260-308.

6 P. B. Eveleth and J. M. Tanner, *Worldwide Variation in Human Growth* (Cambridge, England, 1976).

7 On the Caribbean (Latin America) see G. C. Friedman, "The Heights of Slaves in Tri-

nidad," *Social Science History*, 6 (1982), pp. 482-515; on African heights see D. Eltis, "Nutritional Trends in Africa and the Americas: Heights of Africans, 1819-1839," *Journal of Interdisciplinary History*, 12 (1982), pp. 453-75; B. W. Higman, "Growth in Afro-Caribbean Slave Populations," *American Journal of Physical Anthropology*, 50 (1979), pp. 373-85; on Asian heights see T. Shay, "The Level of Living in Japan, 1885-1938: New Evidence," in J. Komlos, ed., *The Biological Standard of Living and Economic Development on Three Continents: Essays in Anthropometric History* (Chicago, 1994).

[8] R. W. Fogel, S. L. Engerman, and J. Trussell, "Exploring the Uses of Data on Height: The Analysis of Long-Term Trends in Nutrition, Labor Welfare, and Labor Productivity, *Social Science History*, 6 (1982), pp. 401-21.

[9] J. M. Tanner, *Fetus into Man: Physical Growth from Conception to Maturity* (Cambridge, MA, 1978).

[10] For a recent analysis of some of the issues in the American context see R. Margo and G. Villaflor, "The Growth of Wages in Antebellum America: New Evidence," *Journal of Economic History*, 47 (1987), pp. 873-97.

[11] The problems with the construction of wage and income series are manifold. Prices are generally limited to a few staples and are not adequately dispersed geographically. Information on rents is seldom available. Moreover data on wages are typically limited to a few occupations, and are often incredibly stable even in the face of violent fluctuations in commodity prices. Their reliability is therefore brought into question. In addition, daily wages cannot be accurately converted into annual income. The issue is complicated by the fact that the labor force participation rates of women and children changed over time, thereby exerting an influence on family income. Joan Thirsk has warned us recently that "We should remind ourselves of the crude measures used to establish the standard of living index..." in "Horticultural Revolution: A Cautionary Note on Prices," *Journal of Interdisciplinary History*, 14 (1983), pp. 299-302. The literature on the controversy associated with measuring the standard of living is enormous. Among the more recent contributions one might see, for instance, M. Flinn, "English Workers' Living Standards during the Industrial Revolution: A Comment," *Economic History Review*, 2nd ser., 37 (1984), pp. 88-92; G. N. von Tunzelmann, "The Standard of Living Debate and Optimal Economic Growth," in J. Mokyr, ed., *The Economics of the Industrial Revolution* (Totowa, NJ, 1985), pp. 207-26; J. Rule, *The Laboring Classes in Early Industrial England, 1750-1850* (London - New York, 1986).

[12] L. D. Schwarz, "The Standard of Living in the Long Run: London, 1700-1860," *Economic History Review*, 2nd ser., 38 (1985), pp. 24-41; P. Lindert and J. Williamson, "English Workers' Living Standards during the Industrial Revolution," *Economic History Review*, 2nd ser., 36 (1983), pp. 1-25. The study of probate inventories is an especially creative approach to overcoming the limitations of the data. S. McMahon, "Provisions Laid Up for the Family: Toward a History of Diet in New England, 1650-1850," *Historical Methods*, 14 (1981), pp. 4-21; L. G. Carr and L. S. Walsh, "Inventories and the Analysis of Wealth and Consumption Patterns in St. Mary's County, Maryland, 1658-1777," *Historical Methods*, 13 (1980), pp. 81-104.

[13] J. Riley, "Insects and the European Mortality Decline," *American Historical Review*, 91 (1986), pp. 833-58; J. Riley, *The Eighteenth-Century Campaign to Avoid Disease* (London, 1987); J. Riley and G. Alter, "Mortality and Morbidity: Measuring Health Across Time," in R. W. Fogel, ed., *Long-Term Changes in Nutrition and the Standard of Living*, Ninth Congress of the International Economic History Association (Berne, Switzerland, 1986), pp. 97-106; One should not overlook the efforts of anthropologists in this regard, see A. H. Goodman, R. Brooke Thomas, A. C. Swedlund, and G. J. Armelagos, "Biocultural Perspectives on Stress in Prehistoric, Historical, and Contemporary Population Research," *Yearbook of Physical Anthropology*, 31 (1988), pp. 169-202.

[14] J. G. Williamson, "Was the Industrial Revolution Worth It? Disamenities and Death in 19th Century British Towns," *Explorations in Economic History*, 19 (1982), pp. 221-245; J. G. Williamson, "Urban Disamenities, Dark Satanic Mills and the British Standard of Living Debate," *Journal of Economic History*, 44 (1981), pp. 75-84; E. Davin, "The Era of the Common Child," Unpublished Manuscript, Department of History, University of Pittsburgh, 1988.

[15] New evidence is being constantly unearthed, see for instance, H. Freudenberger and

J. B. Pritchett, "New Evidence on the Domestic Slave Trade: Certificates of Good Character," *Journal of Interdisciplinary History*, 21 (1991), pp. 447-77.

16 The psychological dimension of the standard of living has been the one least studied until now from a quantitative perspective.

17 R. Steckel, "Height and Per Capita Income," *Historical Methods*, 16 (1983), pp. 1-7; H. Jan Brinkman, J. W. Drukker, and B. Slot, "Height and Income: A New Method for the Estimation of Historical National Income Series," *Explorations in Economic History*, 25 (1988), pp. 227-64.

18 J. Komlos, "The Secular Trend in the Nutritional Status of the Population of the United Kingdom, 1730-1860," *Economic History Review*, 2nd ser., 46 (1993), pp. 115-144.

19 Fogel, "Nutrition and the Decline in Mortality since 1700."

20 Komlos, "The Height and Weight of West Point Cadets."

21 R. E. Gallman, "The Pace and Pattern of American Economic Growth," in L. E. Davis, R. A. Easterlin, and W. N. Parker, eds., *American Economic Growth: An Economist's History of the United States* (New York, 1972), pp. 15-60.

22 J. Komlos, "Stature and Nutrition in the Habsburg Monarchy: The Standard of Living and Economic Development in the Eighteenth Century," *American Historical Review*, 90 (1985), pp. 1149-61.

23 P. W. Ward and P. C. Ward, "Infant Birth Weight and Nutrition in Industrializing Montreal," *American Historical Review*, 89 (1984), pp. 324-45.

24 "... a wide area of instability is created whenever money becomes the basis of the economy. The disruption of the traditional balance leads to increasing inequality in food consumption, with improvement for some accomplished at the price of the pauperization of a part of the population, especially in the cities. For we know that this disruption often coincides with migration to the cities, which breaks down dietary and culinary habits and changes the traditional channels of supply." M. Aymard, "Toward the History of Nutrition: Some Methodological Remarks," in R. Forster and O. Ranum, *Food and Drink in History. Selections from the Annales Economies, Sociétés, Civilisations* (Baltimore, 1979), pp. 1-16, esp. p. 6. See also T. Cuff, "A Weighty Issue Revisited: New Evidence on Commercial Swine Weights and Pork Production in Mid-Nineteenth Century America," *Agricultural History*, 66 (1992), pp. 55-74.

25 C. L. Pope, "Native Adult Mortality in the U.S.: 1770-1870," in Fogel, ed., *Long-Term Changes in Nutrition and the Standard of Living*, pp. 76-88.

26 While income per capita was undoubtedly rising during the second half of the nineteenth century, as late as 1900 British workers are said to have consumed a mere 2,000 calories a day. D. Oddy, "A Nutritional Analysis of Historical Evidence: The Working-Class Diet, 1880-1914," in D. Oddy and D. Miller, eds., *The Making of the Modern British Diet* (London, 1976), pp. 214-31.

27 Komlos, *Nutrition and Economic Development in the Habsburg Monarchy*, p. 105.

28 J. Komlos, "The Height and Weight of West Point Cadets: Dietary Change in Antebellum America," *Journal of Economic History*, 47 (1987), pp. 897-927.

29 J. Komlos, "Height and Social Status in Eighteenth-Century Germany," *Journal of Interdisciplinary History* (1990), pp. 607-21.

30 P. Riggs, "The Standard of Living in Britain's Celtic Fringe," in Komlos, ed., *The Biological Standard of Living.*

31 J. Komlos, "Toward an Anthropometric History of African-Americans: The Case of the Free Blacks of Antebellum Maryland," in C. Goldin and H. Rockoff, eds., *Strategic Factors in Nineteenth Century American Economic History: A Volume to Honor Robert W. Fogel*, National Bureau of Economic Research, Studies in Income and Wealth, vol. 52 (Chicago, 1992), pp. 297-329.

32 J. Wu, "The Anthropometric History of Pittsburgh and Allegheny County, Pennsylvania, 1890-1950," in Komlos, ed., *The Biological Standard of Living.*

33 W. P. Ward, "Weight at Birth in Vienna, 1865-1930," *Annals of Human Biology*, 14 (1987), pp. 495-506.

34 E. Le Roy Ladurie and N. Bernageau, "Etude sur un Contingent Militaire (1968). Mobilité géographique, délinquance et stature, mises en rapport avec d'autres aspects de la situation des conscrits," *Annales de démographie historique* (1971), pp. 311-37.

35 Komlos, "Height and Social Status in Eighteenth Century Germany."

[36] Floud, Wachter, and Gregory, *Height, Health and History.*

[37] K. Sokoloff and G. Villaflor, "The Early Achievement of Modern Stature in America," *Social Science History*, 6 (1982), pp. 453-81.

[38] Shay, "The Level of Living in Japan, 1885-1938: New Evidence"; J. Mokyr and C. Ó Gráda, "Poor and Getting Poorer? Living Standards in Ireland before the Famine," *Economic History Review*, 2nd ser., 41 (1988), pp. 209-35.

[39] Austria, *Militär-Statistisches Jahrbuch für das Jahr...* (Vienna, various years).

[40] Floud, Wachter, and Gregory, *Height Health and History*; Komlos, "Toward an Anthropometric History of African-Americans."

[41] A. Sen, *Poverty and Famines: An Essay on Entitlement and Deprivation* (Oxford, 1981).

[42] R. H. Steckel, "Slave Height Profiles from Coastwise Manifests," *Explorations in Economic History*, 16 (1979), pp. 363-80; see also J. Trussell and R. H. Steckel, "The Age of Slaves at Menarche and Their First Birth," *Journal of Interdisciplinary History*, 8 (1978), pp. 477-505; R. H. Steckel, "A Peculiar Population: The Nutrition, Health and Mortality of American Slaves from Childhood to Maturity," *Journal of Economic History*, 46 (1986), pp. 721-42; R. H. Steckel, "Birth Weights and Infant Mortality among American Slaves," *Explorations in Economic History*, 23 (1986), pp. 173-98; R. W. Fogel, *Without Consent or Contract. The Rise and Fall of American Slavery* (New York, 1989), pp. 138-47; R. Margo and R. H. Steckel, "The Height of American Slaves. New Evidence on Slave Nutrition and Health," *Social Science History*, 6 (1982), pp. 516-38; R. H. Steckel, "Growth Depression and Recovery: the Remarkable Case of American Slaves," *Annals of Human Biology*, 14 (1987), pp. 111-32.

[43] C. Goldin and R. Margo, "The Poor at Birth: Birth Weights and Infant Mortality at Philadelphia's Almshouse Hospital, 1848-1873," *Explorations in Economic History*, 26 (1989), pp. 360-79.

[44] R. W. Fogel, "The Conquest of High Mortality and Hunger in Europe and America: Timing and Mechanisms," in P. Higonnet, D. S. Landes and H. Rosovsky, *Favorites of Fortune. Technology, Growth, and Economic Development since the Industrial Revolution* (Cambridge, MA, 1991), pp. 33-71; T. Cuff, "The Bio-Mass Index of West Point Cadets," *Historical Methods*, forthcoming in 1994.

[45] K. W. Wachter and J. Trussell, "Estimating Historical Heights," *Journal of the American Statistical Association*, 77 (1982), pp. 279-93.

[46] J. Komlos and J. Han Kim, "On Estimating Trends in Historical Heights," *Historical Methods*, 23 (1990), pp. 116-20.

[47] Floud, Wachter, and Gregory, *Height, Health and History*; Komlos, "The Secular Trend."

[48] C. Tilly, L. A. Tilly, and R. Tilly, "European Economic and Social History in the 1990s," *The Journal of European Economic History*, 20 (1991), pp. 645-71.

# III

# Estimating Trends in Historical Heights

**John Komlos**
**Joo Han Kim**

Recent advances in anthropometric history, that is, 116
the study of human height and weight in a historical context, have greatly contributed to our understanding of secular changes in nutritional status and of the ways in which nutrition was related to economic and demographic processes during the course of the last two centuries.[1] However, for purposes of analysis, one of the major sources of data has been provided by military records and, generally, such records are quite difficult to analyze because most armies invoked a minimum height requirement. This means that the available samples are often though seldom perfectly, truncated because the regulations were not uniformly enforced. Several solutions to the problem of estimating central tendencies in such distributions have been offered, including the *reduced-sample maximum likelihood estimator* (RSMLE), and the *quantile bend estimator* (QBE).[2] They both rely on the well-known fact that the heights of human populations are normally distributed.

## EMPIRICAL PROBLEMS

Since their development, the above estimators have been used in a large number of studies but have proven to be rather unreliable in some cases. At times, it is, therefore, difficult to draw firm inferences from the results. The major problem has been that from one period to the next, the estimates of the mean height of the underlying population have often fluctuated unreasonably either because of sampling error or because outliers exerted an undue influence on the estimates. For instance, Sandberg and Steckel used QBE on five-year moving averages without examining the underlying sample distributions.[3] For quite a few of their estimates, the height of the Swedish population fluctuated by as much as 2 cm from one year to the next; fluctuations of this magni-

Historical Methods 23, 3, pp. 116–120, 1990. Reprinted with permission of the Helen Dwight Reid Educational Foundation. Published by Heldref Publications, 1319 18th Street, N.W., Washington, DC 20036-1802. 

tude are unlikely because such patterns have not been observed in modern populations.[4] In fact, when they used the QBE program to estimate the mean height of the population per decade instead of using the moving-average method, the resultant series was much more stable.[5] Similarly unreasonably large fluctuations in the mean height of British adults and children have also been reported.[6] These irregularities are quite troublesome because the rationale for most height studies is to estimate the trend in height as opposed to its level. The trend itself is what tells us most about fundamental changes in the economic, demographic, or epidemiological environment.

The work of one of the authors on Habsburg military heights confirmed the importance of examining the frequency distributions of heights for outliers and for obvious deviations from normality.[7] For example, using the QBE procedure, the adult height of the male, Hungarian peasant population from which the recruits were drawn was initially estimated to have increased from 165.0 to 170.1 cm between the birth cohorts of the 1730s and the 1740s.[8] Upon examination of the two sample frequency distributions of these decades, it became clear that they both violated the normality assumption, even though the sample sizes were relatively large (figure 1, panel A). With an increase in sample size, however, the two distributions did approach normality (figure 1, panel B), and the estimated change in mean heights (using the QBE procedure) decreased from 5.1 cm to 2.6 cm. (The estimated height in the 1730s became 167.2 and in the 1740s, 169.8 cm; see table 1). Therefore, the rule of thumb in using the QBE program ought to be to inspect the sample frequency distributions first in order to ascertain if the assumption that the obtained height distribution is normal is at all plausible. If the sample frequency distribution obviously deviates from normality, then the use of the QBE procedure is hardly war- 117
ranted, and the sample sizes should be increased prior to further analysis.

However, even this tactic turned out to have drawbacks because occasionally odd distributions persisted even after the sample sizes were repeatedly increased, and constraints of time and of financial resources simply prohibit archival work to continue indefinitely.[9] Thus, in the final sample of more than 150,000 observations, there were only 199 adult men born in Lower Austria in the 1740s, and the change in the height of the soldiers between the 1740s and the 1750s appeared un-

**FIGURE 1**
**Height Distribution of Adult Hungarian Soldiers**

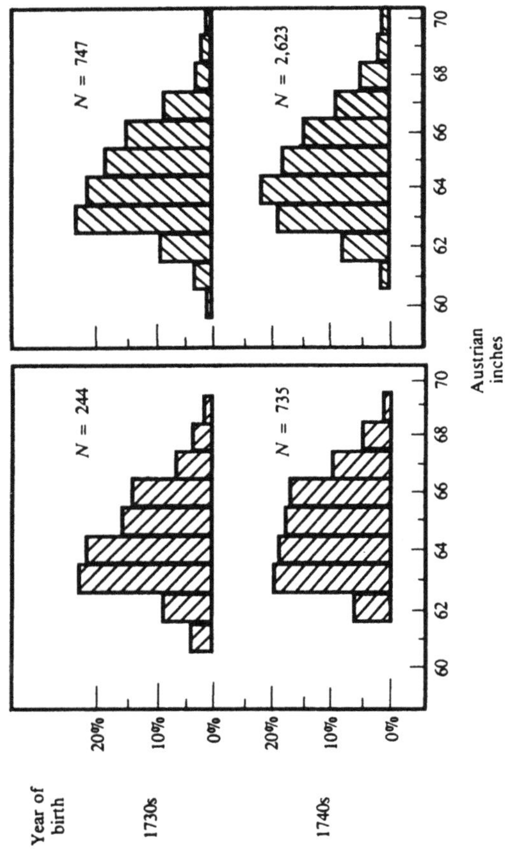

reasonably large: 4.5 cm (table 1 and figure 2). A similarly unreasonable pattern was obtained for the recruits born in the Polish province of Galicia in the 1790s (figure 2). The mean height of that population was estimated to have increased by an improbable 5.7 cm between the 1780s and the 1790s (table 1). Many other such examples could be provided. In such cases, differences in QBEs are not statistically reliable, and there are no statistical procedures to compare two population means using QBEs. Clearly, an alternative approach capable of providing a more reliable way to compare two population means is needed.

**AN ALTERNATIVE APPROACH**

The most frequently discussed issue in anthropometric history is the trend in the mean height of a particular population. Consequently, we explore a method of testing for the sign-of-the-time trend of mean height. Our method is based on a relationship between the mean of the normal distribution and the mean of the truncated normal distribution. During the period to be considered, the truncation point, as well as the degree to which it was enforced, might change. Call the greatest of the minimum height requirments $a$. (In the eighteenth-century Habsburg monarchy, $a$ was 63 Austrian inches or 165.8 cm.) Note that the distribution to the right of 63 inches was not changed by the subsequent lowering of the minimum height requirement. Therefore, the part of the sample above $a$ remains unaffected by administrative changes. If one perfectly truncates all of the eighteenth-century sample at $a$ by excluding all observations below it, one can examine the trend in the most reliable component of the sample to the right of $a$.[10] Call this perfectly truncated sample the *restricted sample*, and its mean $\bar{x}'$.

The object is to determine if a change in the mean of the restricted sample, $\bar{x}'$, has the same sign as a change in the true mean of a population, $\mu$. In the appendix we prove that the mean of a truncated normal distribution, $\mu'$, increases in the mean of the original normal distribution, $\mu$. In other words, if the mean of the population, $\mu$, decreases, then, holding the truncation point $a$ and the standard deviation of the population $\sigma$ constant, the corresponding mean of the truncated distribution, $\mu'$, will also decrease. The reverse is also true: if $\mu'$ increases, then $\mu$ increases. Therefore, while $\mu'$ is not the 118
true mean of the population, a change in $\mu'$ has the

**FIGURE 2**
**Height Distribution of Adult Lower-Austrian and Galician Soldiers**

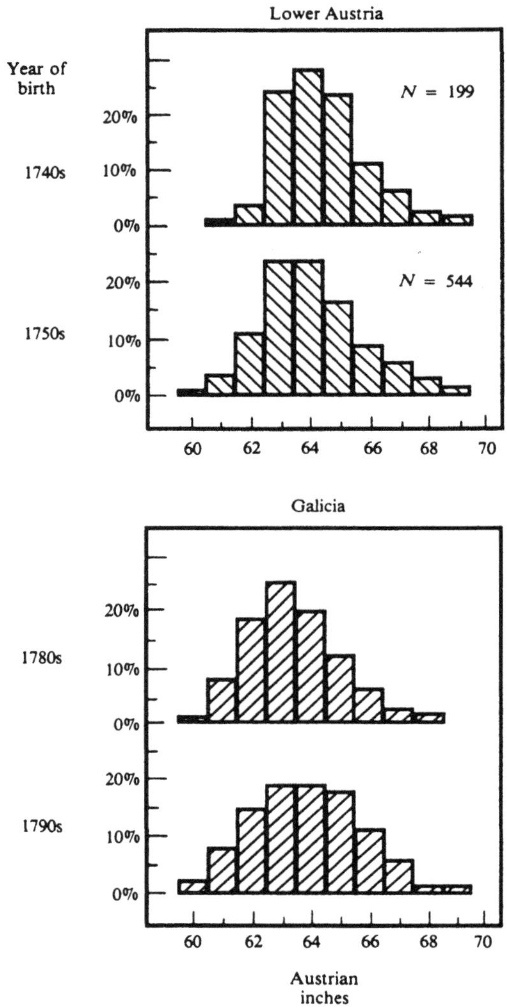

same sign as a change in the true mean of the population $\mu$, as long as the truncation point and the standard deviation are held constant.[11] This result is reassuring because the restricted sample is not contaminated by changes in the minimum height requirements.

With large numbers of observations, the sample mean $\bar{x}'$, by virtue of the central limit theorem, has a normal distribution. So we can test the hypothesis: whether $\mu_1' > \mu_2'$ or $\mu_2' > \mu_1'$. For example, if $\bar{x}_1' > \bar{x}_2'$, and the difference $\bar{x}_1' - \bar{x}_2'$ is statistically significant at the given significance level, then we accept the hypothesis that $\mu_1' > \mu_2'$, from which one can also infer that $\mu_1 > \mu_2$. Because of the large number of observations in many of the samples, even minute changes in $\bar{x}'$ turn out to be statistically significant. For example, even a 0.26 cm change in $\bar{x}'$ between 1740 and 1750 for Hungarian recruits would be statistically significant at the 5 percent level, given that the standard deviation of the restricted sample is about 7 cm.

## DISCUSSION

Our aim has been to derive a statistical procedure that is valid for truncated samples that would be capable of providing a way of determining trends in the height of a population.[12] The mean of the sample restricted to the right of the largest of the minimum height requirements, which is the least contaminated part of the sample, was used as a statistic for the procedure.

The restricted sample can also be analyzed using ordinary least squares (OLS) regression in order to obtain the trend. The intercept of such a regression is obviously upwardly biased, but the coefficients of the dummy variables representing the decade of birth have the right sign, even if not the right magnitude.[13] In addition, the statistic, $\bar{x}'$, together with the theorem in the appendix, can also be used to derive new estimates of the mean height of the population that fluctuate less than the ones obtained with the QBE program.

The strategy is to use one of the QBE (or RSMLE) height estimates as the base for the other estimates. For example, for the Lower Austrian birth cohorts, the 1760s was used as a base (table 1). An estimate of the mean height of the population was obtained for this decade with the QBE program. The result turned out to be 165.6 cm (with a standard deviation of about 7 cm; see table 1). Subsequently, a normal distribution with the above mean and a standard deviation was truncated

**TABLE 1**
**Comparison of Different Ways of Estimating the Mean Adult Height of the Population in the Habsburg Monarchy by Place of Birth**

| Decade of birth | Hungary | | | | Galicia | | | | Lower Austria | | | |
|---|---|---|---|---|---|---|---|---|---|---|---|---|
| | $N$ | $\hat{x}$ | $\bar{x}'$ | $\hat{x}'$ | $N$ | $\hat{x}$ | $\bar{x}'$ | $\hat{x}'$ | $N$ | $\hat{x}$ | $\bar{x}'$ | $\hat{x}'$ |
| 1730 | 746 | 167.2 | 171.3 | 165.8 | | | | | | | | |
| 1740 | 2,620 | 169.8 | 172.1 | 167.8 | 150 | 170.8 | 172.0 | 167.0 | 199 | 169.1 | 170.9 | 165.0 |
| 1750 | 9,044 | 166.3 | 171.5 | 166.3 | 812 | 168.8 | 171.8 | 166.5 | 544 | 164.6 | 170.9 | 165.0 |
| 1760 | 6,589 | 166.5 | 171.1 | 165.2 | 903 | 166.0 | 171.6 | 166.0 | 782 | 165.6 | 171.1 | 165.6 |
| 1770 | 3,780 | 164.1 | 170.1 | 161.9 | 1,552 | 163.5 | 170.8 | 163.6 | 809 | 161.8 | 170.2 | 162.8 |
| 1780 | 5,841 | 163.9 | 170.6 | 163.7 | 1,334 | 163.3 | 170.0 | 160.6 | 1,126 | 163.6 | 169.6 | 160.4 |
| 1790 | 8,760 | 163.0 | 170.6 | 163.7 | 543 | 169.0 | 170.9 | 163.9 | 1,588 | 161.2 | 169.7 | 160.8 |

*Sources:* Kriegsarchiv, Vienna, Austria, Musterlisten, and Standestabellen.
*Note:* Adult recruits are between the ages of twenty-three and forty-five.
1. $N$ = number of observations in the sample.
2. $\hat{x}$ = estimated height using the QBE procedure.
3. $\bar{x}'$ = mean of the restricted sample.
4. $\hat{x}'$ = estimated height calculated using changes in $x'$ (see text).

at 165.8 cm, and the mean of the truncated normal distribution to the right of the truncation point was calculated. Next, in order to arrive at an estimate of the mean height of the population born in the 1740s, we noted that $\bar{x}'$, the mean of the truncated data, was 0.2 cm smaller in the 1740s than in the 1760s (table 1). The change 0.2 cm in $\bar{x}'$ can be regarded as the change in the mean of the truncated normal distribution, $\mu'$, because we have a large number of observations. Holding the truncation point constant, we calculated how much the true population mean height, $\mu$, would have had to change from 165.6 cm in order to obtain a 0.2 cm change in $\mu'$. The estimated mean obtained in this way for the 1740s was 165.0 cm, much more plausible than the 169.1 cm obtained using the QBE program. This procedure was repeated for the other decades of the 119
sample, not only for Lower Austrian soldiers, but also for the other provinces of the monarchy. The results confirm, in the main, the secular trend obtained with the QBE program without some of that program's extreme fluctuations. Thus, the estimated mean height of the Galician population born in the 1790s using the QBE program was 169.0 cm, whereas using the method just outlined, it was 163.9 cm, an intuitively much more plausible result.

This is the case because the estimated trend is less erratic than the one obtained by using the QBE procedure. Hence, our technique is a useful way to confirm the results of the QBE program as well as to overcome the uncertainties associated with some unstable estimates obtained with it. It is true that the verity of our technique is also predicated on the assumption that the population distribution is normal. Yet, as the above examples indicate, our proposed method based on the mean of the restricted sample produces more acceptable results than the results of the QBE program. Consequently, it is incumbent on researchers encountering difficulties with the QBE or the RSMLE routines to make an attempt at applying the technique proposed here in order to illuminate the problem at hand, that is, to ascertain the appropriate trends in historical heights.

## APPENDIX

**Theorem 1**

Suppose that a random variable $X$ has a normal distribution with a mean $\mu$ and a variance $\sigma^2$. Let $Y$ be the truncated random variable obtained by truncating $X$ from below. That is, the probability density function $f_Y(y)$ of $Y$ is given by:

$$f_Y(y) = \begin{cases} \dfrac{f_X(y)}{1 - F_X(y)} & \text{if } y \geq a \\ 0 & \text{if } y < a \end{cases}$$

where $a$ is the truncation point and $f_X$ and $F_X$ are the probability density function and the distribution function of $X$. Assume that $a$ and $\sigma^2$ are known. Then the mean of the distribution of the truncated random variable $Y$, $\mu'$, is an increasing function of the mean of the distribution of the original random variable $X$, $\mu$.

**Proof**

Let $\phi(x)$ and $\Phi(x)$ be the density function and the distribution function of the standard normal distribution respectively. That is,

$$\phi(x) = \frac{1}{\sqrt{2\pi}}e^{-x^2/2}, \quad \Phi(x) = \int_{-\infty}^{x} \phi(t)dt.$$

The mean of the truncated random variable $Y$ is given by:

$$\mu' = a + \sigma\left(\frac{\phi(t)}{1 - \Phi(t)} - t\right)$$

where $t = [a - \mu]/\sigma$. If $g(t) = h(t) - t$, where $h(t) = \phi(t)/[1 - \Phi(t)]$, is a decreasing function in $t$, then $\mu'$ is an increasing function of $\mu$. To show that $g(t)$ is decreasing in $t$, we need to show that the derivative of $g(t)$ is negative for all $t$.

The derivative of $g(t)$ is given by:

$$\frac{dg(t)}{dt} = \frac{dh(t)}{dt} - 1$$

$$= h^2(t) - th(t) - 1$$

$$= \left(h(t) - \frac{t - \sqrt{t^2 + 4}}{2}\right)\left(h(t) - \frac{t + \sqrt{t^2 + 4}}{2}\right).$$

Since $t - \sqrt{t^2 + 4} < 0$ and $h(t) > 0$ for all $t$, the first factor in the above expression is positive for all $t$. The second factor in the 120
derivative can be expressed as follows:

$$h(t) - \frac{t + \sqrt{t^2 + 4}}{2}$$

$$= \frac{e^{-t^2/2}}{\int_t^\infty e^{-x^2/2}} - \frac{t - \sqrt{t^2 + 4}}{2}$$

$$= -\frac{t + \sqrt{t^2 + 4}}{2\int_t^\infty e^{-x^2/2}dx}\left(\int_t^\infty e^{-x^2/2}dx - \frac{2e^{-t^2/2}}{t + \sqrt{t^2 + 4}}\right)$$

Since $t + \sqrt{t^2 + 4} > 0$ for all $t$, it is enough to show that $l(t) > 0$ for all $t$, where,

$$l(t) = \int_t^\infty e^{-x^2/2}dx - \frac{2e^{-t^2/2}}{t + \sqrt{t^2 + 4}}.$$

The derivative of $l(t)$ is given by:

$$\frac{dl(t)}{dt} = c(t)(t - \sqrt{t^2 + 4})$$

where

$$c(t) = \frac{2e^{-t^2/2}}{(t + \sqrt{t^2 + 4})^2\sqrt{t^2 + 4}}.$$

Now $c(t) > 0$ and $t - \sqrt{t^2 + 4} < 0$ for all $t$, so the derivative of $l(t)$ is negative for all $t$, and hence, $l(t)$ is monotonically decreasing in $t$. Therefore,

$$l(t) > \lim_{t \to \infty} l(t) = 0.$$

This completes the proof.

Remark: This theorem can be proved differently by using Iyengar's results about the Mills's ratio.[14]

## NOTES

We are grateful to Roderick Floud and Annabel Gregory for providing us with a copy of the QBE program. The computer simulations were done by Phil Sidel, Joe Salemi, and Albert Pollick, Jr., of the University of Pittsburgh's Social Science Computer Research Institute, whose help was greatly appreciated.

1. Floud, R., K. W. Wachter, and A. Gregory. 1990. *Height, health and history*. Cambridge, England: Cambridge University Press; Fogel, R. W., S. L. Egerman, and J. Trussell. 1982. Exploring the uses of data on height: The analysis of long term trends in nutrition, labor welfare and labor productivity. *Social Science History* 6:401–21; Komlos, J. 1989. *Nutrition and economic development in the Habsburg monarchy: An anthropometric history*. Princeton: Princeton University Press.
2. Schneider, H. 1986. *Truncated and censored samples from normal populations*. New York: Marcel Dekker; Trussell, J., and K. W. Wachter. 1984. Estimating the covariates of historical heights. National Bureau of Economic Research, Working Paper No. 1455; Wachter, K. W., and J. Trussell. 1982. Estimating historical heights. *Journal of the American Statistical Association* 77:279–303; Wachter, K. W. 1981. Graphical estimation of military heights. *Historical Methods* 14:31–42.
3. Sandberg, L. G., and R. H. Steckel. 1988. Overpopulation and malnutrition rediscovered: Hard times in nineteenth century Sweden. *Explorations in Economic History* 25:1–19, figure 1.
4. Tanner, J. 1981. *A history of the study of human growth*. Cambridge, England: Cambridge University Press.
5. Sandberg, L. G., and R. H. Steckel. 1987. Heights and economic history: The Swedish case. *Annals of Human Biology* 14:101–10, table 1.

6. Floud, R., and K. W. Wachter. 1982. Poverty and physical stature, evidence on the standard of living of London boys 1770–1870. *Social Science History* 6:422–52; Floud, Wachter, and Gregory. *Health, health and history*. Estimated changes in height are unreasonable if they are either larger or more erratic than those recorded in reliable nontrucated samples.
7. Original data are given in Austrian inches (1 inch = 2.63 cm).
8. Of course, this strategy assumes that $a$ is known, which is almost always the case. To be sure, the degree to which it was enforced varied, but that would not have affected the distribution to the right of $a$. It is true that an advantage of the QBE program is that it works even if $a$ is unknown. Yet in that case, its use is risky since if $a$ were greater than the true population mean, the QBE procedure would give upwardly biased estimates.
9. Komlos, J. *Nutrition and economic development*.
10. We also tried excluding from the sample all observations below the known minimum height requirement, in other words, truncating the sample perfectly, before running the QBE program. That alternative did not make much difference in the estimates obtained.
11. The relationship between $\mu$ and $\mu'$ was first explored through computer simulations using perfectly normal distributions. A known perfectly normal distribution was truncated and the mean of the truncated distribution was calculated. Next, the whole known distribution was shifted, holding the truncation point and the standard deviation of the normal distribution constant. The new $\mu'$ was calculated and compared with its initial value. The calculation was repeated for larger and larger shifts in the mean of the true distribution. Then the whole procedure was repeated with a slightly different truncation point. The results were invariably the same: the mean of the truncated distribution moved in the same direction as the mean of the true normal distribution.
12. The method can be adopted to compare the mean height of two different populations by using the largest of the two minimum height requirements for the cut-off point $a$.
13. One might also use a linear or quadratic time trend instead of the dummy variables for decade of birth.
14. Iyengar, S. 1986. On a lower bound for the multivariate normal Mills's ratio. *The Annals of Probability* 14:1399–1403.

NB. This article has been indexed under its original pagination, as indicated by the page references in the margin.

# IV

**The Nutritional Status of French Students** The effort of anthropometric historians to unearth the broad patterns of human biological well-being is now too well-known to need detailed reiteration. Less acknowledged, however, is that French historians in the *Annales* tradition were among the first to adopt methods from physical anthropology and from the biological sciences to illuminate historical issues. Until then, the topic of human height was of interest primarily to scholars in sister disciplines.

The real expansion of the field dates from Steckel's and Fogel's exploratory essays, which became the launching manifesto of the discipline on American soil. Now, hundreds of thousands of records, from nearly all continents of the globe, have been examined. All of this research notwithstanding, many important issues remain open. One such question is related to the most amazing discovery to date: that as part of adaptability to socio-economic environments, human physical stature has experienced cycles during the course of the last quarter of a millennium.[1]

John Komlos is Professor and Chair of Economic History, University of Munich. He is the author of *Nutrition and Economic Development in the Eighteenth-Century Habsburg Monarchy: An Anthropometric History* (Princeton, 1989).

I have benefited greatly from conversations with Katherine Lynch, and from comments and suggestions provided by Maurice Aymard, Rondo Cameron, François Crouzet, Paul Hohenberg, Patrick O'Brien, David Weir, Eugene White, and E. Anthony Wrigley, without implicating them in whatever shortcomings might remain. The author thanks the American Philosophical Society for financial support of the archival work, and Claudine Billoux of the archive of the École Polytechnique, Palaiseau, for help in collecting data.



1 Robert W. Fogel, "Nutrition and the Decline in Mortality since 1700: Some Preliminary Findings," in Stanley Engerman and Robert Gallman (eds.), *Long-Term Factors in American Economic Growth* (Chicago, 1986), 439–555; Emmanuel Le Roy Ladurie, Nicole Bernageau, and Y. Pasquet, "Le conscrit et l'ordinateur: Perspectives de recherches sur les archives militaires du XIX$^{e}$ siècle français," *Studi Storici*, X (1969), 260–308; Jacques Houdaille, "La taille des Français au début du XIXe siècle," *Population*, XXV (1970), 1297–1298; Le Roy Ladurie and Bernageau, "Étude sur un Contingent Militaire (1868): Mobilité géographique, délinquance et stature, mises en rapport avec d'autres aspects de la situation des conscrits," *Annales de Démographie Historique*, VII (1971), 311–336; Jean-Paul Aron, Paul Dumont, and Le Roy Ladurie, *Anthropologie du Conscrit Français* (Paris, 1972); M. Demonet, Dumont, and Le Roy Ladurie, "Anthropologie du conscrit: une cartographie

In Europe, a widespread downturn in height has been identified among military men, drawn mostly from the lower classes of Sweden, Great Britain, and the Habsburg monarchy, beginning in the second half of the eighteenth century. On American soil, a similar major downturn in stature started with the birth cohorts of the 1830s. The biological standard of living did not keep pace, on average, with the overall growth in per capita income during the decades that comprised the transition to the industrial age. The scattered and limited evidence that we have on the secular trend in the nutritional status of the upper and middle classes in Britain, Austria, and Germany suggests that they, in contrast to the "common man," did not fare badly at all during these turbulent times. This study is an effort to expand our knowledge of middle-class nutritional experience to France by examining the physical stature of students at the École Polytechnique.[2]

cantonale (1819–1830)," *Annales* (1976), 700–760; Houdaille, "La taille des Parisiens en 1793," *Population*, XXXVIII (1983), 173–177; *idem*, "Les femmes dans les prisons parisiennes sous la Révolution," *Population*, XLII (1987), 369–388; Michiel Alexander van Meerten, "Developpement économique et stature en France, XIX$^{e}$–XX$^{e}$ siècle," *Annales*, (1990), 755–781; James M. Tanner, *A History of the Study of Human Growth* (Cambridge, 1981); Marie-Claude Chamla, "L'Accroissement de la stature en France, de 1880 à 1960," *Bulletin de la Société d'Anthropologie de Paris*, XI (1964), 201–278; Richard H. Steckel, "Slave Height Profiles from Coastwise Manifests," *Explorations in Economic History*, XVI (1979), 363–380; Komlos, *Nutrition and Economic Development in the Eighteenth-Century Habsburg Monarchy: An Anthropometric History* (Princeton, 1989); Roderick Floud, Kenneth Wachter, and Annabel Gregory, *Height, Health, and History: Nutritional Status in the United Kingdom, 1750–1980* (Cambridge, 1990).

2 Komlos, "The Secular Trend in the Nutritional Status of the Population of the United Kingdom, 1730–1860," *Economic History Review*, XLVI (1993), 115–144; *idem*, "A Malthusian Crisis Revisited: The Stature of Runaway Indentured Servants in Colonial America," *Economic History Review*, XLVI (1993), forthcoming; Lars G. Sandberg and Steckel, "Heights and Economic History: The Swedish Case," *Annals of Human Biology*, XIV (1987), 101–110; Fogel, "Nutrition and the Decline in Mortality since 1700," 442; Komlos, "The Height and Weight of West Point Cadets: Dietary Change in Antebellum America," *The Journal of Economic History*, XLVII (1987), 897–927; *idem*, "Toward an Anthropometric History of African-Americans: The Case of the Free Blacks of Antebellum Maryland," in Claudia Goldin and Hugh Rockoff (eds.), *Strategic Factors in Nineteenth-Century American Economic History: A Volume to Honor Robert W. Fogel*, (Chicago, 1992), 297–329; *idem*, "Height and Social Status in Eighteenth-Century Germany," *Journal of Interdisciplinary History*, XX (1990), 607–621; *idem*, "Patterns of Children's Growth in East-Central Europe in the Eighteenth Century," *Annals of Human Biology*, XIII (1986), 33–48; Floud, Wachter and Gregory, *Height, Health, and History*, 177. Other studies of physical stature in France, based primarily on military data, include L. R. Villermé, "Mèmoire sur la Taille de l'homme en France," *Annales d'hygiène publique*, I (1829), 551–559; *idem*, "De la stature et du poids de l'homme," *Annales d'Hygiène Publique et de Médicine Légale*, X (1833), 27–35; A. d'Angeville, *Essai sur la Statistique de la population Française: Considérée sous quelques-uns*

DATA The number of records obtained for the period from 1794 to the incoming class of 1887 is 13,195. Geographically, the north and northeast of France were heavily represented in the student body. The departments of Meurthe, Moselle, Bas Rhin, and Nord each accounted for almost 3 percent, and the Paris region provided 17 percent of the total enrollment during the period under consideration. Slightly more than half of the students came from municipalities with at least 5,000 inhabitants (Table 1).[3]

The height distributions were examined by place of birth, age, and decade of birth for deviations from normality (they are not reported here). Truncation was not observed, even though the distributions were often not quite Gaussian. There was no minimum height requirement, and some students as short as 151 cm, gained acceptance. Many students were said to be poor, but the occupational distribution of the parents leaves no doubt that the overwhelming majority of the students had upper-class backgrounds (Table 1). The entrance examination required one or two years of preparation after completion of high school (lycée), which itself was not free. The attainment of these stringent standards, particularly in mathematics, required considerable financial resources; as a consequence, poor boys were de facto excluded from the school, except for a few on scholarship.[4]

RESULTS The high social status of the students at the École meant that they were decidedly taller than their countrymen as a whole: they were 4 cm taller than the conscripts of 1819–1826. Among the young men called up for military service in 1868,

*de ses rapports physiques et moraux* (Paris, 1969). These data were studied to a limited extent prior to the computer age. Jean Sutter, René Izac, and Tran Ngoc Toan, "L'Évolution de la Taille des Polytechniciens," *Population,* (1958), 373–406. This study was limited to univariate analysis. The authors had the advantage, however, of having gained access to the archives up to the incoming class of 1954, whereas the 100-year rule was applied to my project.

3 Archive, École Polytechnique, Palaiseau, France. Because of the 100-year rule in effect in France, I did not gain access to more recent records even though other researchers were able to do so. The study is based on the complete set of records on all students up to 1887. Administration Générale et Départementale, *Dénombrement quinquennal de la Population du Royaume, 1846* (Paris, 1847). Population growth rates are from Charles H. Pouthas, *La Population Française Pendant La Première Moitié du XIX^e^ Siècle* (Paris, 1956), 32, 70. See also René Le Mée, "Population Agglomérée, Population Éparse au début du XIX^e^ Siècle," *Annales de Démographie Historique,* VII (1971), 455–513.

4 Sutter, Izac, and Toan, "L'Évolution de la Taille des Polytechniciens," 377.

*Table 1* Characteristics of the Whole Sample

DISTRIBUTION OF OBSERVATIONS

| BY BIRTH DECADE | | BY PARENT'S PROFESSION | | BY AGE | |
|---|---|---|---|---|---|
| 1770[a] | 418 | Unknown | 1992 | 13[b] | 7 |
| 1780 | 808 | Upper Class | 9223 | 14 | 13 |
| 1790 | 1,315 | Middle Class | 1338 | 15 | 38 |
| 1800 | 1,049 | Agricultural | 211 | 16 | 220 |
| 1810 | 1,434 | Worker | 431 | 17 | 942 |
| 1820 | 1,472 | | | 18 | 2552 |
| 1830 | 1,181 | | | 19 | 4492 |
| 1840 | 1,417 | | | 20 | 3971 |
| 1850 | 2,285 | | | 21 | 613 |
| 1860 | 1,816 | | | 22 | 155 |
| Total | 13,195 | | | 23 | 99 |
| | | | | 24 | 61 |
| | | | | 25[c] | 23 |
| | | | | Missing | 10 |

| HEIGHT BY AGE | | BY SIZE OF MUNICIPALITY OF RESIDENCE[e] | |
|---|---|---|---|
| 14 | 160.8 | Less than 1,000 | 3,954 |
| 15 | 164.5 | 1–5,000 | 1,794 |
| 16 | 167.2 | 5–10,000 | 1,343 |
| 17 | 169.0 | 10–25,000 | 1,611 |
| 18 | 169.8 | 25–50,000 | 1,004 |
| 19 | 169.8 | 50–100,000 | 606 |
| 20 | 169.7 | 100–Above | 444 |
| 21 | 169.2 | Paris | 1,899 |
| 22 | 169.5 | Missing | 540 |
| 23[d] | 169.1 | | |

[a] Includes two observations from the 1760s.
[b] Age thirteen and below.
[c] Between age twenty-five and above.
[d] Between age twenty-three and twenty-five.
[e] At time of birth.

merely 20 percent reached the average height of the Polytechnique students. A century later, the latter were even taller than the student population at large.[5]

Although they were tall, a general declining tendency in physical stature is evident (Fig. 1). Of the students born in Paris,

5 Aron, Dumont, and Le Roy Ladurie, *Anthropologie du Conscrit Français*, 92, 397, 398. But the Polytechnique students were 8 cm taller than the conscripts of 1815–19 according

*Fig. 1* The Physical Stature of the Students at the École Polytechnique, 1770–1869

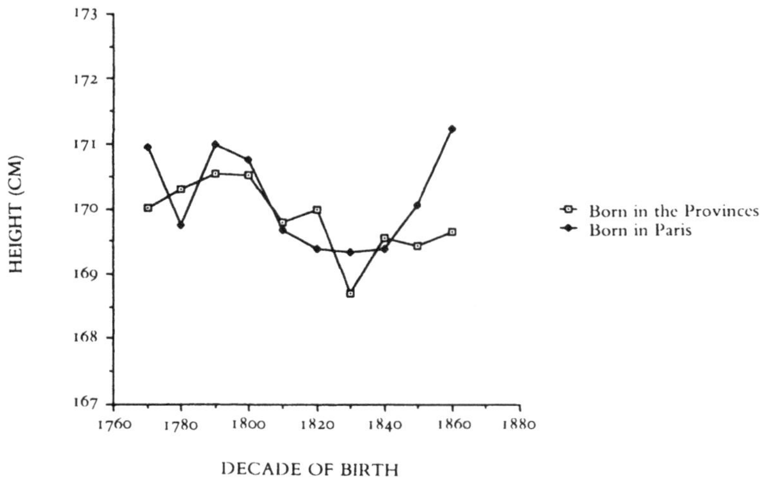


a downward trend in height began with the birth cohorts of the 1810s and remained at the lower level until the 1840s; the recovery started in the 1850s, and a large increase in height was experienced among those born in the 1860s (Table 2). The trend was also examined using the "reduced-sample" method—eliminating students who were shorter than 161 cm. This approach tests for possible discrimination against very short students that might not be evident in casual observations of the frequency distributions. The reduced-sample trend correlates well with the trend obtained from the whole sample (Table 2). In fact, only in the decades of the 1820s and 1830s do the two trends diverge from one another

to d'Hargenvilliers, *Recherches et considerations sur la formation et le recrutement de l'Armeé en France* (Paris, 1817), 1, as cited in Sutter, Izac, and Toan, "L'Évolution de la Taille des Polytechniciens," 399. Le Roy Ladurie and Bernageau, "Étude sur un Contingent Militaire," 335. Compared to the regular soldiers drafted in the 1820s, the students at the Polytechnique (born prior to 1809) were 4.3 cm taller (standard deviation =1.9 cm). The comparison refers to unweighted averages at the departmental level. The Polytechniciens were even taller than students at the Sorbonne (Could their social status also have been higher?).

*Table 2* Trend of Physical Stature by Decade of Birth

| | PLACE OF BIRTH | | | | |
|---|---|---|---|---|---|
| | WHOLE SAMPLE | | | REDUCED SAMPLE[a] | |
| BIRTH DECADE | PARIS | PROVINCES | ALL | PARIS | PROVINCES |
| 1770 | 100.7 | 100.1 | 100.2 | 100.7 | 100.1 |
| 1780 | 100.0 | 100.3 | 100.2 | 100.2 | 100.4 |
| 1790 | 100.7 | 100.4 | 100.5 | 100.6 | 100.4 |
| 1800 | 100.6 | 100.4 | 100.5 | 100.4 | 100.4 |
| 1810 | 100.0 | 100.0 | 100.0 | 100.0 | 99.9 |
| 1820 | 99.8 | 100.1 | 100.1 | 100.1 | 100.0 |
| 1830 | 99.7 | 99.4 | 99.4 | 99.7 | 99.7 |
| 1840 | 99.8 | 99.9 | 99.8 | 99.9 | 99.9 |
| 1850 | 100.2 | 99.8 | 99.8 | 100.0 | 99.8 |
| 1860 | 100.9 | 99.9 | 100.1 | 100.8 | 99.8 |

NOTE Index = 100 for the whole period;

[a]The reduced sample method excludes students shorter than 161 cm. See John Komlos and Joo Han Kim, "Estimating Trends in Historical Heights," *Historical Methods*, XXIII (1990), 116–120.

in any meaningful way. This difference is of no consequence to the overall pattern.[6]

The above trend was standardized for the age-composition of the students. Regression analysis was used to ascertain the trend, holding also the socioeconomic composition of the student body constant. Again, we analyzed both the whole sample and the reduced sample. The results remained unchanged: a marked downturn began among the birth cohorts of the 1810s. There are some notable differences with this method: the amplitude of the cycle is nearly twice as large as the one reported in Figure 1, the trough of the cycle is reached in the 1820s, and a gradual recovery is evident immediately thereafter (Table 3).[7]

The same approach was adopted for the analysis of heights of the students born in the provinces, with broadly similar results:

6 For a discussion of this methodology, see Komlos and Joo Han Kim, "Estimating Trends in Historical Heights," *Historical Methods*, XXIII (1990), 116–120. Students as short as 150 cm attended the school.

7 The trend, not standardized for age, reported in Sutter, Izac, and Toan, "L'Évolution de la Taille des Polytechniciens," 380, 393, is not substantially different from the one described above, even though the decline in height of the first half of the nineteenth century was not noticed by them. The reduced sample regression is not reported here.

*Table 3* Regression Analysis: Dependent Variable Height (cm)

| | BIRTHPLACE | |
|---|---|---|
| | PARIS | PROVINCES |
| Constant | 169.8[a] | 170.6[a] |
| Birth Decade | | |
| 1780[b] | 1.3 | 0.8[a] |
| 1790 | 1.9[a] | 1.1[a] |
| 1800 | 1.6[a] | 0.9[a] |
| 1810 | 0.3 | 0.1 |
| 1820 | −1.0 | 0.3 |
| 1830 | −0.1 | −1.0[a] |
| 1850 | 0.6 | −0.3 |
| 1860 | 1.9[a] | 0.0 |
| Age | | |
| 14 | −15.6[a] | −7.1[a] |
| 15 | −8.3[a] | −4.9[a] |
| 16 | −5.3[a] | −2.7[a] |
| 17 | −1.8[a] | −0.9[a] |
| 19[c] | −0.3 | 0.1 |
| Occupation | | |
| Unknown | −0.1 | −0.3[a] |
| Middle Class | −0.7 | −1.1[a] |
| Menial | — | −0.5 |
| Town | | |
| Ln (size) | | 0.0 |
| $R^2$ | 0.03 | 0.04 |
| F | 4.9[a] | 4.6[a] |
| N | 1,838 | 11,214 |

[a]Significant at 5 percent level.

NOTE Constant refers to eighteen-year-old, upper-class youth born in the 1840s in (1) Paris or (2) Bas Rhin. The coefficients of the dummy variables of the departments are not reported here for the second regression. They are evident in Figure 3.

[b]Includes those born in the 1770s.

[c]Includes those between the ages of nineteen and twenty-five.

the cycle is evident, and the downward trend also begins in the 1810s even when the spatial composition of the sample is kept constant (Fig. 1). In addition, the trends are nearly the same for the whole sample as for the reduced sample (Table 2). The major difference between the Parisian and provincial students is that, in contrast to the upward trend in physical stature among the former in the 1850s and 1860s, the height of the provincial students failed to rebound at all during the period under examination. The range in the variation in height was around 3 cm among the Parisian, and 2 cm among the provincial students.

The pattern of height among the provincial students was explored further by the introduction of a variable for town size in two ways: using the logarithm of town population as an independent variable, and using dummy variables in six size categories (Table 3). Systematic variations were not found with either approach. This result is unusual because, in all samples studied so far, urban men were invariably shorter than rural ones at least until improvements in urban sanitation in the late-nineteenth century.[8]

The occupation of the fathers of the students was found to be a useful explanatory variable. Upper-class students were taller than the rest by between 0.5 and 1.1 cm (Table 3). The sons of artisans and small shopkeepers were the shortest, whereas the sons of military officers were the tallest, both in the capital and in the provinces (Fig. 2).

Students from the north were taller than the ones from the south; those from the northeast tended to be the tallest (Fig. 3). However, a convergence is apparent in the course of the first half of the century. The decline in height tended to be greatest among the students who were born in departments that initially had the tallest populations, and a handful of the shortest departments even increased in height (Fig. 4). The transition matrix shows that about a third of the departments stayed in the same quintile at mid-century as the one they were in in the early nineteenth century, nearly a third moved up a quintile in rankings and about a third moved down (Table 4).[9]

8 The latter are not reported here.

9 The downturn was evident even when the regional composition of the sample was held constant (Table 3). Hence, the downturn was not caused by changes in the regional composition in favor of students from the south.

*Fig. 2* Relative Height by Occupation of Father

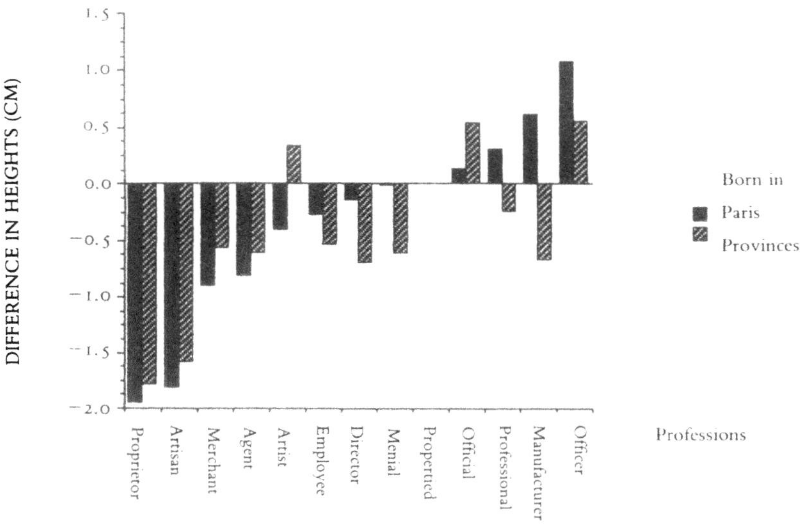


The analysis of the secular trend on a regional basis indicated that the downswing was confined to the northern two-thirds of the country (Fig. 5). It was hardly evident at all on the Mediterranean littoral (Zone 9) or in the narrow zone of departments extending from Isère to Gironde (Zone 8). Yet, even in these two zones there was a reduction in stature among the birth cohorts of the 1830s. Zones 5, 6, and 7 follow the overall provincial pattern. Zones 3 and 4, east and southwest of Paris, deviated from the overall pattern in that a negative trend did not appear until the 1830s. When the trend did appear, it became quite pronounced. In Zone 2, encompassing Brittany, one of the poorer parts of France, the decline began a decade earlier than average (in the 1800s), and the recovery was evident during the final decades of our period. Zone 1 is similar to Zone 2 in that the recovery of the 1850s and 1860s has more similarity to Paris than to the provincial average. In sum, nutritional status declined everywhere in the 1830s. The decline began in the 1810s in the central belt extending from Savoy to Vendée, earlier still in Brittany, but later

*Fig. 3* Spatial Distribution of Heights in France, 1770–1869

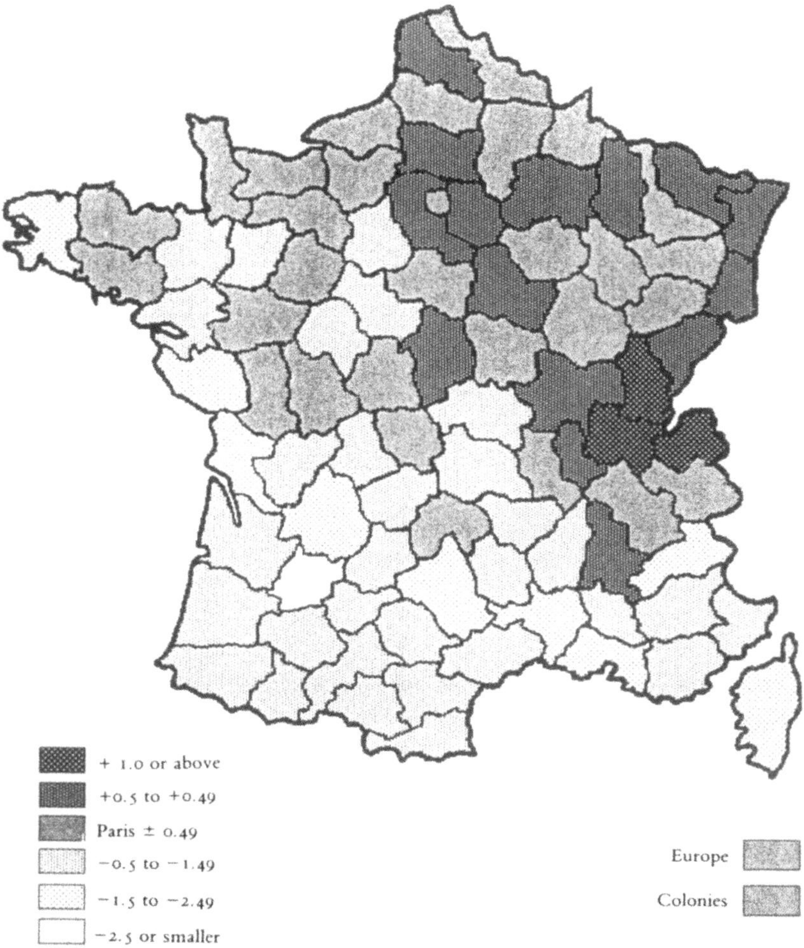


in the regions around Paris. In the south, a diminution in height was not evident except for the birth cohorts of the 1830s.[10]

International comparisons of upper-class youth indicate that

10 These trends are standardized for varying shares of the sample within a region, originating from different departments.

*Fig. 4* Change in Height Between 1810–69 and 1770–1809 By Provinces

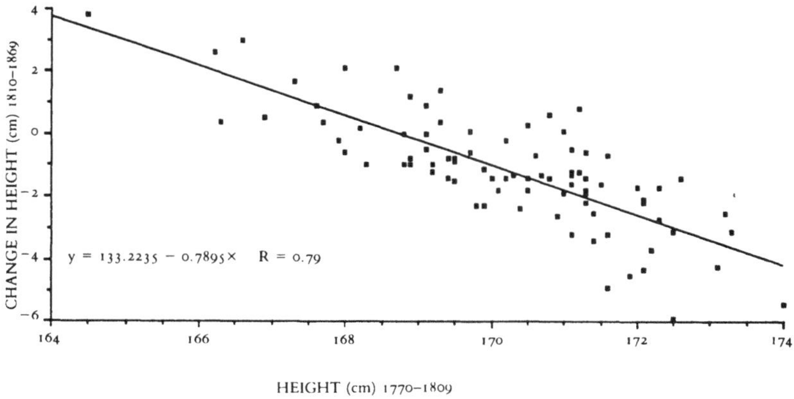


the English were the tallest in the nineteenth century, Americans were second, followed by the French, and then the Germans (Table 5). There is no evidence of a downturn in nutritional status among the English elite; on the contrary, the students attending the Sandhurst military academy were gaining in height during the period under consideration. They were the only ones to do so: all other evidence so far examined has indicated that the decline in physical stature was a widespread phenomenon of the nineteenth century in both Europe and America. The unique aspect of the French pattern is its early start. The American downturn, as well as the downturn among British soldiers, began in the late 1830s and, among the Swedish soldiers, it began in the 1840s. In contrast, the French decline began two decades earlier among the birth cohorts of the 1810s, and its amplitude was 1–2 cm greater than that of the diminution in the stature of the West Point cadets (Table 2).[11]

11 Perhaps the Sandhurst cadets were taller because they came from a higher social stratum than the Polytechniciens. Perhaps their superiority was due to truncation bias. Floud, Wachter, and Gregory suggest that their height "distributions are very close to normal," but they did not present the actual evidence, *Height, Health and History*, 174, 177, 149. Komlos, "Toward an Anthropometric History of African-Americans"; *idem*, "The Height and Weight of West Point Cadets," 901; Sandberg and Steckel, "Heights and Economic History: The Swedish Case."

*Table 4* Transition Matrix of Rank Order of Heights by Provinces, Quintiles

NUMBER OF OBSERVATIONS

| RANK | | 1810–1869 | | | | |
|---|---|---|---|---|---|---|
| | | 1 | 2 | 3 | 4 | 5 |
| 1770–1809 | 1/ | 7 | 3 | 2 | 2 | 4 |
| | 2/ | 5 | 7 | 4 | 1 | 1 |
| | 3/ | 2 | 4 | 5 | 4 | 2 |
| | 4/ | 2 | 3 | 3 | 6 | 3 |
| | 5/ | 2 | 1 | 3 | 4 | 7 |

PERCENTAGES

| RANK | | 1810–1869 | | | | |
|---|---|---|---|---|---|---|
| | | 1 | 2 | 3 | 4 | 5 |
| 1770–1809 | 1/ | 39 | 17 | 11 | 11 | 22 |
| | 2/ | 28 | 39 | 22 | 6 | 6 |
| | 3/ | 12 | 24 | 29 | 24 | 12 |
| | 4/ | 12 | 18 | 28 | 35 | 18 |
| | 5/ | 12 | 6 | 18 | 24 | 41 |

SUMMARY OF CHANGE

| QUINTILES | MOVED | | |
|---|---|---|---|
| | UP | | DOWN |
| 0 | | 32 | |
| 1 | 16 | | 14 |
| 2 | 8 | | 5 |
| 3 | 3 | | 3 |
| 4 | 2 | | 4 |
| Total | 29 | | 26 |

The students at the École were tall by the standards of the time. This finding is consistent with the well-documented strong positive correlation between socioeconomic status and physical stature. The convergence in height at the provincial level has also been found to be a general concomitant of the expansion and integration of markets. The spatial distribution of heights of the French elite is broadly familiar: like the general population, students from the economically more developed north, with a more

*Fig. 5* Regional Trends in Physical Stature

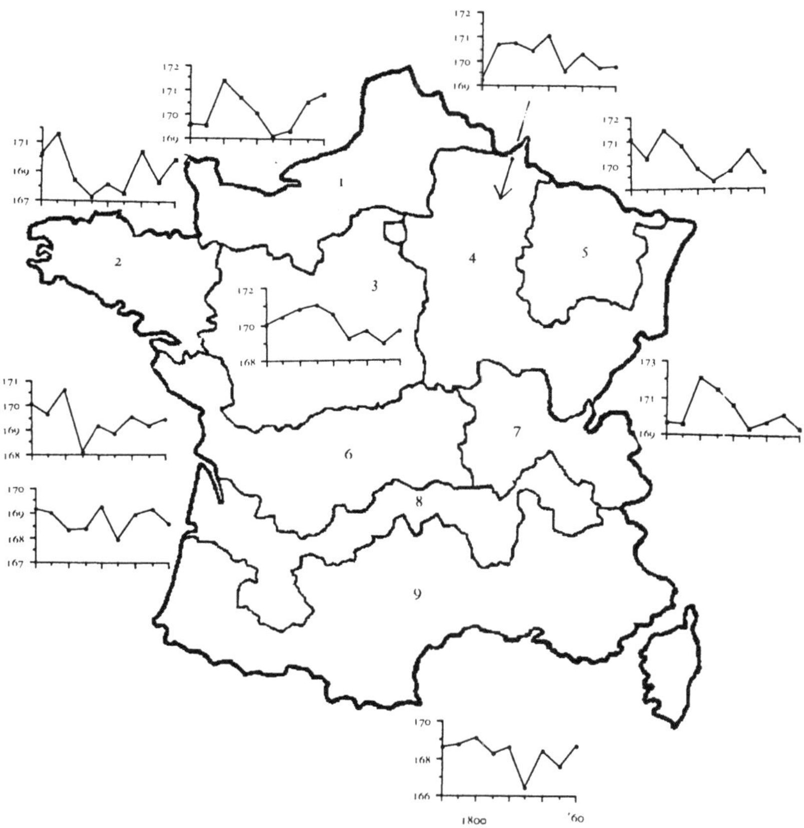


literate population, also tended to be taller than students from the south. In the north, wheat cultivation was more prevalent, the productivity of the soil was higher, and cereal prices were lower than in the southern part of the country. One would expect these factors to have influenced nutritional status positively.[12]

12 Komlos, "Height and Social Status in Eighteenth-Century Germany"; *idem, Nutrition and Economic Development in the Eighteenth-Century,* 97; Sutter, Izac, and Toan, "L'Évolution de la Taille des Polytechniciens," 394; Aron, Dumont, and Le Roy Ladurie, *Anthropologie du Conscrit Français,* 92; Alain Bideau, "Les Structures Anthropologiques," in Jacques

*Table 5* Height of Students at Elite Schools, International Comparisons

| | BORN BETWEEN 1770 AND 1789 | | | |
|---|---|---|---|---|
| | MIDDLE CLASS | | UPPER CLASS | |
| AGE | FRANCE | GERMANY | FRANCE | GERMANY |
| 17 | 168.1 | 164.8 | 168.8 | 164.6 |
| 18 | 171.4 | 166.6 | 170.6 | 168.5 |
| 19 | 170.3 | 167.0 | | |

| | BORN BETWEEN 1840 AND 1869 | | |
|---|---|---|---|
| AGE | FRANCE | ENGLAND | U.S. |
| 17 | 169.3 | 171.5 | 169.4 |
| 18 | 169.4 | 172.6 | 170.5 |
| 19 | 169.6 | 173.4 | 171.0 |

SOURCE See Table 1; John Komlos, "The Height and Weight of West Point Cadets: Dietary Change in Antebellum America," *Journal of Economic History*, XLVII (1987), 901; *idem*, "Height and Social Status in Eighteenth-Century Germany," *Journal of Interdisciplinary History*, XX (1990), 613; Roderick Floud, Kenneth Wachter, and Annabel Gregory, *Height, Health, and History: Nutritional Status in the United Kingdom, 1750–1980* (New York, 1990), 177.

The nineteenth-century cycle in physical stature among the students conforms, but also deviates to some extent, to the ones found in other samples. In the north, the downturn in the cycle began at about the same time as in Great Britain and the United States. In Brittany and the center of the country, two of the poorer areas of France, as well as in Paris, it began a decade or two earlier. In the Mediterranean basin and immediately north of it, the downturn was confined to the decade of the 1830s. The average decline in height at the departmental level was 1.2 cm.

The decline in nutritional status of the students at the beginning of the nineteenth century does not correlate well with the trend in the crude death rate (CDR), inasmuch as the latter was improving rapidly during the first two decades of the century,

Dupaquier (ed.), *Histoire de la Population Française: De 1789 à 1914* (Paris, 1988), III, 268–278; François Furet and Jacques Ozouf, *Lire et Écrire: L'alpahétisation des français de Calvin à Jules Ferry* (Paris, 1977), Figure 4; *Atlas Historique de la France Contemporaine, 1800–1965* (Paris, 1966), 55–57.

just as heights were starting to diminish. After the 1820s, however, the correspondence between the two variables is somewhat better. On the whole, there was little or no improvement in the CDR between 1820 and 1890. In fact, there were some reversals as well, such as during the cholera epidemic of the early 1830s, and during the decades of the 1850s and 1860s when the infant mortality rate rose. Thus, the decline in nutritional status, indicated by the trends in the physical stature of the students, is at least not as incongruent with the mortality trend as it would have been had the CDR continued to improve after the 1820s.[13]

At this stage of the research, it is unclear why physical stature would have been increasing among those born just before or at the turn of the nineteenth century, when wars raged, and harvest failures abounded. It is also puzzling that such prolonged and marked nutritional stress was experienced by the French elite early in the nineteenth century. The 1810s were not the best of times: war continued, and, in addition, the birth cohorts of that decade would have been subject to the major harvest failure of 1816. Yet, it would seem that, as in England, the upper classes in France were sufficiently wealthy to be able to isolate themselves from such transient phenomenon, as well as from the unhealthy environmental conditions of early industrialization. Food prices increased between 1826 and 1833 and again in the 1840s, but the elite certainly had enough food to eat. It is anomalous that physical stature among men of military age, per capita gross national product, and per capita calorie intake are all estimated to have increased during the first half of the century. Real wages declined after the 1820s and stagnated for a long period thereafter: only after 1880 did real purchasing power increase. Only with the increase in real wages did meat consumption increase among the general population.

Is it possible that the offspring of the social elite, attending the prestigious École Polytechnique, would have been affected by the rising cost of living that was experienced by workers? Or, is it more likely that the social composition of the students became

13 Jacques Vallin, "Mortality in Europe from 1720 to 1914: Long-Term Trends and Changes in Patterns by Age and Sex," in Roger Schofield, David Reher, and Alain Bideau (eds.), *The Decline of Mortality in Europe* (Oxford, 1991), 40, 50; E. Anthony Wrigley, "The Fall of Marital Fertility in Nineteenth-Century France: Exemplar or Exception? (Part I)," *European Journal of Population*, I (1985), 36.

increasingly heterogeneous after the beginning of the July Monarchy with its "citizen king"? Could infant feeding practices have changed in elite families, such as the age of weaning, or the use of wet nurses? Or, did the incidence of childhood diseases become more severe? Did spatial mobility increase, so that the differences in nutritional status between town and country became blurred? These hypotheses are worth considering in future research.[14]

Why did the French lead the nineteenth-century reversal in physical stature? Urbanization and industrialization were not proceeding at a faster rate, and population pressure was not more pronounced than in England. Although many questions remain, it is evident—and one of the noteworthy contributions of anthropometric history—that the marked improvements in the standard of living (as conventionally conceived) were not generally translated into an improvement in the biological standard of living until the end of the nineteenth century.

14 John D. Post, *The Last Great Subsistence Crisis in the Western World* (Baltimore, 1977); Hugh Clout, *Agriculture in France on the Eve of the Railway Age* (Totowa, N.J., 1980), 42; bread riots returned to France during the harvest failures of the 1840s. George Grantham, "Scale and Organization in French Farming, 1840–1880," in William N. Parker and Eric L. Jones (eds.), *European Peasants and Their Markets: Essays in Agrarian Economic History* (Princeton, 1975), 310; David R. Weir, "Parental Consumption Decisions and Child Health during the Early French Fertility Decline, 1790–1914," *Journal of Economic History,* LIII (1993), 259–274; Maurice Aymard, "Toward the History of Nutrition: Some Methodological Remarks," in Robert Forster and Orest Ranum (eds.), *Food and Drink in History: Selections from the Annales* (Baltimore, 1979), 2; Maurice Levy-Leboyer and François Bourguignon, *l'Économie Française au XIX^e siècle: Analyse macro-économique* (Paris, 1985), 19; J. Toutain, "La consommation alimentaire en France de 1789 à 1964," *Économies et Sociétés,* V (1971), 1909–2049.

# V

# *A Malthusian episode revisited: the height of British and Irish servants in colonial America*[1]

Recent research into the secular changes in the nutritional status of various European populations has demonstrated with remarkable consistency the beginning of a critical phase during the second half of the eighteenth century. The demographic expansion brought with it the ancient threat of food shortages, widespread hunger, and occasionally even famine. That the population would actually be able to break out of the partial homeostatic equilibrium that had prevailed since time immemorial was not at all clear. During the last similar episode of rapid growth in the sixteenth century it had not yet been able to do so.[2] However, this time was to be different: a general subsistence crisis of major proportions was ultimately avoided.[3] Much progress had been made during the previous century and a half in practically all branches of the economy, including transport, agriculture, and technology.[4] New sources of food imports had become available, and previously unknown foods had been introduced from the New World. Merchants were wealthier, and governments were better organized. All these factors, and others too, helped Europeans break through the Malthusian ceiling. Population growth, and therefore economic activity, could continue, freed from the limitations imposed until then by the supply of nutrients.[5] The industrial revolution was under way.[6]

Thus, in the eighteenth century the age-old contest between population expansion and food supply was ultimately decided in the latter's favour. Yet malnutrition became widespread, even endemic, during the course of the century. As a consequence, the lower classes suffered a diminution in their nutritional status. The notion of a general decline in food consumption

[1] I am deeply indebted to Peter Coclanis for bringing the availability of these data to my attention, to Paul Mageli for helping to collect them, to Philip Sidel of the University of Pittsburgh's Social Science Computer Research Institute for assistance in processing them, to Tim Cuff for help in verification and computation, and to the University of Pittsburgh's Office of Research for its financial support. The work on the project was begun while the author was at the University of North Carolina's Carolina Population Center at Chapel Hill, supported by an N.I.C.H.D. National Research Award No. 2732HD07168 from the Center for Population Research.

[2] For the agrarian cycles of the early modern period see, for instance, Le Roy Ladurie, *Peasants of Languedoc*; on demographic cycles see Wrigley and Schofield, *Population history*; on pre-industrial famines see Appleby, *Famine in Tudor and Stuart England*.

[3] For the difficulties experienced by the French see Hufton, *Poor of eighteenth century France*.

[4] Mokyr, *Lever of riches*.

[5] Thomas, 'Escaping from constraints'.

[6] Komlos, *Nutrition and economic development*; for a recent assessment of the issues see Mokyr, 'Is there still life in the pessimist case?'.

during the second half of the eighteenth century was already evident to an earlier generation of scholars using conventional indicators.[7] However, that evidence was not integrated well into a conceptualization of the industrial revolution, except insofar as its relevance to the debate over the changes in the standard of living was perceived. Recent research has added support and amplification to the notion of a decline in average availability of food per caput by using physical stature as a proxy for calorie and protein intake.[8]

The anthropometric evidence unearthed thus far indicates that from the middle of the eighteenth century average height in several European countries declined and continued to do so until the beginning of the next century. This was the case in east-central Europe, in Sweden, and in Great Britain and Ireland.[9] In fact, the slum boys of London in *c.* 1800 were so malnourished that they were shorter than boys living in practically any society today.[10]

A weakness in the evidence is our reliance primarily on military sources for information about eighteenth-century heights. This causes difficulties of interpretation, because of doubts about the representativeness of soldiers.[11] In addition, since European armies generally rejected a large number of men with heights below the minimum requirement, the samples obtained are inherently biased. In principle, inferences can be drawn from such truncated samples, since the heights of populations are known to be normally distributed; in practice, however, the estimating procedures are difficult, and may yield ambiguous results.[12]

One other sample of adult civilians is provided by convicts sent to Australia, and begins with the birth cohorts of the 1770s, nicely complementing the sample under consideration.[13] It is clearly useful to corroborate the trend in the height of soldiers with a civilian sample.

[7] Abel, *Agrarkrisen und Agrarkonjunktur*, p. 238; Sandgruber, *Die Anfänge der Konsumgesellschaft*, p. 115. Crafts, *British economic growth*, estimates that per caput agricultural output declined towards the end of the eighteenth century.

[8] Anthropometric measures enable one to quantify nutritional status, that is, the quantity of nutrients consumed minus the amount of calories and proteins claimed by work and disease. Nutritional status has a major impact on stature at a particular age, on the tempo of growth, on the timing and intensity of the adolescent growth spurt, and on the age at which final height is reached: Fogel, 'Nutrition and the decline in mortality'; Tanner, *Foetus into man*.

[9] Sandberg and Steckel, 'Heights and economic history'; Komlos, 'Secular trend'; Nicholas and Steckel, 'Heights and health of English workers'.

[10] Floud and Wachter, 'Poverty and physical stature'. In contrast, the middle classes do not seem to have felt the Malthusian pressure: Komlos, 'Height and social status'.

[11] Moreover, the estimated decline in the average height among the soldiers coincided with the enormous expansion of all European armies as a consequence of the Napoleonic wars; one cannot be absolutely sure that the decline was in no way related to recruiting those elements of the lower classes that would not have been considered for the military during peacetime.

[12] The statistical procedures estimate the amount of shortfall in the sample, i.e., the extent to which the sample distribution deviates from an expected Gaussian one: Wachter, 'Graphical estimation of military heights'. For the difficulties of estimation see Komlos and Kim, 'Estimating trends in historical heights'. As an example of the difficulties, note that British heights in the eighteenth century were increasing according to Floud, Wachter, and Gregory, *Height, health and history*, but decreasing according to Komlos, 'Secular trend'.

[13] Nicholas and Steckel, 'Heights and health of English workers'. The other non-military datasets found thus far pertain to children; while one of the two does support the trends deduced on the basis of the soldiers' heights, the other also suffers from the difficulties associated with truncated samples,

I

The advertisements for runaway servants found in colonial American newspapers were placed by owners of the contracts, in the hope of apprehending those who broke their obligations and fled to regain their freedom.[14] The notices began to appear in the early eighteenth century, in the early newspapers in the colonies, but seldom contained precise information on stature until mid century, although descriptive references such as tall, medium, or short were used even earlier.[15] The evidence diminishes soon after the revolution, when the importance of indentured service as a major component of the labour market declined rapidly.[16] The evidence the advertisements provide on female stature is the only such known source for the period.

The question of representativeness arises with the advertisements just as it does with military samples. What segments of British society were likely to become servants? The consensus is that they were drawn from a rather broad range of the population. Paupers and, of course, members of the upper middle class were underrepresented. Otherwise, the occupational distribution, as well as the rate of literacy of the servants, was quite similar to that of average Englishmen.[17] As many as one-third originated from London and its vicinity.[18] Thus, the indentured servants can be considered representative of the bottom 'half'. of the income distribution, excluding the destitute.[19]

In addition to indentured servants, the sources pertain also to convict servants transported to America for 7- to 14-year terms. From 1718, when this policy became a part of the English criminal justice system, until the American Revolution, about 50,000 persons, overwhelmingly men, mainly English with one-third from Ireland and just a sprinkling from Scotland, were banished to the colonies of North America.[20] The convicts were habitual felons most frequently guilty of theft.[21] That fully half of them

and the results based on it are currently disputed: Floud, Wachter, and Gregory, *Height, health and history*; Komlos, *Nutrition and economic development*, p. 85.

[14] *Early American newspapers*.

[15] I have not counted the percentage of advertisements which mentioned stature. A referee pointed out that until the 1740s practically no numeric height values were given in the advertisements. According to his tally the proportion jumped to 10 per cent in the 1740s and reached 42 per cent by 1752.

[16] German redemptioners continued to come to America 'long after the Revolution': McCormac, *White servitude*, p. 76.

[17] Galenson, *White servitude*, p. 78.

[18] Bailyn, *The peopling of British North America*, p. 12.

[19] These are precisely the ones most likely to have experienced difficulties in protecting themselves from the nutritional stress accompanying a decline in real wages.

[20] Banishment 'beyond the seas' was practised earlier but not until 1718 did it become institutionalized, although in Scotland it started later. Ekirch calculates that 13,000 came from Ireland, 700 from Scotland, and the rest of the 50,000 from England: Ekirch, *Bound for America*, pp. 1, 27. The present sample contains 1.7 per cent of the convicts shipped to the colonies, and approximately reflects the ethnic composition of all convicts (table 1).

[21] The most dangerous criminals were executed, and the petty thieves, beggars, and prostitutes were not threatening enough to be transported. Hence those transported were grand larcenists who were not shown mercy because of the lack of extenuating circumstances. The Irish, however, included a large number of vagabonds. Many of the English convicts, perhaps more than half, came from London and its surroundings, where the crime rate was well above that of rural areas, although many of these were immigrants to the city: Ekirch, *Bound for America*, pp. 28-31, 43, 47.

V



were either unskilled or did not report an occupation indicates that their social position was dissimilar from those who crossed the ocean of their own accord. The group as a whole might be characterized as well below the middle class in both income and social status.[22]

Thus the advertisements pertain to two distinct, yet overlapping segments of the lower classes, the very portion of society from which the military must also have recruited heavily. Yet each of the three samples, the soldiers, the indentured servants, and the convict servants, illuminates the experience of the lower classes from a unique perspective. A consistent pattern among these groups would greatly reinforce confidence in our ability to generalize about the trend in the biological well being of the lower classes.

Admittedly, the evidence gleaned from the newspaper advertisements also has shortcomings. Runaways may not be representative of all indentured servants,[23] and the advertisements may not be representative of all runaways.[24] A priori, one can hardly rule out the possibility that the stronger, hence taller, servants might have been more likely to take the risks of running away.[25] At this time there is no independent way of estimating this possible source of bias. Yet it is reassuring that one student of servitude concluded that 'in most respects, runaways differed little from the general convict population'.[26]

A more serious potential source of error is the fact that the references to heights in the advertisements were based on the recollection of the owners and not on actual measurements. This, in itself, might lead to heaping on

[22] Convicts did include a few well-to-do, who probably do not feature in this sample, because they would have purchased their freedom upon landing: ibid., pp. 52, 54, 59, 68, 119.

[23] Runaways must have been the most daring of their peers, since those who were caught faced retribution, in addition to the extra time they had to serve in order to compensate their master for their absence and for the expenses incurred in their apprehension: ibid., p. 196. Of course we do not know whether the height of the runaways can be considered representative of the height of all indentured servants. This question cannot be answered with certitude, but an analogy might help to formulate at least a conjecture. A sample of runaway slaves was collected as a part of this project. Their height can be compared with that of other segments of the slave population. American-born runaway slaves who were in custody of a gaoler at the time the advertisements were placed, and whose height estimates should, therefore, be considered the most accurate, were about 66.5 inches tall. This is below the height of slaves recorded on shipping manifests. Since African-American runaways were not taller than slaves on average, it is perhaps reasonable to suppose that indentured servants would not have been either: Steckel, 'Slave height profiles'.

[24] In addition to the price of advertisements, the inconvenience of placing them may have deterred some owners, particularly those who were not living near towns with newspapers. Some may have merely posted notices in public places. Many masters viewed advertising as a last resort: Ekirch, *Bound for America*, p. 194. Between 6 and 10 per cent of the servants are said to have run away: Salinger, '*To serve well and faithfully*', pp. 103, 113.

[25] It is also conceivable that owners may have been more likely to advertise for those servants who were most valuable, and therefore more skilled than average. This does not appear to have been the case, however, because the share of skilled runaways was less than that of the general indentured population: 'Skilled servants appear to have run away less often than unskilled servants, since the former were concentrated in the city and urban servants ran away at a lesser rate than rural servants': Salinger, '*To serve well and faithfully*', p. 113. It is also possible that height may have been mentioned in the advertisements for those servants whose height was a distinguishing feature. But with the exception of the 1720s the height distributions are quite normal, although there was considerable heaping on even numbers. An advertisement for a week cost 3s. in the *Pennsylvania Gazette* of the 1770s: Quimby, *Apprenticeship in colonial Philadelphia*, p. 86.

[26] Except that males were more likely to run away. Males made up 80 per cent of the convict servants, but 95 per cent of the runaways: ibid., p. 195.

round numbers, but is unlikely seriously to bias the analysis: the master had an incentive to recall the height of runaways accurately, since their return could be influenced by it.[27] A few of the runaways were caught in flight and were gaoled. The gaolkeepers placed advertisements seeking the owner of the runaways in their custody, and the heights mentioned in those notices were less subject to errors caused by faulty memory. Even if these notices had not been based on exact measurements, they would have been more accurate than the masters' advertisements, inasmuch as they were based mostly on direct observation.[28]

These potential shortcomings of the sample would tend to influence the intercept of the regressions, without biasing the other coefficients, as long as they were uncorrelated with the independent variables. This is important, because the main purpose of this study is to estimate the trend in height. As long as the degree of bias, if any, did not change over time, this potential source of error would not affect the estimated trend.[29] In the absence of contrary evidence, it seems reasonable to make this assumption.

Another issue worth considering is that after arrival the younger servants could have benefited to some degree from the better nutritional opportunities available in the New World. Hence their height could exceed to some extent that of their countrymen left behind. This is not a problem if servants fled soon after commencing their indentures. Thus, it is reassuring that according to one estimate the average length of time convicts served prior to attempting to flee was nine and a half months. Most did so within two months after their contract took effect.[30] Hence, this does not seem to be a serious source of bias, and particularly not among adults on whose height most of the inferences are based.[31]

The main dataset collected consists of some 3,800 observations on male indentured servants and some 700 records on convict servants for whom numeric height information is mentioned in the advertisements (table 1).[32] This is supplemented by 949 records on men with height descriptors, such as 'tall, medium or short' mentioned in the notices. For female runaways fewer than 100 records provided numeric height information.[33] Most of the advertisements collected were placed in Maryland and Pennsylvania, with New York occupying third place (table 1). Convicts were on average 28.8

[27] However, one cannot rule out the possibility that there was some 'learning by doing' over time in such a way as to improve the accuracy of the masters' estimates.

[28] Some of the advertisements placed by the gaolers referred to servants who had run away from gaol, and were therefore not directly observed at the time the advertisement was placed.

[29] In other words, if for some reason in one period advertisements had been placed only for tall servants, and in another period only for short ones, the bias would still affect the results obtained.

[30] Ekirch, *Bound for America*, p. 242.

[31] Moreover, this potential source of bias would affect only the intercept, not the other coefficients of the regression, as long as the length of time spent in the colonies prior to running away did not vary substantially over time. Since there is no reason for thinking that it did, the analysis of the trend should remain unaffected.

[32] If a range of height was specified in the advertisements, the interpolated height was used. Multiple advertisements were eliminated to the extent possible by matching names and other characteristics of the individual. If the height was given as 'near six feet', then 5 feet 11 inches was assigned to the observation; if the reference was to 'above six feet', then 6 feet 1 inches was assigned.

[33] Notices were also placed for lost persons, for suspected felons wanted for crimes committed in America, for runaway apprentices, and for soldiers and sailors who deserted.

Table 1. *Characteristics of the male runaway sample*

| | *Convicts* | *Indentured servants* |
|---|---|---|
| *Advertisement appearing in* | | |
| Connecticut | — | 25 |
| District of Columbia | — | — |
| Delaware | — | — |
| Georgia | 1 | 16 |
| Kentucky | — | — |
| Massachusetts | 6 | 46 |
| Maryland | 481 | 591 |
| N. Carolina | — | — |
| New York | 38 | 412 |
| Ohio | — | — |
| Pennsylvania | 271 | 3,007 |
| Rhode Island | 6 | 13 |
| S. Carolina | 3 | 41 |
| Tennessee | — | — |
| Virginia | 26 | 316 |
| Total | 832 | 4,467 |
| *Profession* | | |
| Specified | 269 | 1,444 |
| Unspecified | 541 | 2,985 |
| Farmer | 22 | 38 |
| *Aged* | | |
| 9-13 | — | 3 |
| 14-9 | 33 | 581 |
| 20-4 | 155 | 1,123 |
| 25-9 | 166 | 636 |
| 30-4 | 96 | 363 |
| 35-9 | 45 | 182 |
| 40-4 | 49 | 102 |
| 45-9 | 20 | 54 |
| 50 and over | 21 | 51 |
| Age unknown | 247 | 1,372 |
| Of urban origin | 17 | 60 |
| In gaol[a] | — | 319 |
| *Country of birth* | | |
| England | 327 | 855 |
| Ireland | 123 | 1,644 |
| Scotland | 15 | 171 |
| Germany | 5 | 105 |
| Holland | 3 | 216 |
| Elsewhere in Europe | 9 | 56 |
| Unknown | 350 | 1,420 |
| *Birth decade*[b] | | |
| 1710s | 43 | 155 |
| 1720s | 105 | 424 |
| 1730s | 284 | 918 |
| 1740s | 335 | 1,887 |
| 1750s | 65 | 1,073 |
| 1760s | — | 10 |

*Notes*:
[a] due to a shortcoming in the sample design, convict servants in custody of a gaoler were not recorded as such
[b] if age is not given, it is assumed to be equal to average age.
*Source*: Advertisements in colonial American newspapers. Includes also those for whom only height descriptors are available in the advertisements.

years old at the time they ran away, while indentured servants, at 25.2 years, were younger. Age of the runaways tended to diminish over time (table 2).[34]

Table 2. *Mean age of runaways[a] by date of advertisement*

| | *Convicts* | | *Indentured servants* | |
|---|---|---|---|---|
| | *N* | *age* | *N* | *age* |
| 1720 | — | — | 8 | 26.0 |
| 1730 | 1 | 35.0 | 14 | 24.7 |
| 1740 | 6 | 31.3 | 81 | 27.0 |
| 1750 | 86 | 30.0 | 432 | 25.6 |
| 1760 | 288 | 29.2 | 1,066 | 25.3 |
| 1770 | 181 | 27.5 | 1,494 | 25.0 |
| Total | 562 | 28.8 | 3,095 | 25.2 |

*Notes*:
N = number of observations
[a] limited to English and Irish servants and to those whose birthplace is unknown.
*Source*: see tab. 1.

## II

Regression models using dummy variables for age and birth decade categories were used for the main part of the analysis (tables 3 and 4).[35] The average height of those adult English runaway indentured servants who were not from a large town, such as London, was about 67.0 inches (tables 3, 5, and 6).[36] Thus, they were almost an inch shorter than the contemporary American average, but well above previous estimates based on military evidence.[37] The average height of those who were caught in flight and gaoled was not significantly different from the rest of the sample (table 3).[38]

[34] The mean age of convicts declined from about 30 to 27.5 years between the 1750s and the 1770s, and the age of indentured servants declined by 0.6 years in the same period.

[35] The results are robust with respect to various specifications. The *F*-statistics of all regressions are significant at the 5 per cent level. The $R^2$ statistics indicate that typically about 10 per cent of the variation in height is explained by the regressions. This is not unusual in height studies because there is a lot of unexplained variation in physical stature. The $R^2$ of the 'English' regression (not reported here) becomes 0.2 if the convicts and indentured servants are combined, but falls to 0.05 when the heights of the servants of unknown provenance are analysed.

[36] Not enough Welshmen were found for independent analysis, and they were included among the English. One should note also that in a handful of cases the reference to an indentured servant as an 'Englishman' could have referred to ethnic origin rather than birthplace. This caveat does not apply to convicts.

[37] The first estimate of British heights, made on the basis of the Royal Marines, was 64.7 inches, or 3.3 inches below American heights. The more recent estimate put the average adult height of the British men born in the 1740s and 1750s at 65.4 inches. This estimate is still too low, insofar as it is also based at least partly on sailors' height. Because the social status of indentured servants was not likely to have been very different from that of English soldiers, the difference between this civilian estimate and earlier military estimates of height is probably not due to variations in the social composition in the two sources. Rather, the reason that previous estimates were downwardly biased is that they were based entirely or partly on the height of Royal Marines, and tall men were systematically excluded from the navy, because they were at a disadvantage in the cramped quarters of a ship. The soldiers in the American army who were born in England were estimated to have been about 66.7 inches tall, close to the height of indentured servants: Sokoloff and Villaflor, 'Modern stature in America', pp. 457, 468; Floud, Wachter, and Gregory, *Height, health and history*, p. 148. The runaway indentured servants were about as tall as the American slaves: Margo and Steckel, 'Height of American slaves'.

[38] The height of servants in gaol differed by only 0.1 inch (table 3). Among adults, the difference was greater: those in custody of a gaoler were reported to have been about 0.5 inch shorter, but there

V



Table 3. *Regression equation. Dependent variable: height of indentured servants*

| | Country of birth | | |
|---|---|---|---|
| | *unknown* | *England* | *Ireland* |
| Constant | 67.2[a] | 66.9[a] | 66.9[a] |
| Aged 14 | −17.2[a] | −12.1[a] | −13.5[a] |
| 15 | −13.8[a] | −3.9[a] | −6.2[a] |
| 16 | −5.2[a] | −4.5[a] | −5.7[a] |
| 17 | −3.8[a] | −5.0[a] | −3.3[a] |
| 18 | −1.6[a] | −2.3[a] | −2.2[a] |
| 19 | −1.5[a] | −2.6[a] | −1.0[a] |
| 20-2 | −0.6[a] | −0.5 | 0.0 |
| Age unknown | −0.2 | 0.1 | −0.1 |
| Birth decade | | | |
| 1710s | 0.6[a] | 0.6 | −0.1 |
| 1720s | 0.8 | 0.6[b] | 0.2 |
| 1730s | −0.1 | 0.0 | 0.1 |
| 1750s | 0.1[a] | 0.0 | 0.3 |
| Of urban origin | | −0.9[a] | −1.3 |
| Knowledge of English | | | |
| Speaks | 0.8 | | |
| Accented | 0.1 | | |
| Little | −1.2 | | |
| Skilled craftsmen[b] | 0.1 | 0.0 | 0.1 |
| Farmer | 1.4 | 1.0 | 0.2 |
| In gaol | 0.0 | −0.1 | 0.1 |
| $R^2$ | 0.21 | 0.12 | 0.13 |
| $F$ | 22.8[a] | 8.4[a] | 15.2[a] |
| N | 1,360 | 850 | 1,628 |

*Notes*:
[a] significant at the 5 per cent level
[b] significant at the 10 per cent level
Constant represents the height of an adult runaway born in the 1740s, with no occupation mentioned in the advertisement, who was not from a large city, and who was not in custody of a gaoler at the time of the advertisement.
*Source*: see tab. 1.

An examination of the trend in physical stature over the first half of the eighteenth century indicates that the nutritional status of Englishmen began to decline with the birth cohorts of the 1730s.[39] This was consistently the case in several age categories, and among the convicts as well as the indentured servants (tables 3, 4, 6, and 7). The trend was also negative among indentured servants of unknown provenance, a large number of whom must have been English.[40] According to the regression estimates, the

are only 28 such observations on adults in the sample. Moreover, the sign of the coefficient of the 'in gaol' dummy variable among the Irish turns out to be positive. Consequently, the masters' height estimates, though based on memory and not on actual measurements, do not appear to have differed systematically from gaolkeepers' observations. (The comparison is, of course, made on two different groups of individuals.)

[39] A polynomial in time was also tried in place of the decadal dummy variables. There was no meaningful change in the results.

[40] An attempt was made to use surnames as a proxy for birthplace. The regression confined to those with English surnames was similar to the one reported in table 3 in the 'unknown' column.

Table 4. *Regression equation. Dependent variable: height of convicts*

| | Country of birth | | | | | | |
|---|---|---|---|---|---|---|---|
| | *unknown* | | *England* | | | *Ireland* | |
| | *(A)* | *(B)* | *(A)* | *(B)* | *(C)* | *(A)* | *(C)* |
| Constant | 67.1[a] | 66.9[a] | 67.4[a] | 67.3[a] | 67.4[a] | 66.7[a] | 66.9[a] |
| Age group | | | | | | | |
| 14-6 | −7.7[a] | | −12.6[a] | | −12.3 | | |
| 17-9 | −3.7[a] | | −2.4[a] | | −2.0 | −3.8[a] | −3.9[a] |
| 20-2 | −0.8 | −0.5 | 0.9[b] | 1.0 | 1.1[a] | −0.1 | −0.3 |
| Age unknown | 0.2 | | 0.0 | | | 0.7 | |
| Birth decade | | | | | | | |
| 1710s | 2.1[a] | 1.9[a] | −0.3 | −0.2 | −0.3 | 2.1 | 2.4[b] |
| 1720s | 0.2 | 0.5 | 0.5 | 0.6 | 0.5 | 0.3 | −0.1 |
| 1730s | 0.1 | 0.6 | 0.3 | 0.4 | 0.4 | 0.8 | 0.3 |
| 1750s | 0.6 | 0.1 | −0.2 | −0.4 | −0.7 | 0.3 | 0.2 |
| Knowledge of English | | | | | | | |
| Speaks | −0.6 | −1.8 | | | | | |
| Accented | −1.2 | 0.4 | | | | | |
| Little | −0.8 | −0.5 | | | | | |
| $R^2$ | 0.07 | 0.02 | 0.32 | 0.00 | 0.40 | 0.07 | 0.09 |
| *F* | 3.5[a] | 1.6[b] | 19.6[a] | 1.2[b] | 23.7[a] | 2.3[a] | 2.4[a] |
| N | 334 | 200 | 319 | 228 | 241 | 122 | 89 |

*Notes*:
[a] significant at the 5 per cent level
[b] significant at the 10 per cent level
Constant represents the height of an adult born in the 1740s.
(A) whole sample
(B) restricted to those aged 20 and over
(C) excludes those whose ages were unknown
*Source*: see tab. 1.

decline between the 1720s and the 1740s was between 0.5 and 0.6 inches.[41] The decline in physical stature was synchronous with the acceleration in the English population expansion in the 1730s.[42]

In contrast, the height of the Irish did not decline, indicating that they did not experience similar nutritional stress during the period.[43] Subsequently, Irishmen tended to be taller than Englishmen, but not in our sample

[41] Among the indentured servants the coefficient is significant at the 10 per cent level. Partial *F*-tests indicate that the trend among the indentured servants is significantly different from zero at the 10 per cent level. The same is true when the indentured servants are combined with the convict servants. These regressions are not reported here.

[42] Komlos, 'Industrial revolution in England'.

[43] One could argue that the decline in English heights is a statistical artefact caused by the fact that masters learned to estimate the height of the runaways more accurately after an initial period of learning by doing. A cursory examination of the height distributions does indicate that they are far from normal in the early period, and become much more nearly so for the post-1730 birth cohorts. In other words, there appears to be an improvement in the accuracy of the observations. There are two reasons why I would not draw from this the conclusion that the downward trend is caused by this factor: (1) The pattern of height distributions is the same among both the Irish and the English, but only the English height declines; (2) The number of observations in the 69-inch cell, a category that one would expect to be least affected by such biases, fell from 15.5 per cent prior to 1730 to 10 per cent after that date.

Table 5. *Height of adult convicts and indentured servants by place of birth*

| | *Convicts* | | *Indentured servants* | | | |
|---|---|---|---|---|---|---|
| | *aged 23+* | | *aged 23+* | | *aged 28+* | |
| *Country of birth* | *N* | *height* | *N* | *height* | *N* | *height* |
| England | 181 | 67.6 | 408 | 67.0 | 220 | 67.2 |
| Ireland | 69 | 67.2 | 588 | 67.0 | 286 | 67.0 |
| Scotland | 9 | 68.8 | 67 | 67.5 | | |
| Germany | | | 52 | 65.6 | | |
| Holland | | | 78 | 66.1 | | |
| Elsewhere in Europe | | | 22 | 67.7 | | |
| Unknown | 164 | 67.3 | 467 | 67.3 | 255 | 67.2 |

*Source*: see tab. 1.

(table 5).[44] Apparently the Irish height advantage developed during the second half of the eighteenth century.[45]

Adult English convicts appear to have been taller than indentured servants by about 0.6 inches (tables 3-6).[46] The difference among the Irish was either less striking or nonexistent (tables 4-8). Another result worthy of note is that English farmers were markedly taller than average (table 3). This is

Table 6. *Height of runaway servants born in England*

| | *Convicts* | | | | | | *Indentured servants* | | | | | |
|---|---|---|---|---|---|---|---|---|---|---|---|---|
| | *aged 20-2* | | *aged 23-50* | | *age unknown* | | *aged 20-2* | | *aged 23-50* | | *age unknown* | |
| *Birth decade* | *N* | *height* | *N* | *height* | *N* | *height* | *N* | *height* | *N* | *height* | *N* | *height* |
| 1710s | — | — | — | — | — | — | — | — | 31 | 67.4 | 4 | 67.0 |
| 1720s | — | — | 40 | 67.9 | 19 | 68.1 | — | — | 69 | 67.4 | 30 | 67.4 |
| 1730s | 8 | 69.1 | 76 | 67.7 | 32 | 67.2 | 21 | 66.2 | 107 | 66.8 | 57 | 66.6 |
| 1740s | 22 | 68.3 | 56 | 67.3 | 26 | 67.4 | 34 | 66.5 | 155 | 66.8 | 103 | 67.2 |
| 1750s | 17 | 67.9 | 4 | 67.3 | — | — | 63 | 66.3 | 46 | 67.2 | 48 | 66.9 |
| All | 47 | 68.3 | 176 | 67.6 | 77 | 67.5 | 118 | 66.4 | 408 | 67.0 | 242 | 67.0 |

*Source*: see tab. 1.

[44] Mokyr and ÓGráda, 'Poor and getting poorer?' Another study which found no difference between Irish and English soldiers was Steegman, 'Eighteenth century British military stature'.

[45] At the end of the eighteenth century Irish soldiers were about the same height as Englishmen of the northern border counties such as Westmorland, but taller than those of the midlands and the London area: Komlos, 'Secular trend'. By the end of the century Irish convicts shipped to Australia were, however, taller than English ones: Nicholas and Steckel, 'Heights and health of English workers'.

[46] The difference was significant at the 5 per cent level. The regression is not reported here. This finding might be interpreted as being consistent with the notion that taller men had a comparative advantage in criminal activity. It is also just possible that the convicts originated from a higher social class than the indentured servants. Convicts imprisoned in Glasgow and born in the early nineteenth century in rural Scotland were 67.3 inches tall, Irish criminals 66.5 inches, and Englishmen 66.2 inches: Riggs, 'Standard of living in Scotland'.

Table 7. *Height of runaway servants born in Ireland*

| | *Convicts* | | | | *Indentured servants* | | | | | |
|---|---|---|---|---|---|---|---|---|---|---|
| | *aged 23-50* | | *age unknown* | | *aged 20-22* | | *aged 23-50* | | *age unknown* | |
| *Birth decade* | *N* | *height* | *N* | *height* | *N* | *height* | *N* | *height* | *N* | *height* |
| 1710s | 6 | 69.3 | — | — | — | — | 33 | 66.5 | 3 | 70.7 |
| 1720s | 10 | 66.9 | 6 | 68.3 | 11 | 66.6 | 74 | 66.9 | 45 | 67.4 |
| 1730s | 28 | 67.2 | 18 | 68.3 | 63 | 67.5 | 175 | 67.2 | 63 | 65.9 |
| 1740s | 24 | 66.8 | 9 | 66.6 | 141 | 66.9 | 245 | 66.8 | 276 | 66.9 |
| 1750s | — | — | — | — | 157 | 67.0 | 61 | 67.4 | 43 | 67.2 |
| All | 68 | 67.2 | 33 | 67.8 | 372 | 67.1 | 588 | 67.0 | 430 | 66.9 |

*Source*: see tab. 1.

Table 8. *Height of runaway servants whose country of birth is unknown*

| | *Convicts* | | | | | | *Indentured servants* | | | | | |
|---|---|---|---|---|---|---|---|---|---|---|---|---|
| | *aged 20-22* | | *aged 23-50* | | *age unknown* | | *aged 20-22* | | *aged 23-50* | | *age unknown* | |
| *Birth decade* | *N* | *height* | *N* | *height* | *N* | *height* | *N* | *height* | *N* | *height* | *N* | *height* |
| 1710s | — | — | 12 | 68.8 | — | — | — | — | 21 | 68.2 | 7 | 68.1 |
| 1720s | — | — | 27 | 67.4 | 33 | 67.0 | 14 | 67.8 | 59 | 68.0 | 45 | 67.5 |
| 1730s | — | — | 57 | 67.4 | 48 | 67.3 | 15 | 66.0 | 128 | 67.2 | 99 | 66.7 |
| 1740s | 22 | 66.3 | 64 | 66.9 | 40 | 67.6 | 82 | 66.4 | 217 | 66.9 | 286 | 67.2 |
| 1750s | 10 | 66.8 | 4 | 66.5 | — | — | 84 | 66.8 | 42 | 67.9 | 88 | 66.7 |
| All | 32 | 66.5 | 164 | 67.3 | 121 | 67.3 | 195 | 66.7 | 467 | 67.3 | 525 | 67.1 |

*Source*: see tab. 1.

yet another example of the by now well-established pattern that, at least until the end of the nineteenth century, propinquity to the source of nutrients was beneficial to nutritional status.[47] In other words, urban men were always shorter than rural men; among the servants the difference was almost a full inch (table 3).

Could some of the younger servants have benefited from the nutritional advantages of the New World and reached a greater final height than if they had remained in Europe? Those who were aged 28 or older at the time they took flight can be compared with those who were 23 or older.[48] The former group would certainly have been too old to benefit from the American

[47] Komlos, *Nutrition and economic development*, p. 97.

[48] The average age of male indentured servants leaving England in the eighteenth century varied between 18 and 24 years, but skilled servants were somewhat older: Galenson, *White servitude*, pp. 30, 59. Because at that time growth ceased around the age of 23, a substantial proportion, perhaps even most of the servants were too old on arrival to benefit substantially from the abundant nutrients of the New World. Since the length of the indenture of adults was about four years, and since the servants were less likely to run away during the last year of indenture, the height of those aged over 28 at the time they ran away would not have been influenced by American nutritional circumstances.

V



nutritional environment, while those between the ages of 23 and 28 conceivably might still have gained some benefit. Yet the heights of the indentured servants in the two groups were virtually identical, suggesting that catch-up growth did not have a substantial effect on estimated adult heights (table 5).[49]

The paucity of evidence limits the range of conclusions that can be made in this sample of female indentured servants on the basis of the 87 observations found. Their terminal height was, on average, 62.5 inches, about the same centile of modern heights as the men's stature.[50]

It remains only to examine the part of the sample of advertisements which did not include numeric values for the height of the runaway, but such descriptions as 'short', 'average', or 'tall'. This part of the sample is presumed to be less accurate than the numeric data, but using them allows one to extend the evidence backward by a decade, however tentatively. The number of observations in the three categories was weighted by assigning the following numerical values to the descriptors: 65.0, 67.5, and 70.5 inches. In this way the average height becomes 67.0 inches (table 9). There is some variability in the height estimates over time, but no definite trend is discernible. The diminution in stature thus appears not to have begun prior to the 1720s.[51] A similar approach was used for weight descriptors. 'Light' body build was assumed to be 110 lbs., 'average' 140 lbs., and 'heavy' 160 lbs.[52] Two results are worthy of mention: the Irish appear to have been heavier than the English, and their weight increased between the 1710s and 1720s (table 9). The latter pattern is consistent with the notion that nutritional status did not decline before the 1720s. These tentative results are presented here only for illustrative purposes.

## III

The evidence is suggestive rather than providing incontrovertible proof of the inferences drawn from it. Physical stature of individuals varies considerably, even if the records are based on exact measurements. The evidence is based mainly on subjective estimates made from memory; the sample sizes are fairly small, and consequently some of the coefficients of

[49] A different way of looking at this problem is to examine the height of the few servants who ran away from their ship immediately after arrival on these shores. (Those who were clearly sailors were not considered.) They would not have benefited from American nutritional standards, and if the hypothesis is to be accepted they would be shorter than average. The handful of available observations fails to support this conjecture for those in their early twenties: those who ran away directly from their ship upon arrival were not systematically shorter than average, with the exception of the teenagers. (Teenagers (N = 7) were 2.3 inches below average stature; in contrast, adults and young adults were 0.5 inches taller than average.) An increase in nutrient intake would have had a larger impact on teenagers than on those in their twenties. Although the number of observations is much too small for statistically significant results, the evidence lends little credence to the notion of a strong catch-up growth affecting the terminal height of adult indentured servants in America.

[50] The height of those aged over 20 was as follows: Irish: 62.2; English: 63.1; unknown: 62.6 inches.

[51] This is also the implication of the few numerical observations available for the birth cohorts of the 1710s.

[52] The weight of an 18-year-old cadet was 128 lbs. in the West Point Military Academy in antebellum America: Komlos, 'Height and weight'.

Table 9. *Estimated height[a] and weight[b] of adult servants based on descriptors*

| | Country of birth | | | | | |
|---|---|---|---|---|---|---|
| | *England* | | *Ireland* | | *unknown* | |
| *Birth decade* | *N* | *height* | *N* | *height* | *N* | *height* |
| 1700s | 31 | 67.2 | 40 | 67.6 | 76 | 67.0 |
| 1710s | 83 | 66.8 | 134 | 67.0 | 124 | 67.2 |
| 1720s | 53 | 67.1 | 132 | 66.7 | 67 | 66.5 |
| All | 167 | 67.0 | 306 | 67.0 | 267 | 67.0 |
| | *N* | *weight* | *N* | *weight* | *N* | *weight* |
| 1710s | 22 | 132.7 | 50 | 140.0 | 32 | 130.3 |
| 1720s | 19 | 135.3 | 59 | 142.0 | 27 | 133.3 |
| All | 41 | 133.9 | 109 | 141.1 | 59 | 131.7 |

*Notes:*
[a] assuming the following numerical equivalents for the height descriptors: short = 65.0 ins.; middle = 67.5 ins.; tall = 70.5 ins.
[b] assuming the following numerical equivalents for the weight descriptors: light = 110 lbs.; average = 140 lbs.; heavy = 160 lbs.
*Source*: see tab. 1.

the regression estimates are not statistically significant. Hence more work still needs to be done on the nutritional history of this period.

Yet even if we acknowledge these limitations of the evidence, it appears that the height of the lower segments of the English population was probably beginning to decline *prior* to the beginning of the industrial revolution, and contemporaneously with the acceleration in population growth.[53] In other words, population growth had started once again to press on available food supplies.[54] That the decline in height continued after the end of the period under scrutiny is supported by the height of convicts transported to Australia. The birth cohorts of the 1770s among these convicts were on average 66.5 inches tall, i.e., 1.1 inches shorter than the English runaway convicts in colonial America.[55]

The height of the Irish servants does not seem to have declined during the first half of the eighteenth century.[56] This result is not anomalous, provided one notes that the Irish population stagnated before 1750, and only began to increase thereafter, and in addition, that the export of

[53] Wrigley and Schofield, *Population history*.

[54] Admittedly we do not know the nutritional significance of the half-inch difference in height because the functional relationship between stature and nutrition is not known. At this point we must be content with specifying the direction of change. One might note in this regard that the decline in American heights of about twice this magnitude during the decades prior to the Civil War was accompanied by a decline in consumption of about 300 calories per person per day: Komlos, 'Height and weight'. Note that changes in the distribution of income play a role in determining the impact of average calorie intake on average physical stature.

[55] Nicholas and Steckel, 'Heights and health of English workers'.

[56] Unfortunately the regional composition of the sample within Ireland and within England could not be held constant.

V



nutrients from Ireland to England did not reach meaningful proportions until mid century.[57] However, military evidence indicates that during the second half of the century the stature of the Irish also diminished.[58]

The relationship between industrialization and the well being of the societies experiencing it remains controversial. The inferences in this comment, all the caveats notwithstanding, could transform the debate in its essentials. If the biological standard of living began to decline just prior to the industrial revolution, then the latter process could not have been responsible for the onset of deterioration.

[57] Daultrey, Dickson, and Ó Gráda, 'Eighteenth-century Irish population'; Thomas, 'Escaping from constraints'.

[58] Nicholas and Steckel, 'Heights and health of English workers'; Komlos, 'Secular trend'.

**Footnote references**

Abel, W., *Agrarkrisen und Agrarkonjunktur: eine Geschichte der Land- und Ernährungswirtschaft Mitteleuropas seit dem hohen Mittelalter* (1935; repr., Hamburg, 1966).

Appleby, A., *Famine in Tudor and Stuart England* (Stanford, 1978).

Bailyn, B., *The peopling of British North America: an introduction* (New York, 1986).

Crafts, N. F. R., *British economic growth during the industrial revolution* (Oxford, 1985).

Daultrey, S., Dickson, D., and Ó Gráda, C., 'Eighteenth-century Irish population: new perspectives from old sources', *J. Econ. Hist.*, 41 (1981), pp. 601-28.

*Early American newspapers, 1704-1820* (New York, 1979).

Ekirch, A. R., *Bound for America: the transportation of British convicts to the colonies, 1718-1775* (Oxford, 1987).

Floud, R. and Wachter, K., 'Poverty and physical stature: evidence on the standard of living of London boys, 1770-1870', *Soc. Sci. Hist.*, 6 (1982), pp. 422-52.

Floud, R., Wachter, K., and Gregory, A., *Height, health and history: nutritional status in the United Kingdom, 1750-1980* (Cambridge, 1990).

Fogel, R. W., 'Nutrition and the decline in mortality since 1700: some preliminary findings', in S. Engerman and R. Gallman, eds., *Long term factors in American economic growth* (Chicago, 1987; National Bureau of Economic Research, Studies in Income & Wealth, vol. 51), pp. 439-555.

Galenson, D. W., *White servitude in colonial America: an economic analysis* (Cambridge, 1981).

Hufton, O. H., *The poor of eighteenth century France, 1750-1789* (Oxford, 1974).

Komlos, J., 'The height and weight of West Point cadets: dietary change in antebellum America', *J. Econ. Hist.*, 47 (1987), pp. 897-927.

Komlos, J., *Nutrition and economic development in the eighteenth-century Habsburg monarchy: an anthropometric history* (Princeton, 1989).

Komlos, J., 'Height and social status in eighteenth-century Germany', *J. Interdisc. Hist.*, 20 (1990), pp. 607-21.

Komlos, J., 'Nutrition, population growth and the industrial revolution in England', *Soc. Sci. Hist.*, 14 (1990), pp. 69-91.

Komlos, J., 'The secular trend in the biological standard of living in the United Kingdom, 1730-1860', *Econ. Hist. Rev.*, XLVI (1993), pp. 115-44.

Komlos, J. H. and Kim, J. H., 'Estimating trends in historical heights', *Hist. Meth.*, 23 (1990), pp. 113-20.

Le Roy Ladurie, E., *The peasants of Languedoc* (Urbana, Ill., 1974).

Margo, R. and Steckel, R. H., 'The height of American slaves: new evidence on slave nutrition and health', *Soc. Sci. Hist.*, 6 (1982), pp. 516-38.

McCormac, E. I., *White servitude in Maryland, 1634-1820* (Baltimore, 1904).

Mokyr, J. and Ó Gráda, C., 'Poor and getting poorer? Living standards in Ireland before the Famine', *Econ. Hist. Rev.*, 2nd ser., XLI (1988), pp. 209-35.

Mokyr, J., *The lever of riches: technological creativity and economic progress* (New York, 1990).

Mokyr, J., 'Is there still life in the pessimist case? Consumption during the industrial revolution, 1790-1850', *J. Econ. Hist.*, 48 (1988), pp. 69-92.
Nicholas, S. and Steckel, R. H., 'Heights and health of English workers during the early years of industrialisation, 1770-1815', *J. Econ. Hist.*, 51 (1991), pp. 937-57.
Quimby, I. M. G., *Apprenticeship in colonial Philadelphia* (New York, 1985).
Riggs, P., 'The standard of living in Scotland, 1800-1850', in J. Komlos, ed., *The biological standard of living and economic development on three continents: essays in anthropometric history* (Chicago, forthcoming, 1994).
Salinger, S. V., *'To serve well and faithfully': labor and indentured servants in Pennsylvania, 1682-1800* (Cambridge, 1987).
Sandberg, L. G. and Steckel, R. H., 'Heights and economic history: the Swedish case', *Annals Human Biol.*, 14 (1987), pp. 101-10.
Sandgruber, R., *Die Anfänge der Konsumgesellschaft, Konsumgüterverbrauch, Lebensstandard und Alltagskultur in Österreich im 18. und 19. Jahrhundert* (Vienna, 1982).
Sokoloff, K. and Villaflor, G. C., 'The early achievement of modern stature in America', *Soc. Sci. Hist.*, 6 (1982), pp. 453-81.
Steckel, R. H., 'Slave height profiles from coastwise manifests', *Exp. Econ. Hist.*, 16 (1979), pp. 86-114.
Steegman, A. T., Jr., 'Eighteenth century British military stature: growth cessation, selective recruiting, secular trends, nutrition at birth, cold and occupation', *Human Biol.*, 57 (1985), pp. 77-95.
Tanner, J. M., *Foetus into man: physical growth from conception to maturity* (Cambridge, Mass., 1990).
Thomas, B., 'Escaping from constraints: the industrial revolution in a Malthusian context', *J. Interdisc. Hist.*, 15 (1985), pp. 729-53.
Wachter, K. W., 'Graphical estimation of military heights', *Hist. Meth.*, 14 (1981), pp. 31-42.
Wrigley, E. A. and Schofield, R. S., *The population history of England, 1541-1871: a reconstruction* (Cambridge, 1981).

# VI

# *The secular trend in the biological standard of living in the United Kingdom, 1730-1860*[1]

Records collected by the army and the marines as well as by the Marine Society, a charitable institution, enable one to explore the trend in the height of the British population during the course of a century that included the beginning of the industrial revolution and its immediate aftermath.[2] The standard of living in this period has remained a controversial issue among historians, and it is in this connection that the evidence about to be examined assumes its significance.[3] Within the last decade we have been frequently reminded of the importance of anthropometric history for understanding the well-being of societies of the past.

Anthropometric measures, such as height, weight, and biomass help us to assess the biological standard of living because they enable us to quantify nutritional status. Medical research has established beyond doubt that the net cumulative nutritional intake of a population has a major influence on its average height. Height at a particular age, as well as adult height, can be thought of as an historical record of the net cumulative nutritional status during the growing years, indicating the intake of nutrients minus the claims of basal metabolism, of energy expenditure, and of disease. The higher the nutritional status, the more calories and protein remain available for physical growth. Therefore, human stature can be used as a proxy measure of the biological well-being of a population, an important component of that complex variable called the standard of living.[4]

In conventional terms the standard of living, regarded as synonymous with a material standard, has been measured by such indices as per caput gross national product, real per caput income, or hourly wages. The

[1] The author is indebted to Roderick Floud and Annabel Gregory for providing him with a copy of the QBE computer program, and to them and the University of Essex's ESRC Data Archive for making available for analysis the datasets upon which this study is based. I am grateful to Marcus Hein of the Technical University of Vienna for undertaking the calculations and to Phil Sidel of the University of Pittsburgh's Social Science Computer Research Institute for facilitating the transfer of data. I also appreciate comments from Kenneth Wachter and from members of the Economic History Workshop of the Universities of Toronto and Pittsburgh.

[2] The goal of the Marine Society was to train poor boys for life at sea. For a detailed description of the Marine Society dataset, see Floud and Wachter, 'Poverty and physical stature'.

[3] The standard of living debate has a well-known literature; see for example, Crafts, 'English workers' real wages', pp. 139-44; and, for more recent studies, Mokyr, 'Consumption during the industrial revolution'; Brown, 'England and the standard of living'; Mokyr and Ó Gráda, 'Living standards in Ireland'.

[4] For further details see Komlos, 'Height and weight of West Point cadets'; Fogel, 'Nutrition and the decline in mortality'.

construction of these indexes is fraught with numerous uncertainties. It is hardly surprising then, that the debate over the standard of living conventionally conceived, particularly during the industrial revolution, has reached an impasse, in spite of noteworthy efforts to overcome the limitations of the data.[5] Yet, the standard of living concept can be broadened to encompass the biological dimension. Income, the basis of the conventional measure, is one of the determinants of food consumption, and therefore has implications for biological well-being. Biological and economic measures are thus interrelated.

The height records relating to the UK are useful in considering the impact of industrialization on the lower segments of society.[6] Floud, Wachter, and Gregory have done pathbreaking work by collecting and examining these data.[7] Without their early statistical contribution, including work coauthored with Trussell, anthropometric history would possibly not have taken off at all in the European context, where most of the records originate from truncated data sets.[8] For example, the imposition of a minimum height requirement for admission into the army, the marines, and the Marine Society means that the samples drawn from their records are distorted.[9] It is extremely difficult to draw reliable inferences as part of the underlying distribution is missing, and therefore the samples are inherently biased. In fact, in their pioneering contributions Floud, Wachter, and Trussell stressed that great care must be exercised in analysing such distributions. The statistical problem is complicated, moreover, by the fact that the height requirements changed over time, and were not perfectly enforced. The issues associated with correcting such biased samples are quite complex and have been explored in detail elsewhere.[10] Here we need only stress that the sample frequency distributions must be carefully examined before reliable conclusions can be reached.[11]

While the underlying population's height distribution can be estimated from a truncated sample either by the Quantile Bend Estimator (QBE) method or by the Reduced Sample Maximum Likelihood Estimator (RSMLE) method (a contribution of Floud, Wachter, and Trussell), these procedures are not without problems.[12] Errors may arise for many reasons. It must be assumed, for instance, that the sample frequency distributions are part of the normal family; if this assumption is violated, however, the estimates can be quite inaccurate.[13] Unusual sample distributions can lead to large variations

[5] Schwarz, 'The standard of living in the long run'.
[6] For the importance of these data see Tanner, *Study of human growth*.
[7] Floud, Wachter, and Gregory, *Height, health and history*.
[8] Floud and Wachter, 'Poverty and physical stature'; Wachter, 'Graphical estimation'; Wachter and Trussell, 'Estimating historical heights'.
[9] Floud, Wachter, and Gregory, *Height, health and history*, p. 138. In their own words, the problem 'shows itself in haphazard scatter on the graphs'.
[10] Komlos and Kim, 'Estimating trends', pp. 116-20.
[11] Komlos, *Nutrition and economic development*, pp. 52-3.
[12] Wachter, 'Graphical estimation'; Wachter and Trussell, 'Estimating historical heights'.
[13] The sensitivity of the estimation technique is explored using simulations in Wachter and Trussell, 'Estimating historical heights', pp. 284-6. However, the difficulties of archival sampling coupled with the heterogeneous composition of military units render the empirical problem more complicated than this sensitivity analysis acknowledges. See the nature of actual unstable estimates in Komlos and Kim, 'Estimating trends'.

in the estimates. The validity of the normality assumption must therefore be validated by examining each height distribution prior to using the estimation technique.[14] If the normality assumption is violated, further estimations, using the QBE or RMSLE procedures should be reserved for cases not affecting the inferences being made.[15]

Floud, Wachter, and Gregory conclude that heights, and hence nutritional status,

> rose from the middle of the eighteenth century and into the late 1820s, though with a possible check for those born just before and thus growing up in the Napoleonic wars. Average male heights then began to fall for those experiencing childhood in the hungry thirties and continued to fall for the birth cohorts of the 1840s and early 1850s,

but a recovery began after the 1860s.[16] In their view, therefore, the biological standard of living started to improve at the start of the industrial revolution and the improvement was first interrupted by a factor exogenous to the economy, namely the French wars. The rise in economic productivity following shortly after the cessation of hostilities did not filter down to the lower classes for a very long time. In their words,

> Thus the era of the early Industrial Revolution seems on this evidence to have led to an improving standard of living for the working population, while the years of further industrialization and urbanization, after the Irish famine and the repeal of the Corn Laws, stand out as years of retrogression.[17]

Thus they arrive at an optimistic conclusion regarding the effect of the industrial revolution on the standard of living for the eighteenth century, but adopt a more pessimistic view for the nineteenth.[18] This paper will reconsider this assessment by re-examining the secular trend in physical stature first from the heights of poverty-stricken boys, and then from those of soldiers.

[14] The sample distributions might depart from normality, if, *inter alia*, the sample size is small, or if the changes in the minimum height requirement bring about the superimposition of two different parts of a normal distribution. Moreover, if the minimum height requirement is to the right of the population's true mean, the procedure results in an upwardly biased estimate of the true mean, and the bias increases the further the minimum height requirement is from the true mean.

[15] There is no evidence in their book that they examined the frequency distributions of the children. However, they did examine the height distributions of the soldiers and recognized that 'aberrant cases are certainly not absent, [though] they are relatively rare'. They decided not to discard these cases on methodological grounds. They felt it was more 'important to maintain consistency of procedures, and to avoid adjustment of "special cases" and ad hoc judgments, even though . . . some estimates could certainly be improved using detailed inspection of the particular raw distribution'; Floud, Wachter, and Gregory, *Height, health and history*, p. 138. It seems odd to overlook information in order to follow a particular algorithm in analysis. Instead, one might argue that the computer ought not be allowed to take precedence and it is precisely creative human intervention, subjective as it might be, that is needed in order to obtain consistent results. Otherwise there is the danger that inferences might turn on these aberrant cases.

[16] Ibid., pp. 28 and 275.

[17] Ibid., pp. 151-2.

[18] They do make an exception in their optimistic assessment of the eighteenth-century pattern: 'the immiseration of the later 1790s is clearly seen among those born and passing through childhood at the time. The rapid increase comes very soon after the war.' Ibid., p. 153.

# I

The boys accepted into the Marine Society were extremely malnourished, so much so that there are very few populations in the world today whose mean height does not exceed that of these 'Oliver Twists'.[19] This finding is hardly debatable: contemporary peasant boys of the Habsburg lands, for instance, and even American slaves, were markedly taller.[20] However, a number of factors call into question the reliability of the estimated trend.

The trend, based on moving averages, is influenced by large fluctuations in the individual estimates (table 1). The variations are too large to be plausible, and the trends differ between the various ages. For example, the increase in height of the 1770s is confined to the 15- and 16-year-old boys.[21] Among the 13- and 14-year-olds the temporary increase appears later, in the 1780s, at a time when nutritional status is alleged to have fallen. Thus in the 1770s the pattern is not consistent between all age groups, as one might expect if an increase in nutritional status had actually taken place. Would improvements in nutrition have been age-specific? Is a blip in the estimates a reflection of an upward trend or of the instability of the QBE computer procedure?

Table 1. *Floud, Wachter, and Gregory's estimates ($\hat{x}$) of the height of London boys (cm)*

| | *Age* | | | |
|---|---|---|---|---|
| *Year of birth* | *13* | *14* | *15* | *16* |
| 1768 | n.a. | 135.5 | 141.8 | 147.4 |
| 1778 | 129.5 | 134.6 | 146.5 | 150.5 |
| 1780 | | 137.1 | 140.8 | 145.1 |
| 1781 | 128.7 | | | |
| 1786 | 133.8 | | | |
| 1797 | 132.0 | 135.7 | 136.7 | 139.7 |
| 1802 | | 135.3 | 139.3 | 151.1 |
| 1805 | 131.5 | | | |
| 1807 | | | 148.6 | 155.2 |
| 1812 | | 144.2 | | |
| 1815 | 130.5 | | | |
| 1820 | 137.3 | | | |

*Note*: Dates are approximate; age is at last birthday; the estimates have been adjusted for the minimum height requirements by the QBE method.
*Source*: Floud, Wachter, and Gregory, *Height, health and history*, pp. 167-9.

The latter possibility is made more likely by the fact that in 1786 and then again in 1792 the minimum height requirement for admission into the Society changed; both changes affected the birth cohorts of the 1770s. Could these administrative changes result in sample frequency distributions sufficiently abnormal to mislead the computer program? If that were the

[19] Eveleth and Tanner, *Human growth*.
[20] Steckel, 'Growth, depression, and recovery'; Komlos, 'Patterns of children's growth'.
[21] Here, as elsewhere, unless otherwise noted, reference to a particular year is meant to refer to the birth cohorts of that year.

case, one would expect the younger boys to be affected at a later date, a pattern that is, indeed, visible. For example, the change in the minimum height requirement of 1792 (fully effective in 1793) first affected the birth cohorts of 1777 among 16-year-olds, whereas among the 14-year-olds it first had an impact on those born in 1779. The delay in the increase in stature of the 14-year-old boys (in 1780, rather than in 1778 among the 15- and 16-year-olds) is, therefore, consistent with the notion that the rise in the height estimates is related to the changes in the minimum height requirement rather than to changes in nutritional status.[22]

Moreover, the estimated increases in height are often unreasonably abrupt and unacceptably large.[23] For example, an increase of 15.5 cm. in the stature of 16-year-old lower-class boys within just one decade around the turn of the nineteenth century appears to be outside normal biological limits (table 1). Are there precedents for such large shifts in the age-by-height profile (in the absence of large doses of protein supplements)? If nutritional circumstances improved to such an extent, why did they not simultaneously affect children of all ages?

Table 2. *Simple mean height of boys in the sample (cm)*

| | *Age* | | | | | | | |
|---|---|---|---|---|---|---|---|---|
| | *13* | | *14* | | *15* | | *16* | |
| *Year of birth* | *N* | *x̄* | *N* | *x̄* | *N* | *x̄* | *N* | *x̄* |
| 1776 | | | | | 162 | 142.4 | 45 | 146.6 |
| 1777 | | | 252 | 137.7 | 67 | 142.0 | 41 | 148.9[a] |
| 1778 | 149 | 132.7 | 109 | 136.4 | 320 | 144.8[a] | 72 | 147.3 |
| 1779 | 45 | 132.5 | 482 | 138.5[a] | 94 | 142.6 | 63 | 145.3 |
| 1780 | 343 | 133.1[a] | 163 | 137.8 | | | | |
| 1781 | 179 | 131.8 | | | | | | |

*Notes*: *N* is the number of observations; x̄ is the simple mean of the sample uncorrected for the minimum height requirement
[a] indicates birth cohorts recruited in 1793
*Source*: 'Long term changes in nutrition' (SN 2131), ESRC Data Archive, Univ. of Essex.

The above reservations are supported by examining the annual raw mean height of boys born in the late 1770s. There is no evidence of increasing stature other than a single increment among the boys recruited in 1793 (table 2). That is why the increase affected boys of different ages with a successive one year lag. Consequently, the apparent increase in stature of the 1770s ought not be considered a reflection of changes in nutritional status. Instead, it was caused either by the unusual admission practices of 1793, or by the

[22] A practically identical pattern is obtained for the alleged stupendous improvement in nutritional status in the early nineteenth century. It appears later among the younger boys, thereby indicating that it might also be associated with the changes in the minimum height requirement instituted between 1814 and 1824.
[23] Floud, Wachter, and Gregory, *Height, health and history*, pp. 116, 289.

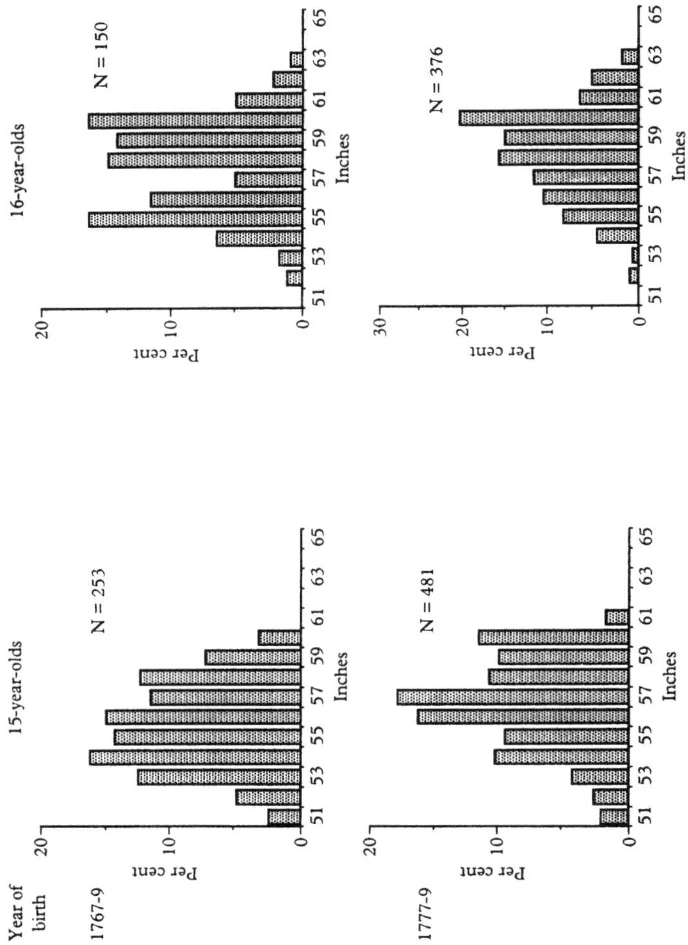
Year of birth
1767-9
1777-9
15-year-olds
16-year-olds
N = 253
N = 150
N = 481
N = 376
Per cent
Inches

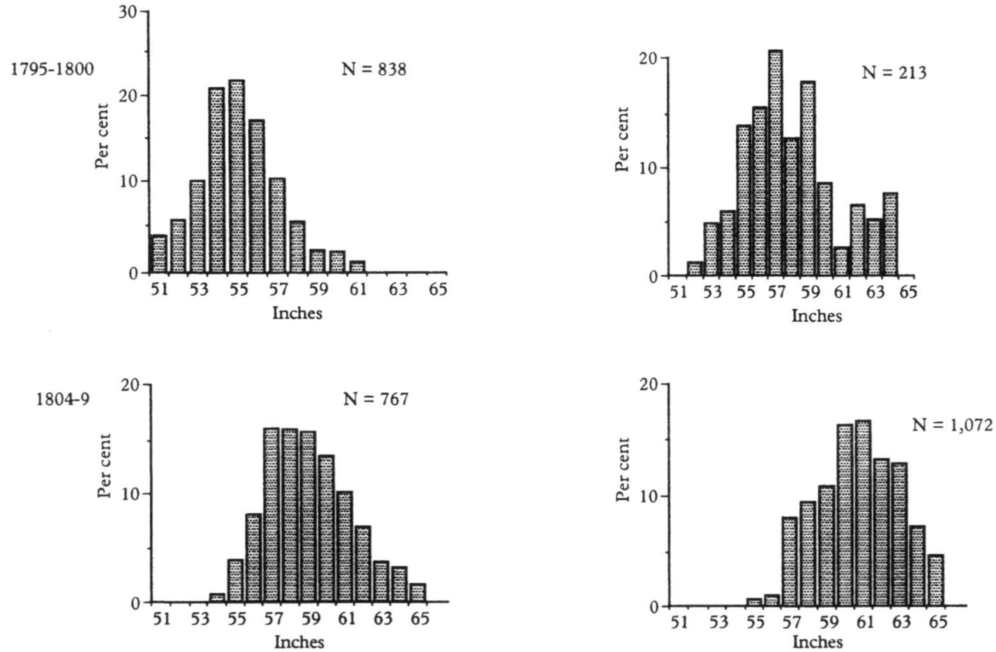


Figure 1. *Height distributions used by Floud, Wachter, and Gregory to estimate the height of 15- and 16-year-old Marine Society recruits, using the QBE program (selected years)*

*Source*: Long-term changes in nutrition (SN 2131), ESRC Data Archive, University of Essex

fact that the 'offer curve' of indigent boys shifted dramatically that year.[24] For these reasons, the children recruited in 1793 should be analysed separately from the rest of the sample.

In addition to the problem of the 1793 recruits, an examination of some of the height distributions on which Floud, Wachter, and Gregory base their QBE estimates (table 1), and ultimately their conclusions, confirms that their results are not grounded in nutritional reality, but merely reflect instabilities associated with the fact that the QBE computer program was led astray by the unusual frequency distributions (figure 1). The height distributions of the 15- and 16-year-old boys are often bimodal, a sign that two distributions with different means or with different truncation points were superimposed on one another.[25] Moreover, these distributions are also frequently truncated not only on the left, but also on the right, albeit not perfectly. None of the height distributions of the older boys appears sufficiently normal to warrant the use of the QBE program, which was not designed to work on frequency distributions that deviate so much from the ideal normal one.[26]

In order to provide a robust estimate of the secular trend in the stature of the Marine Society boys I have, therefore, taken into consideration three peculiarities of the sample: the right-hand truncation of the height distribution among older boys; the generally abnormal frequency distributions; and the unusually tall recruits of 1793. Moreover, in order to eliminate the influence of at least some of the changes in recruiting practices, the analysis is divided into three sub-periods which were more homogeneous in terms of labour market conditions and recruiting practices than the period as a whole: before 1818, between 1818 and 1853, and after 1853.

Floud, Wachter, and Gregory estimated the height of the boys for short time intervals, in order to minimize pooling over different recruitment practices. My strategy, in contrast, is to attempt to minimize such error by recognizing peculiarities of the sample, as in the case of 1793, and also to use several techniques of analysis in conjunction, keeping in mind the shifting nature of the minimum height requirements. Further, in contrast to their strategy of estimating heights for two-, three-, or five-year periods, I chose the decade as a more appropriate period of analysis, because under normal circumstances the height of a population does not change appreciably during such short time intervals. The frequency distributions of the boys' heights by decade indicate that while the sample distributions seldom resemble an ideal normal one, the distributions of the younger boys tend to

[24] Floud, Wachter, and Gregory recognize that the offer curve could have shifted over time: 'Unemployment was certainly a major factor underlying the recruitment by the Marine Society . . . of boys to serve on naval and merchant ships'. However, they did not notice the peculiar nature of recruitment in 1793. One possible reason for the unusual sample of 1793 is that in that year the established strength of the navy was almost quadrupled and the Marine Society's recruitment was perhaps related to this change. Ibid., pp. 55, 69, 117, and 152.

[25] Other factors that could have contributed include heaping on even numbers or measurement error.

[26] At times the right-hand truncation could have been a deliberate policy of the Society due to the fact that tall sailors were less desirable; it could also have come about due to the fact that taller boys had more possibilities of employment, including military service, than shorter ones. (It was easier for them to mislead military recruiting officers about their age.) The reason for the right-hand truncation hardly matters for the purposes of analysis. What does matter is that, as a consequence, such data should not be analysed with the use of the QBE program.

be better than those of the older ones (figures 2 and 3), among whom right-hand truncation is often evident. Consequently, the QBE program is inappropriate for estimating the height of the older boys, but might be used for the younger ones.

The QBE estimates ($\hat{x}$) of 13- and 14-year-old boys show a diminution in their height during the second half of the eighteenth century (table 3). Because they were very short, even a small further decline in stature is worthy of note, as they must already have been close to their genetic minimum height. The QBE estimates also do not suggest that nutritional status improved during the first decade of the nineteenth century.[27] On the contrary, the estimated height ($\hat{x}$) was still below the mid-century maximum at both ages. The trend in $\hat{x}$, the raw mean of the sample, shows an essentially similar declining trend. Even the trend of the unadjusted mean height fails to confirm a strong post-Napoleonic nutritional upswing.

These results are confirmed by a more reliable method of estimating the trend in truncated height distributions which need not be confined to the younger boys. The trend in the mean of the so-called 'reduced sample' is an unbiased estimate of the true trend of a population's height.[28] The reduced sample is restricted to the part of the distribution above the largest of the minimum height requirements of the period (54 inches for the Marine Society boys recruited before 1818),[29] thereby eliminating the disturbing effects of shifts in the point of truncation. The subsequent sample becomes smaller, but it also becomes 'cleaner', in the sense that those data which pollute the sample and detract from, rather than contribute to, its information content are discarded.[30] The trend in the simple mean of the reduced sample ($\bar{x}'$) is more accurate than the one obtained from the QBE estimates, because it is less sensitive to abnormalities in the sample distributions due to shifts in the truncation point.[31] To be sure, $\bar{x}'$ is a biased estimate of the true mean of the population, but its purpose is not to estimate actual heights. Rather, the changes in $\bar{x}'$ are used as indicators of the sign of the true trend in height.[32]

Among the 13- and 14-year-old Marine Society recruits the trend in $\bar{x}'$ is generally negative during the second half of the eighteenth century, and this

[27] In order to eliminate the problems associated with changes in recruiting practices after 1818, the calculations in table 3 have been restricted to those boys who entered the Society prior to that date.

[28] Komlos and Kim, 'Estimating trends'.

[29] Floud, Wachter, and Gregory, *Height, health and history*, p. 164.

[30] In the presence of right-hand truncation, however, it is still a problematical measure if the truncation point, or the degree of its enforcement, shifts over time. If that is the case, the sample might be further reduced to include only that range which was admissible throughout the period under consideration. This was one of the procedures adopted to calculate the index for the 16-year-old boys (table 4). In this case, the range for which $\bar{x}'$ was calculated was restricted to between 54 and 60 inches. See n. 48 for caveats on using this method.

[31] Hence this part of the analysis is restricted to boys whose height exceeded 54 inches, the greatest of the minimum height requirements prior to 1818. The sample height distribution is truncated perfectly at 54 inches, i.e., anyone below this height is eliminated from the sample. The remaining portion of the distributions is unaffected by the changes in the minimum height requirement, but some anomalies nonetheless remain even within the reduced sample.

[32] Floud, Wachter, and Gregory's strategy was first to estimate actual heights at a moment in time and then to deduce the trend from these estimates. In contrast, the reduced-sample method does not estimate actual heights but merely the direction of change, but it does so more reliably.

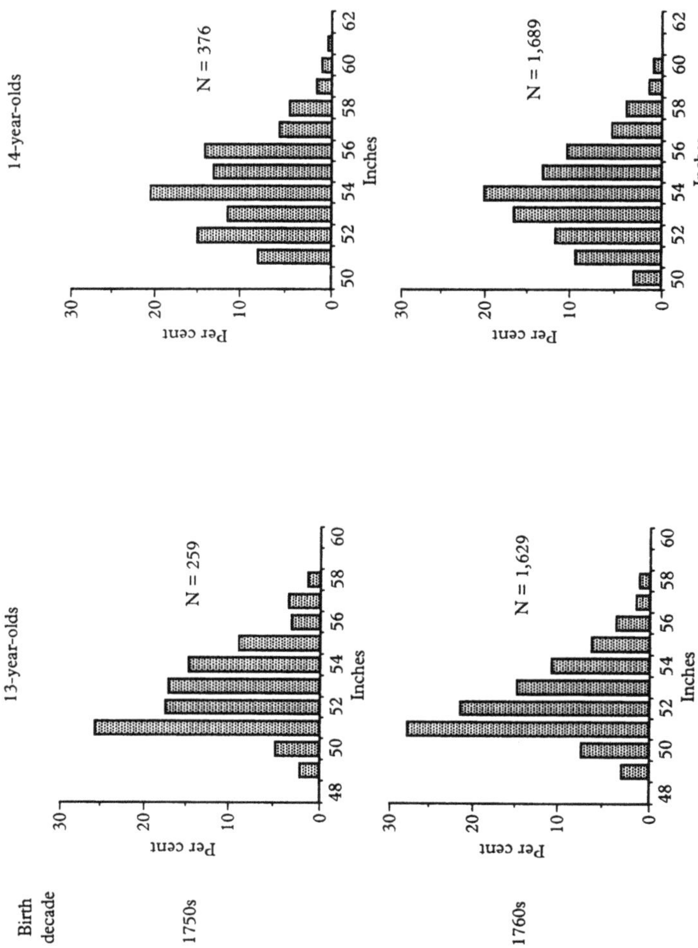
Birth decade
13-year-olds
14-year-olds
1750s
1760s
N = 259
N = 376
N = 1,629
N = 1,689
Per cent
Inches

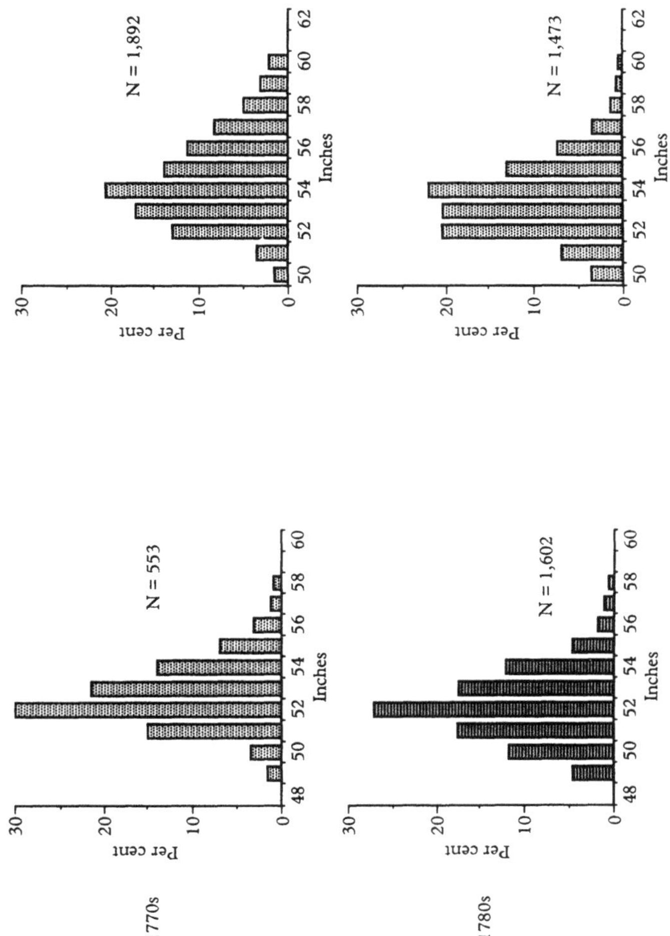


**Figure 2.** *Height distributions of 13- and 14-year-old boys recruited by the Marine Society, 1750-1789*

*Source:* see fig. 1

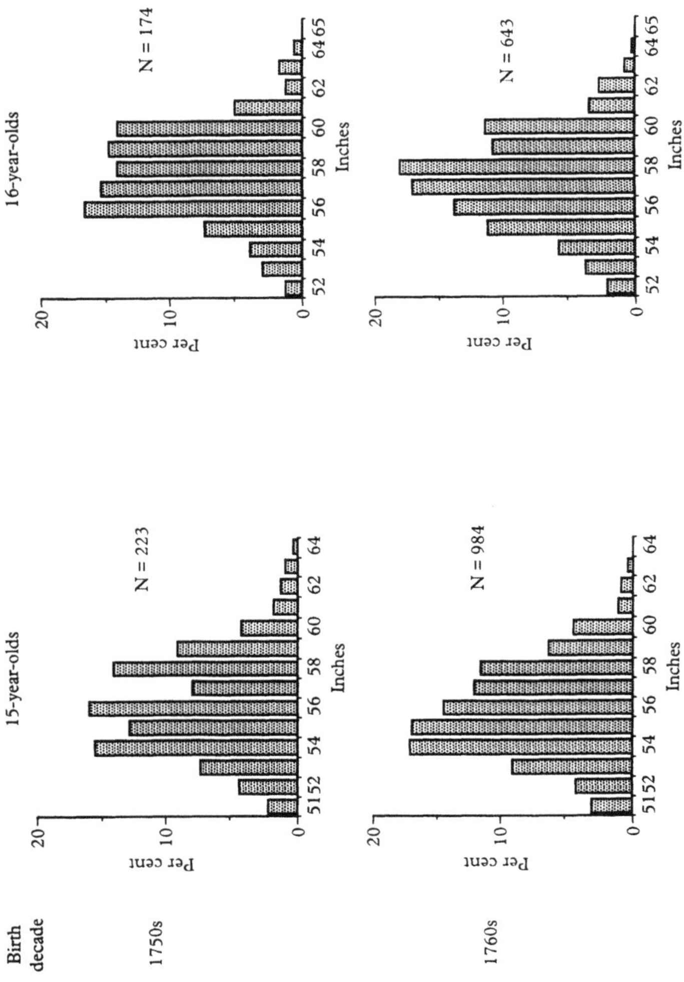
Birth decade
1750s
1760s
15-year-olds
16-year-olds
N = 223
N = 984
N = 174
N = 643
Per cent
Inches
20
10
0
51 52
54
56
58
60
62
64
64 65

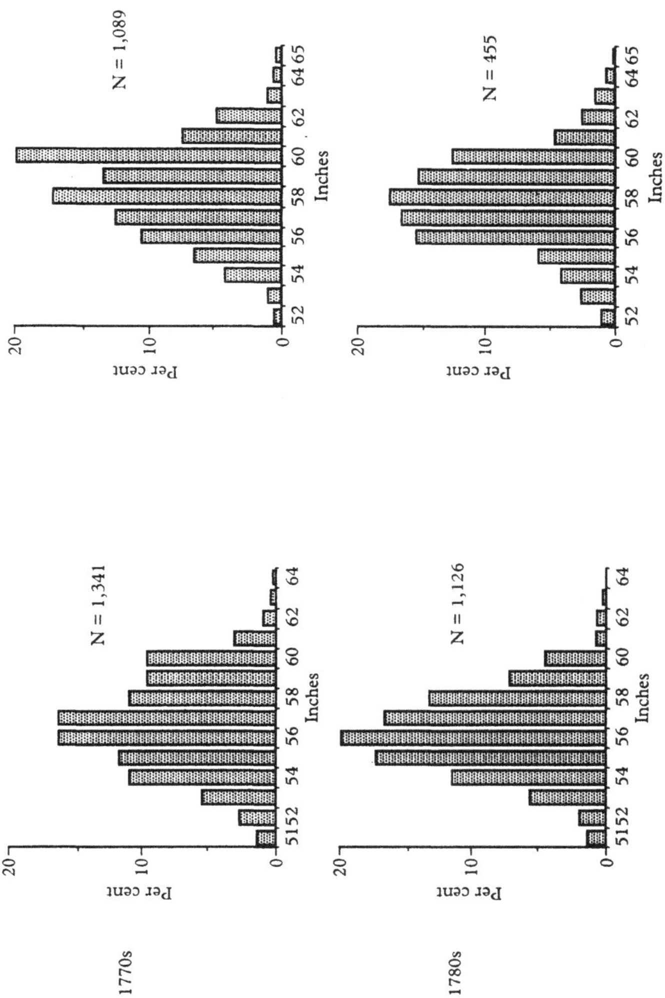


Figure 3. *Height distributions of 15- and 16-year-old boys recruited by the Marine Society, 1750-1789*

*Source:* see fig. 1

Table 3. *Height of poor London boys, various indicators (cm)*

| | Age | | | | | |
|---|---|---|---|---|---|---|
| | *13* | | | *14* | | |
| *Birth decade* | *N* | $\bar{x}$ | $\hat{x}$ | *N* | $\bar{x}$ | $\hat{x}$ |
| 1750-9 | 259 | 133.8 | 131.0 | 376 | 138.1 | 138.2 |
| 1760-9 | 1,629 | 133.0 | 128.7 | 1,689 | 137.2 | 135.8 |
| 1770-9 | 553 | 133.8 | 130.9 | 1,892 | 138.4 | 135.6 |
| 1780-9 | 1,602 | 132.5 | 130.5 | 1,473 | 135.9 | 135.3 |
| 1790-9 | 3,739 | 129.7 | 130.9 | 895 | 134.7 | 133.1 |
| 1800-9[a] | 628 | 133.3 | 130.8 | 210 | 139.1 | 132.0 |

*Notes*: $\bar{x}$ is the simple mean of the sample; $\hat{x}$ is the estimated height of the population obtained with the QBE program after correcting for the minimum height requirement
[a] Estimates for 1800-9 are based on those accepted in the Marine Society before 1818.
*Source*: see tab. 2

Table 4. *Estimated trend of the height of boys (1750-9 = 100)*

| | Age | | | | | |
|---|---|---|---|---|---|---|
| *Birth decade* | *13* | *14* | *15* | *16* | *Average*[a] | *Average*[b] |
| 1750-9 | 100.0 | 100.0 | 100.0 | 100.0 | 100.0 | 100.0 |
| 1760-9 | 100.1 | 99.7 | 99.1 | 99.8 | 99.7 | 99.2 |
| 1770-9 | 99.9 | 99.9 | 100.2 | 100.5 | 100.1 | 99.3 |
| 1780-9 | 99.4 | 98.8 | 99.6 | 100.1 | 99.7 | 98.7 |
| 1790-9 | 99.5 | 98.7 | 98.1 | 99.3 | 99.1 | 98.4 |
| 1800-9 | 99.2 | 99.5 | 99.8 | 99.7 | 99.5 | 99.8 |
| 1810-9 | 98.7 | 99.2 | 99.5 | 99.8 | 99.3 | 100.0 |
| 1820-9 | 98.9 | 99.4 | 99.3 | 100.0 | 99.4 | 100.3 |
| 1830-9 | 98.5 | 99.2 | 99.2 | 98.5 | 98.8 | 100.2 |
| 1840-9 | 97.8 | 98.6 | 98.7 | 97.7 | 98.2 | 98.7 |
| 1850-9 | | 98.7 | 98.6 | 97.1 | 98.1 | |

*Note*: Index is based on the $\bar{x}'$, the simple mean of the reduced sample; see text for definition (boys recruited in 1793 were excluded). The reduced sample includes boys whose height exceeded the following minimum height requirements: prior to recruiting year 1818: 54 inches; between 1818 and 1853: 57 inches; after 1853: 59 inches. Because truncation points shifted, the series was spliced together at the decade of overlap. The splicing was done by recalculating the index for the decade of overlap by using the larger of the two minimum height requirements, and the later index was subsequently shifted by the resultant factor. The index of the trend of 16-year-old boys is based on the reduced sample of those between 54 and 60 inches, because the distribution of this age group is truncated on both sides. The properties of the $\bar{x}'$ statistic for doubly truncated distributions have not been investigated. The indices for the older boys have been included here (with reservation) only because they do not influence the conclusions substantially. For a discussion of the minimum height requirements see Floud, Wachter, and Gregory, *Height, health and history*, p. 164.
[a] simple average of the four previous cols.
[b] calculated from the regression coefficients in tab. 5; see text.
*Source*: see tab. 2.

persists with minor deviations into the first decade of the next century (tables 3 and 4; figure 4, panel A). The recovery among the 14-year-olds after the turn of the century is insufficient to enable them to reach their earlier peak. All in all, there is hardly a sign of a nutritional upswing in these age groups at any time during the eighteenth century, and after a weak and temporary rebound following the cessation of hostilities, the decline in stature persisted well into the nineteenth century.

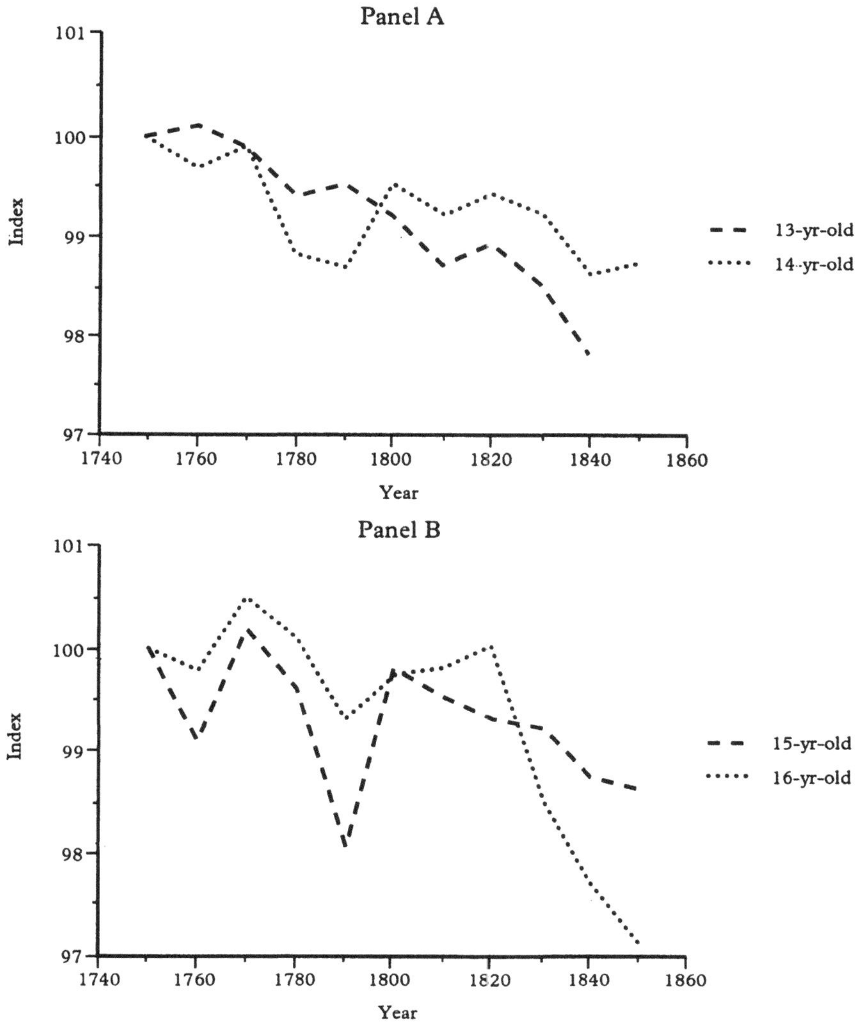


Figure 4. *Estimated trend in the height of boys (1750=100)*

*Source*: tab. 4

The indices of the height of the older boys are more erratic than those of the younger ones, an indication that sampling problems persist even in the restricted part of the sample (table 4). There is a slight upward blip in the index of older boys in the 1770s, but thereafter the trend is essentially negative; the recovery after 1800 is weak and temporary (figure 4, panel B). The subsequent decline correlates well with the pattern found for the younger boys.

It is true that the use of the $\bar{x}'$ statistic sacrifices information. Consequently, the drawback of the $\bar{x}'$ method is that a large part of the sample is discarded. Among the 13-year-olds, for instance, it is clear from the histograms that the 54 inch cut-off point was seldom enforced. However, it is reassuring that the QBE method can be used for the 13-year-olds, because their height distribution was not truncated on the right hand side, and for these boys, the QBE method is essentially in agreement with the $\bar{x}'$ statistic.

It would be a mistake to attach too much significance to the slight deviations from the main trend on the part of the older boys, because these are most probably caused by biases remaining in the sample, produced either by the degree to which tall boys were recruited, or by changes in employment opportunities, which would have influenced the applicant pool drawn upon by the Society. Because these effects would have been less significant among the younger boys, the trend in their height is probably a better estimator of the true secular trend in nutritional status of lower-class boys than the trend in the height of the older boys. Nonetheless, the simple average of the indices of height, even if one includes the older age groups (table 4), confirms the downward trend of the late eighteenth century, with the trough reached in the 1790s. The turn-of-the-century recovery appears to have been small and temporary. The downturn after the 1820s is pronounced, and is consistent among the individual series as well.

The trend can also be estimated using OLS regression analysis on the restricted sample (table 5). This estimate has the advantage that it controls for various socio-economic attributes of the boys, an important factor which was not considered above.[33] The intercepts of the regressions on the restricted sample are obviously biased upward, and the other coefficients, too, are not of the right magnitude, but their sign, and their relative magnitude, are, in fact, correct. The trend obtained from the regression equations (table 4, final column), controlling for the socioeconomic characteristics of the boys, is negative from the 1750s until the second decade of the nineteenth century.[34] The mid eighteenth-century level is not exceeded until the 1820s, and a fairly sharp decline is registered in the 1840s.

Most of the coefficients have the expected sign. Literacy, as indicated by the ability to write, correlates positively with height, while inoculation against smallpox correlates negatively. The former is an indication that literacy was acquired by children whose wealthier than average parents could provide at least a rudimentary form of schooling and additional nutrients.

Boys who were brought to the Society by an indigent relative did not differ much in height from other boys until the mid nineteenth century, when they were shorter. Boys residing in London were appreciably shorter than rural boys throughout the period. This is an indication of the harsh

[33] However, this method has the disadvantage that it does not control for right-hand truncation, which might be the reason why the trend increases more substantially in the early nineteenth century than it does in the other indices.

[34] The trend from the regression estimates was obtained by setting the value of 1750s (140.5 + 0.03) equal to 100, and dividing the other estimates by this value. The regressions were spliced together at the decade of overlap (1790s and 1830s).

Table 5. *Regression equations. Dependent variable: height of boys (cm), reduced samples*

| | *Recruiting dates* | | |
|---|---|---|---|
| | *Before 1818* | *1818-1853* | *After 1853* |
| *Intercept* | 140.5[a] | 147.8[a] | 153.4[a] |
| Age 14 | 2.0[a] | 0.8 | |
| 15 | 5.1[a] | 3.2[b] | 2.3[a] |
| 16 | 8.8[a] | 7.3[a] | 3.3[a] |
| Literate | 0.7[a] | 1.4[a] | 0.3 |
| Smallpox | −0.9[a] | −1.8[a] | |
| Birth decade | | | |
| 1750-9 | 0.3 | | |
| 1760-9 | −0.8[a] | | |
| 1770-9 | −0.7 | | |
| 1780-9 | −1.5[a] | | |
| 1790-9 | −2.0[a] | | |
| 1810-9 | | 0.4[b] | |
| 1820-9 | | 0.7[a] | |
| 1830-9 | | 0.6[a] | |
| 1840-9 | | | −2.3[a] |
| Poor relative | −0.2 | 0.1 | −1.1[b] |
| Residence | | | |
| London | −0.7[a] | −1.1[a] | −0.4 |
| Occupation | | | |
| Pauper | −1.2[b] | −0.7 | 1.2 |
| Agriculture | 0.2 | −0.2 | −1.2 |
| Recruited | | | |
| in 1793 | 1.9[b] | | |
| N | 1445 | 3297 | 423 |
| $R^2$ | 0.3 | 0.3 | 0.2 |
| F | 35.6[a] | 96.2[a] | 9.9[a] |

*Notes*:
[a] Significant at the 5% level;
[b] significant at the 10% level.
Intercept refers to a boy unable to write, not inoculated against smallpox, whose nearest relative was neither destitute nor unknown (including the designation 'other'), whose occupation or whose nearest relative's occupation was not pauper or in workhouse or employed in the agricultural sector, and who was not recruited in 1793; moreover, in the first two regressions the intercept pertains to a 13-year-old boy born between 1800 and 1809, while the intercept of the third regression pertains to a 14-year-old boy born in the 1830s. The first regression is limited to boys above 54 inches, the second to those above 57 inches, and the third to those above 59 inches.
*Source*: see tab. 2.

epidemiological environment and the higher food prices in cities.[35] Paupers were shorter than average, particularly in the eighteenth century.[36] Having an agricultural occupation has a negative but small coefficient in the mid nineteenth-century component of the sample. Finally, boys recruited in 1793 were 1.9 cm. taller than average.

[35] Floud and Wachter discuss the socioeconomic correlates of height in 'Poverty and physical stature', pp. 439-40. Their results are in essential agreement with the ones reported here.
[36] Includes those who were in a workhouse or whose nearest relative was in one.

## II

The difficulties of analysing the military data are similar to those already discussed. Floud, Wachter, and Gregory's height estimates are again quite unstable, suggesting unreliability.[37] Their strategy was to smooth out the variations using moving averages. This is quite risky because the procedure still bases the analysis on potentially questionable estimates. One of the main reasons for the erratic behaviour of the estimates is that the records relate to two distinct groups: the army, and the Royal Marines. Floud, Wachter, and Gregory estimated the mean height of the population using either the QBE or the RSMLE procedures after combining the two samples.[38] However, the frequency distributions of the two samples indicate quite clearly that the pooling is not warranted, because the recruits to the two branches were selected on different principles (figure 5). The height distributions of the Royal Marines recruits are clearly well to the left of Army recruits. The marines could be smaller because at sea height was not an advantage, but on land the length of the stride, for instance, depended on it. Thus in practice the minimum as well as the maximum height requirements in the two branches differed substantially. The two samples should not be merged, because there is no reason to suppose that the composite distribution would reflect the true height distribution of the population. Its shape would be influenced by such factors as the number of men recruited into each branch and by the relative sample sizes drawn from them, as well as by the differences between the truncation points.

Moreover, the height distributions seldom appear normal in either sample,[39] and among the Marines right-hand truncation is also evident at

[37] For instance, the height of adult soldiers is estimated to have risen by an implausible 3 cm between 1757 and 1762. They do realize that the 'scatter . . . reflects the uncertainties in the size of the correction for under-representation of short recruits', but have not eliminated dubious estimates: Floud, Wachter, and Gregory, *Height, health and history*, p. 139.

[38] They did experiment with a large number of combinations of the two samples by giving different weights to them. However, these procedures did not seem to improve the estimates, and finally the decision was made to merge the two samples, with one exception, i.e. after *c.* 1840 they analysed the army records without the Marines. They realize just how crucial these decisions were but apparently did not try to analyse the two series separately for the pre-1840 period. Once the decision to merge the two datasets was made their task became practically hopeless. Their justification is straightforward: 'many recruits were apparently indifferent between units and . . . the height standards essentially operated to sort men to units dependent upon their height and the height demanded by the units'; ibid., pp. 139, 150. Unfortunately, this hypothesis is inaccurate, as can be ascertained from a perusal of the sample distributions (figure 5). For instance, the distributions of the 1750s for the army peak at 68 inches, while those of the Royal Marines peak between 64 and 66 inches. Their composite could not be a normal distribution. This is the case because their hypothesis was tantamount to assuming that anyone who applied to one of the branches but did not qualify on account of height requirements, would apply and be accepted into the other branch of the military. This might not have been the case for two reasons: first, some men might have been willing to serve only in one branch of the military, and second, the number of recruits in each branch was set independently and fluctuated considerably from year to year: ibid., pp. 45, 69, 139, and 150. This meant that the two branches represented different shares of their respective portions of the underlying height distributions of the male population. The difficulty is compounded by the fact that the number of observations in the two branches of the military in the sample varied greatly from period to period. For instance, in the 1780s, the number in the army sample is 200 per cent of the number in the Marine sample. In the 1740s, however, the corresponding share is only 10 per cent.

[39] Another reason why the army sample has unusual frequency distributions is that various units, such as the Guards or the Military Train, with different minimum height requirements, were merged

times (figure 5).[40] Therefore, the use of neither the QBE nor the RSMLE program is warranted. The secular trend of the height of the army recruits can, however, be estimated using the reduced sample ($\bar{x}'$) (the Marine sample is unsuitable because it is doubly truncated).[41] The filters used were the largest of the minimum height requirements (68 inches before 1780 and 67 inches thereafter). The two sub-periods were allowed to overlap, enabling the two resultant indices to be spliced together. The indices were first calculated for army recruits in three age categories: 18 to 19; 20 to 23; and 24 to 29 (table 6). Subsequently, these three indices were averaged, using the number of observations in each series as weights. The indices for recruits born in Scotland, England (including Wales), and Ireland were calculated separately.

In general, the resultant trends in the stature of soldiers averaged across all ages are quite similar in the three groups, and also bear close resemblance to the one found for the children. Contrary to the pattern posited by Floud, Wachter, and Gregory, the trend found is negative in the eighteenth century throughout the British Isles (figure 6). The post-Napoleonic upturn is evident among the birth cohorts of the 1810s. However, the peak of the 1740s is not reached, and a pronounced downturn follows in the 1830s or 1840s. This latter downturn does agree with the findings reported by Floud and his coauthors.[42]

Deviations from this general tendency are minor. Because human stature is not likely to fluctuate abruptly from one decade to another, single deviations from the main trend ought not be given much credence.[43] That the downswing in nutritional status during the second half of the eighteenth century was milder in Ireland could indicate the biological advantages of isolation from the main impact of industrialization and urbanization. Furthermore, the nineteenth-century downturn began in Ireland and in Scotland in the 1830s, a little earlier than in England (figure 6, panel B). This might be indicative of Malthusian population pressure and of the

into one sample; ibid., p. 113. In addition, the sample would be influenced by the vagaries of labour market conditions.

[40] This is clearly visible among the birth cohorts of the 1730s, 1770s, 1790s, and 1800s, for instance; in order to crowd as many men as possible into a limited space, it was desirable to impose an upper bound, even if a flexible one, on the height of men accepted into the Marines. Apparently the height standards for the Marines were not published; ibid., p. 133.

[41] This is the case, because only a small range of the height distribution was ever allowed to enter the Marines. Relative shifts in frequency within this small range may not occur at all, or be difficult to ascertain.

[42] Ibid., pp. 28, 151, 275.

[43] Nonetheless, it is worth noting that the decline of the 1750s is sharpest among the Irish soldiers. The sample size of the Irish-born in the 1740s is very small, but if the finding is dependable it may indicate that the upsurge in Irish agricultural exports to England after the mid century had a relatively large impact on Irish nutritional status. Among the English soldiers a rebound is visible among the birth cohorts of the 1760s (index = 99.9). Too much significance ought not to be accorded to this deviation from the general negative trend, as it is likely to be a statistical artefact for several reasons. The rebound is not evident either among the Scottish or the Irish recruits. It is also not apparent among the English 18- and 19-year-old soldiers (index = 99.7). It is evident among the adult soldiers, but this may not be a reliable indication of nutritional trends, because of the small number of observations, and because of the unusual frequency distribution of the birth cohort of the 1760s (figure 5). The finding is more likely to be due to changes in recruiting practices than to actual changes in nutritional status.

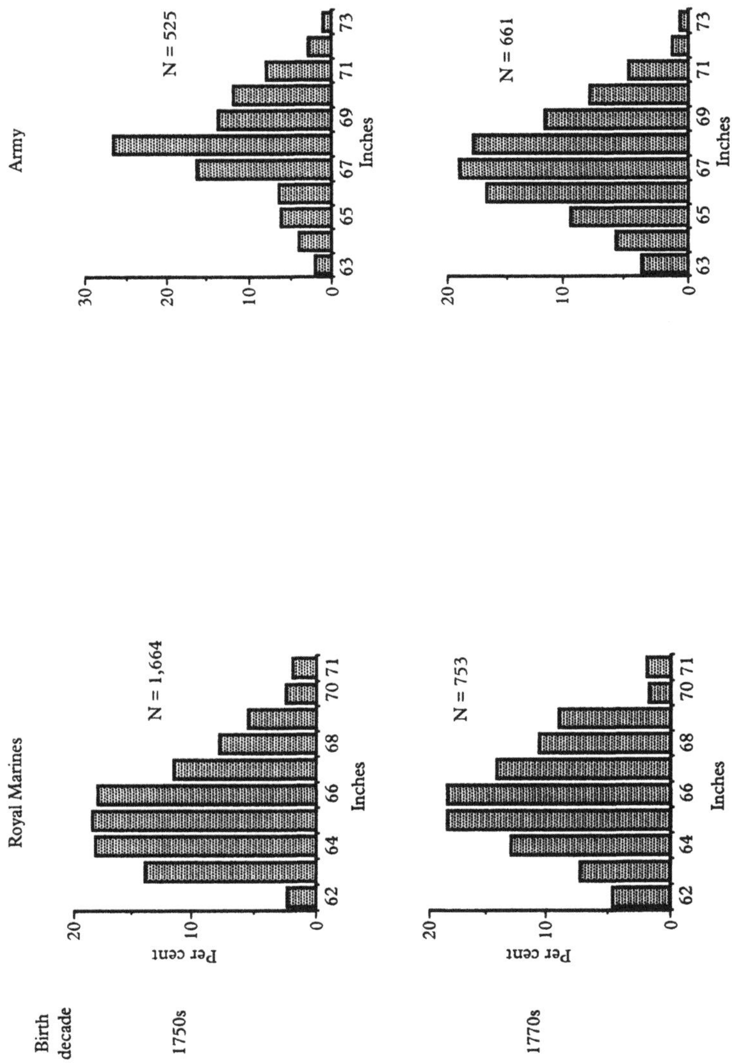
Birth decade
Royal Marines
Army
1750s
1770s
Per cent
Inches
N = 1,664
N = 525
N = 753
N = 661

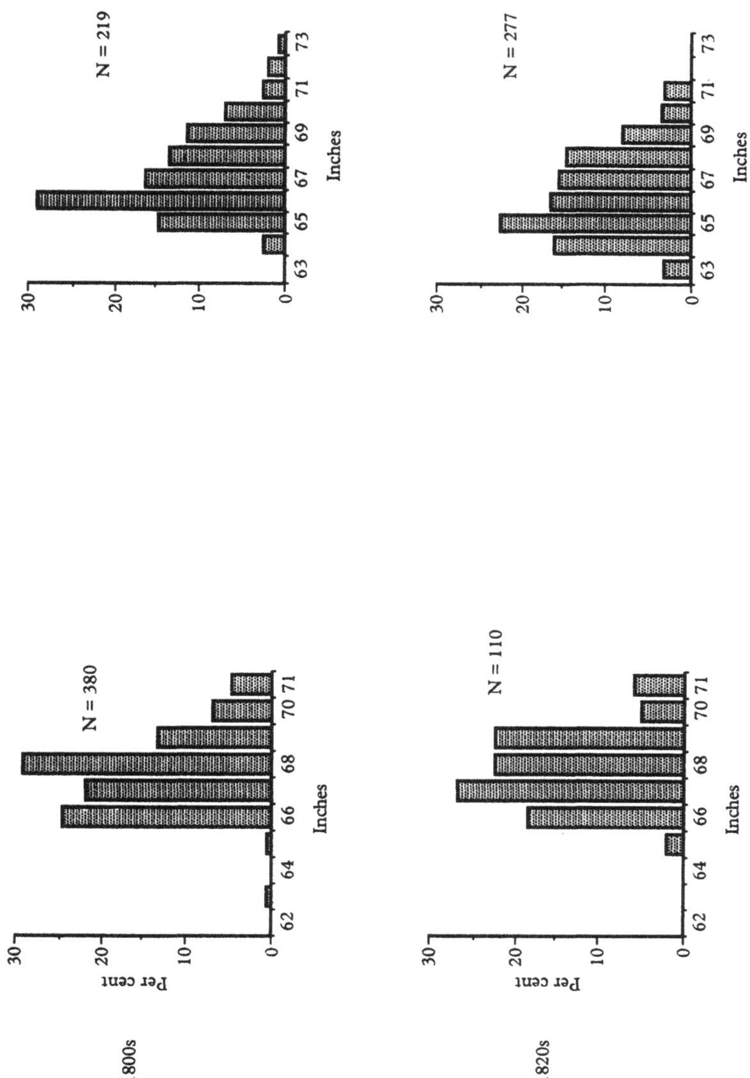


*Figure 5. Height distributions of adult English soldiers (selected decades)*

*Source:* see fig. 1

Table 6. *Indices of the height of soldiers in the British army by age group, 1740-1869*

| | England[a] | | | | | | Scotland | | | | | |
|---|---|---|---|---|---|---|---|---|---|---|---|---|
| | 18-19 | | 20-23 | | 24-29 | | 18-19 | | 20-23 | | 24-29 | |
| | N | Index | N | Index | N | Index | N | Index | N | Index | N | Index |
| 1740-9 | 18 | 100.0 | 138 | 100.0 | 86 | 100.0 | 32 | 100.1 | 133 | 100.1 | 45 | 100.0 |
| 1750-9 | 219 | 99.7 | 504 | 99.2 | 335 | 99.6 | 195 | 100.1 | 320 | 99.7 | 72 | 99.8 |
| 1760-9 | 345 | 99.7 | 806 | 99.9 | 121 | 100.2 | 353 | 99.8 | 570 | 99.8 | 71 | 99.7 |
| 1770-9 | 211 | 99.6 | 319 | 99.4 | 293 | 99.4 | 180 | 99.6 | 141 | 99.5 | 84 | 99.4 |
| 1780-9 | 323 | 99.1 | 571 | 99.0 | 421 | 99.2 | 202 | 98.7 | 164 | 99.6 | 76 | 99.5 |
| 1790-9 | 348 | 98.9 | 489 | 98.8 | 176 | 99.2 | 121 | 99.0 | 64 | 99.0 | 39 | 99.8 |
| 1800-9 | 321 | 98.6 | 343 | 98.8 | 115 | 99.5 | 151 | 98.9 | 131 | 98.8 | 26 | 99.2 |
| 1810-9 | 626 | 99.0 | 761 | 99.2 | 62 | 99.3 | 213 | 99.5 | 161 | 99.7 | 4 | 101.1 |
| 1820-9 | 575 | 99.1 | 400 | 98.9 | 119 | 99.1 | 148 | 100.0 | 141 | 100.1 | 43 | 99.6 |
| 1830-9 | 919 | 99.0 | 1087 | 99.1 | 340 | 99.2 | 56 | 98.9 | 92 | 99.2 | 42 | 99.2 |
| 1840-9 | 339 | 98.7 | 573 | 98.9 | 244 | 99.2 | 46 | 99.0 | 83 | 98.8 | 122 | 99.8 |
| 1850-9 | 850 | 98.5 | 618 | 98.6 | 115 | 98.7 | 107 | 99.1 | 115 | 98.6 | 37 | 99.3 |
| 1860-9 | 104 | 98.4 | | | | | | | | | | |

| | Ireland | | | | | | Average[b] | | |
|---|---|---|---|---|---|---|---|---|---|
| | 18-19 | | 20-23 | | 24-29 | | Eng.[a] | Scot. | Ire. |
| | N | Index | N | Index | N | Index | | | |
| 1740-9 | 33 | 100.2 | 82 | 99.8 | 14 | 100.5 | 100.0 | 100.1 | 100.0 |
| 1750-9 | 56 | 99.7 | 174 | 99.2 | 78 | 99.3 | 99.4 | 99.9 | 99.3 |
| 1760-9 | 127 | 99.4 | 117 | 99.6 | 92 | 99.2 | 99.9 | 99.8 | 99.4 |
| 1770-9 | 163 | 99.6 | 317 | 99.3 | 157 | 99.5 | 99.5 | 99.5 | 99.4 |
| 1780-9 | 134 | 99.5 | 229 | 99.0 | 171 | 99.2 | 99.1 | 99.2 | 99.2 |
| 1790-9 | 334 | 99.1 | 381 | 99.0 | 197 | 99.2 | 98.9 | 99.1 | 99.1 |
| 1800-9 | 457 | 99.1 | 421 | 99.3 | 59 | 99.3 | 98.8 | 98.9 | 99.2 |
| 1810-9 | 507 | 99.6 | 457 | 99.4 | 23 | 100.1 | 99.1 | 99.6 | 99.5 |
| 1820-9 | 170 | 99.4 | 154 | 99.5 | 21 | 99.0 | 99.0 | 100.0 | 99.4 |
| 1830-9 | 211 | 99.0 | 287 | 99.0 | 192 | 99.0 | 99.1 | 99.1 | 99.0 |
| 1840-9 | 291 | 99.0 | 451 | 99.1 | 150 | 98.9 | 98.9 | 99.3 | 99.0 |
| 1850-9 | 212 | 99.3 | 199 | 99.0 | 43 | 99.5 | 98.6 | 98.9 | 99.2 |

*Notes*: The base for the index is the mean height of English soldiers born in the 1740s, of the particular age group listed. Index is based on the reduced sample. Filters used: 1740-79: 68 inches; 1780-1869: 67 inches. The two segments were allowed to overlap for the decade of the 1770s and the latter part was adjusted to coincide with the earlier segment.
[a] includes Wales
[b] of the three age groups
*Source*: see tab. 2.

impending nutritional crisis of the 1840s. Further, only among the Irish does one find an increase in stature in the 1850s. As one might expect, the substantial decline in population after the famine of the 1840s meant that the nutritional status of those born subsequently could improve. Thus the secular trend in Irish nutritional status deviates in a few, but important, respects from the British one. Yet between 1770 and 1829 the correlation between the indices is particularly good.

The Scottish secular trend is closer to the English one than is the Irish. This is to be expected as Scotland was integrated earlier into the English labour and product markets, and into its disease pool. The one major difference in the Scottish pattern is that the post-Napoleonic rebound in height is much more pronounced there, perhaps even unreasonably so.

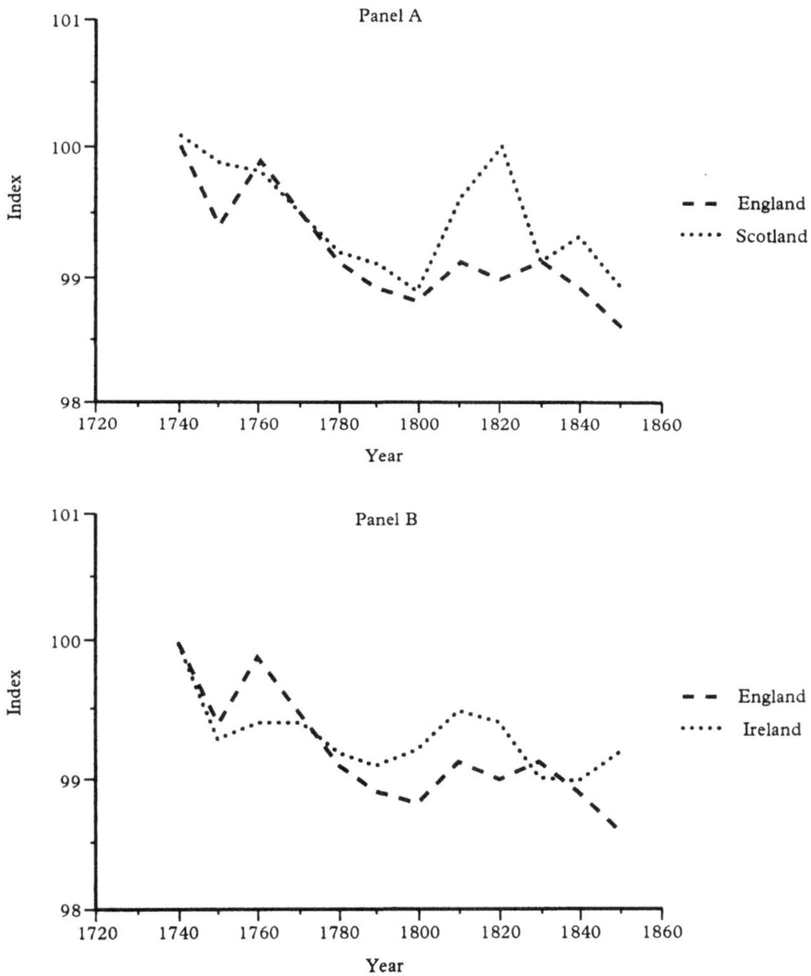


Figure 6. *Estimated trend in the height of soldiers in the army, 1740-1859 (average of all ages) (100 = the English national average, 1740-9; see tab. 6)* *(Panel C overleaf)*

*Source*: tabs. 6 and 7

Irish as well as Scottish soldiers tended to be taller than English ones after the 1780s. Similar patterns have been found in many other datasets as well: at the initial stages of economic development, agricultural regions, remote from major markets, tended to enjoy a higher nutritional status than more developed areas.[44] This cross-sectional pattern is practically identical to the one found by Floud and his collaborators.[45]

[44] Komlos, *Nutrition and economic development*, p. 96.
[45] Floud, Wachter, and Gregory, *Height, health and history*, pp. 200-6.

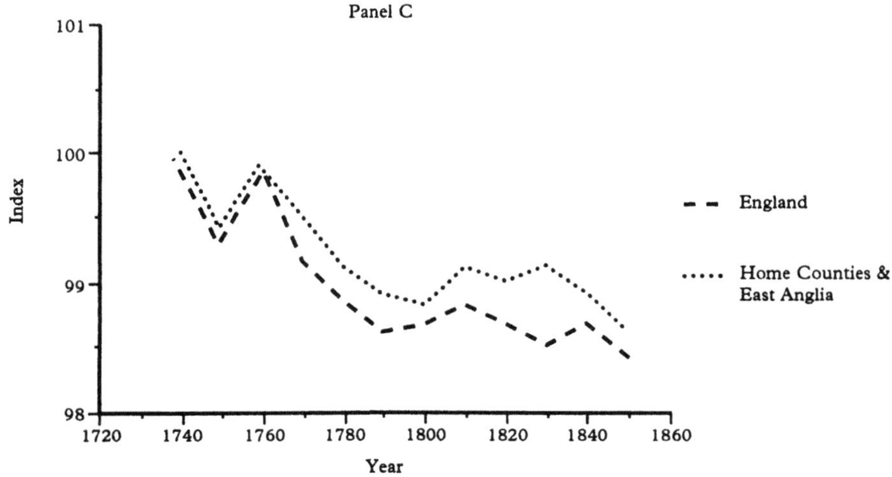


Figure 6. *(continued)*

Table 7. *Index of the height of English[a] soldiers by region of birth (100 = the English national average, 1740-9; see tab. 6)*

| | (1) *Home Counties*[b] | (2) *North of England*[c] | (3) *Midlands West*[d] | (4) *East* | (5) *South west* | (6) *East Anglia* | (7) *Home Counties and East Anglia ((1) + (6))* | (8) *Midlands ((3) + (4))* | (9) *North and SW ((2) + (5))* |
|---|---|---|---|---|---|---|---|---|---|
| 1750-9 | 99.3 | 100.0 | 99.5 | 99.6 | 99.5 | 99.5 | 99.4 | 99.6 | 99.8 |
| 1760-9 | 99.8 | 100.0 | 100.1 | 99.7 | 100.1 | 100.1 | 100.0 | 99.9 | 100.1 |
| 1770-9 | 99.2 | 99.5 | 99.5 | 99.5 | 99.4 | 99.1 | 99.2 | 99.5 | 99.5 |
| 1780-9 | 98.8 | 99.0 | 99.1 | 99.2 | 99.0 | 98.9 | 98.9 | 99.2 | 99.0 |
| 1790-9 | 98.6 | 98.9 | 99.0 | 98.7 | 99.2 | 98.6 | 98.6 | 98.9 | 99.1 |
| 1800-9 | 98.7 | 98.4 | 98.8 | 99.0 | 98.6 | 98.6 | 98.7 | 98.9 | 98.5 |
| 1810-9 | 99.1 | 99.6 | 99.2 | 98.9 | 99.3 | 98.5 | 98.8 | 99.2 | 99.5 |
| 1820-9 | 98.7 | 99.6 | 99.4 | 99.1 | 99.4 | 98.6 | 98.7 | 99.3 | 99.5 |
| 1830-9 | 98.4 | 99.3 | 99.0 | 98.9 | 98.9 | 98.6 | 98.5 | 99.0 | 99.1 |
| 1840-9 | 98.6 | 98.7 | 99.0 | 98.8 | 99.1 | 98.7 | 98.7 | 98.9 | 98.9 |
| 1850-9 | 98.5 | 98.9 | 98.5 | 98.5 | 98.8 | 98.3 | 98.4 | 98.5 | 98.9 |

*Notes*:
[a] excluding Welsh soldiers
[b] does not include London itself
[c] Northumberland, Westmorland, Cumberland, Durham, North and East Ridings of Yorkshire, Cheshire
[d] includes the West Riding of Yorkshire
*Source*: see tab. 2

Within England itself regional patterns were generally similar to the national average: there was a decline in the eighteenth century, and a small and temporary upturn in the early nineteenth (table 7). The trends in the height of the recruits born in the Home Counties and in East Anglia resemble one another sufficiently to merge the two into a single series (column (7) in table 7).[46] The nutritional status of these recruits (born in the south east) was well below the English average after the 1760s, and the nutritional gap widened over time (figure 6, panel C).

In contrast, the recruits born in the remote regions, such as the north and the south west, tended to be taller than average.[47] Since the trends in these two areas were similar they were subsequently merged (table 7). A similar comparison of the heights of recruits born in the east and west midlands showed that they were also similar to one another. Finally these three areas were compared to each other. The recruits born in the south east were consistently shorter than those born in either the midlands, or the south west and the north. The nutritional status in the midlands was similar to the English average.

Finally, it is useful to note that the height of recruits born in the regions of England most remote from the London market (north and south west) was quite similar to the Irish and Scottish average, with the latter perhaps still enjoying a slight advantage after the 1780s (figure 7, panels B and C). Thus the frequently noted Irish and Scottish height advantage is true only when comparison is made with those Englishmen born in the more commercialized, industrialized, and urbanized midlands and south east (figure 7, panel A).[48]

## III

The results of this study underline vividly the difficulties associated with analysing height data whose frequency distributions are truncated. A clear lesson is that the QBE and the RSMLE programs should be used only after careful consideration of the frequency distributions. They should be used only when the distributions are at least close to normal. Otherwise it is easy to be misled by the unstable estimates produced when using irregular frequency distributions. If the sample distribution is not Gaussian, the results obtained can be quite inaccurate, and unsuitable for estimating the trend in height. In contrast, in this study a more stable, hence more reliable method was used to ascertain the trend in height, the simple mean of the reduced sample.

The cross-sectional pattern shows that in general the further a soldier was born from London and the industrializing areas, such as the midlands, the

[46] Two of the regions, Wales and the city of London, did not have sufficient recruits for meaningful analysis and were disregarded.

[47] The West Riding of Yorkshire was excluded from this group. Because of its industrial character it was classified with the midlands. The cross-sectional pattern found by Floud, Wachter, and Gregory was quite similar to the ones posited here; *Height, health and history*, p. 201.

[48] This discussion of the military recruits has been limited to soldiers in the army. The Royal Marines data are much more difficult to analyse because of the probable truncation on both sides of the frequency distribution (figure 5). It is possible to undertake an analysis that is limited to that range of the height distribution which was accepted into the Royal Marines during a particular time interval, as was done for the 16-year-old boys in the Marine Society. That is, one can restrict the sample to those above the greatest of the minimum height requirements, and below the least of the maximum height requirements. Such an analysis produces similar results to that based on army recruits but the properties of the $\bar{x}'$ statistic for doubly truncated distributions have not been investigated, which presents problems of interpretation, and the results are therefore not reported here.

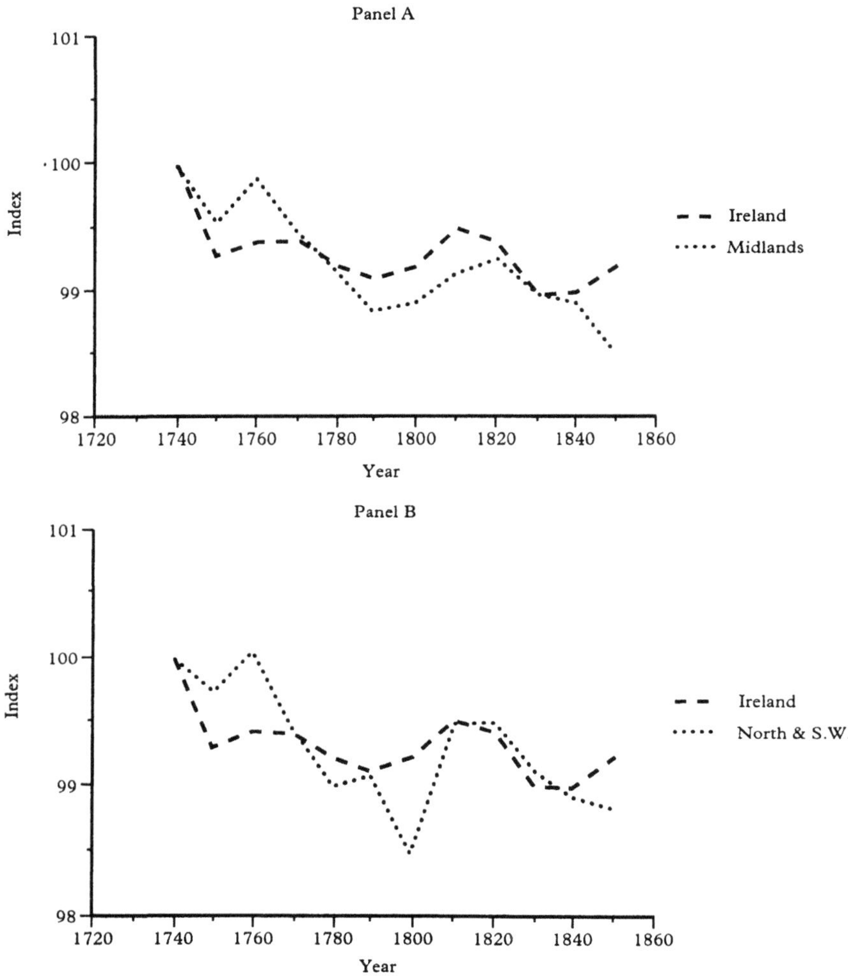


Figure 7. *Estimated trend in the height of soldiers in the army, 1740-1859 (average of all ages) (100 = the English national average, 1740-9; see tab. 6)*
*Source*: tabs. 6 and 7

higher was his nutritional status. In this respect my results are in agreement with those published by Floud, Wachter, and Gregory; we also agree on the nineteenth-century trend in nutritional status.

We differ fundamentally, however, over the direction of the eighteenth-century trend. Floud and his coauthors believe that height increased. This is extremely odd as their own evidence shows a monotonically declining trend among the military recruits of all age groups between *c.* 1762 and

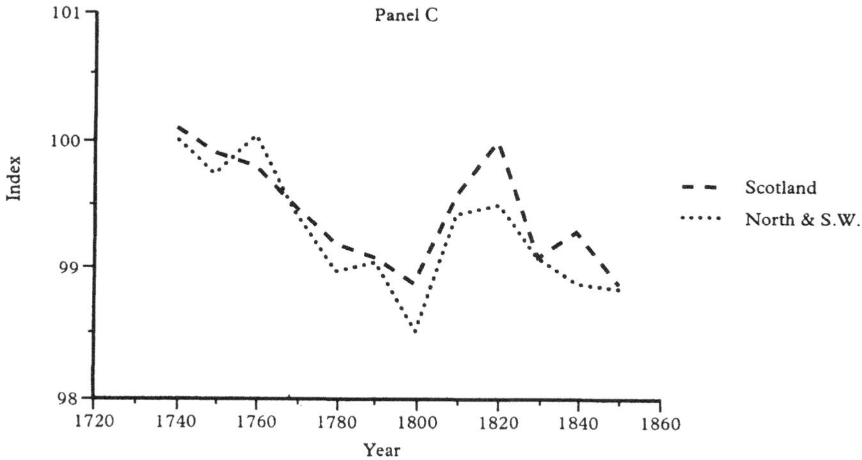


Figure 7. *(continued)*

1800,[49] although the tendency gets lost in their smoothing procedure. If it had not been for this technique our results would have differed only for one or two turn-of-the-century decades and even for those we would have differed only in the magnitude of the change, not in its direction. Given the different statistical techniques we used, the similarity between our raw results is actually quite remarkable. The same could not be said for our analysis of the trend in the height of the Marine Society boys. In that respect our results are irreconcilable.

The key issue is not what would have resulted if a different smoothing procedure had been used or if the Royal Marine data had not been merged with those of the infantry. It is rather that, in contrast with the findings of Floud and his colleagues, my statistical approach suggests that the nutritional status of the British and Irish populations was deteriorating during the second half of the eighteenth century.[50] This finding has just been corroborated by an independent investigation of English convicts transported to Australia.[51] In other words, contrary to the claims of Floud and his collaborators, there are hardly any signs of an improving biological standard of living among the lower segments of society during the first century of the industrial revolution. Conditions improved somewhat after the end of the Napoleonic wars, but the eighteenth-century maximum was not reached. After a weak and temporary recovery a downward tendency is found, beginning with the birth cohorts of the 1830s. The improvements posited by Floud, Wachter, and Gregory for the period prior to 1830 do not appear

[49] Floud, Wachter, and Gregory, *Height, health and history*, pp. 140-9.
[50] Moreover, it is possible that in England the beginning of this deterioration in nutritional status preceded the period under consideration; Komlos, 'A Malthusian episode revisited'.
[51] Nicholas and Steckel, 'Heights and health of workers'.

to have been grounded in changes in nutritional status, but were an artefact of the kind of statistical analysis and sampling strategies they used.

The secular trend in the nutritional status of the British and Irish populations thus turns out to be very similar to that experienced by the contemporary Swedish and Habsburg ones: a decline at the end of the eighteenth century, followed by a recovery, albeit partial, after 1815.[52] The imminent threat of a Malthusian crisis was not confined to continental Europe, but was also evident in the UK.[53] The results obtained are therefore even more pessimistic than those of Floud, Wachter, and Gregory. In spite of the remarkable growth in GDP as a consequence of the industrial revolution, the lowest segments of society apparently saw little or no improvement in their biological standard of living in the first century of the most momentous recorded expansion in industrial productivity. In an aggregate sense, society was saving and investing more out of the increments generated by industrialization than it was consuming. It is also probable that the distribution of income was becoming more skewed. Thus a rising average for the whole society might obscure a decline in the biological standard of living among the lower classes.

The decline in the biological well-being of the population of the UK supports those who point to widespread poverty in the early stages of the industrial revolution. Examining the course of consumption during the early nineteenth century, Mokyr has argued recently that the idea that 'before 1850 the majority of Britons benefited little from the Industrial Revolution is not yet dead'. 'If real wages were rising so fast', he asks, 'why . . . did consumers not consume more?'[54] Why, indeed! The 'endemic' nature of food riots throughout England during the last few decades of the eighteenth century also supports this more pessimistic view.[55] The present study confirms Mokyr's scepticism of the accuracy of the real wage estimates, and suggests that, at least in the dimension of welfare here considered, matters were actually deteriorating in Great Britain as well as throughout the continent.

The fact that the decline in nutritional status extended across Europe, coupled with the fact that its beginning antedated the industrial expansion not only in England, but even in such relatively backward countries as Hungary and Sweden, points to the demographic explosion and not to the industrial revolution itself as the original cause of the adverse developments.[56]

[52] Komlos, 'Patterns of children's growth'; *idem*, *Nutrition and economic development*; Sandberg and Steckel, 'Heights and economic history'. The episode also bears resemblance to the experience of the American population during the Antebellum period when rapid population growth and urbanization was accompanied by both an industrial expansion and a fall in nutritional status; Komlos, 'The height and weight of West Point cadets'. For a similar pattern in Canada see Ward and Ward, 'Infant birth weight and nutrition'.

[53] To be sure, the decline in stature among the 'Oliver Twists' of England was less than the one experienced in east-central Europe. This is understandable, insofar as the Marine Society boys were so short that they must have been close to their biological minimum height and hence a further decline in their stature was hardly possible.

[54] Mokyr, 'Consumption in the industrial revolution', p. 90.

[55] Rose, 'Price riots', p. 283. See also Jackson, 'Growth and deceleration'; Schwarz, 'Standard of living'.

[56] Komlos, 'A Malthusian episode revisited'.

Europe was entering a critical phase after the middle of the eighteenth century. Demographic expansion was bringing with it the ancient threat of food shortages, widespread hunger, and, as in the case of the French Revolution, political instability. It was not at all clear that these threats could be overcome. During the last similar episode of expansion in the sixteenth century there had not been a happy ending. However, this time it was to be different. Much progress had been made during the preceding century-and-a-half in practically all branches of the economy, including transportation and agricultural production, and in the related area of technological know-how. New sources of food imports were available, and previously unknown foods were introduced from the New World. Merchants were wealthier, and governments were better organized. These and other factors allowed societies to break through the Malthusian ceiling, which was determined largely by biological needs.

In the eighteenth century this age-old contest between the rate at which the population was expanding and the growth of food production was decided at the last minute in the latter's favour. Although malnutrition was becoming widespread, fewer people fell below the biological minimum than during earlier periods of rapid demographic expansion; hence, mortality rates did not increase as, for instance, they had in the seventeenth century. Consequently, a general and major subsistence crisis was ultimately avoided. Population growth, and therefore economic activity, was freed from the constraint imposed by the supply of calories. The industrial revolution was under way.

Industrialization did not immediately bring about a general improvement in well-being, but it did enable the population to survive, by creating additional income that could be traded for nutrients even if not on the most propitious terms.

What was the role of the industrial revolution in this? An answer might be ventured by asking the hypothetical question, what would have happened to the British, or even the European population in the absence of the industrial revolution? The most plausible projection is that without a rise in labour productivity, diminishing returns to labour in the agricultural sector would have reduced the nutritional status of the population until Malthusian positive checks operated, as they had in the past. Of course, the crude death rate would have risen during the subsistence crisis, and many would have perished. The increase in mortality, however, would have had a moderating effect on population growth, and a homeostasis would have been established at a lower population density. For those who survived the biological standard of living would have been greater than that obtained with industrialization. How can one balance the pains of the casualties with the gains of those who remained alive after such a crisis? Perhaps that imponderable question cannot be answered. Yet the ultimate culprit was clearly the demographic expansion; the industrial revolution did at least allow the population to survive. The skewing of the distribution of income is yet another issue which prevented the lower classes from enjoying a greater share of the fruits generated by the industrial revolution.

In conclusion I agree with Floud, Wachter, and Gregory that the anthropometric indices illuminate the British and Irish experience of the industrial revolution. I also agree that, in the main, the above results 'suggest that conventional measures of living standards such as real wages or the level of gross domestic product per capita, which show relatively steady upward trend movements, have misled scholars concerned with the impact of industrialization and economic development on the British people'.[57]

[57] Floud, Wachter, and Gregory, *Height, health and history*, p. 154.

**Footnote references**

Brown, J. C., 'Condition of England and the standard of living: cotton textiles in the northwest, 1806-1850', *J. Econ. Hist.*, L (1990), pp. 591-614.

Crafts, N. F. R., 'English workers' real wages during the industrial revolution: some remaining problems', *J. Econ. Hist.*, XLV (1985), pp. 139-44.

Eveleth, P. B. and Tanner, J. M., *Worldwide variation in human growth* (Cambridge, 1976).

Floud, R. and Wachter, K. W., 'Poverty and physical stature: evidence on the standard of living of London boys, 1770-1870', *Soc. Sci. Hist.*, 6 (1982), pp. 422-52.

Floud, R., Wachter, K. W., and Gregory, A., *Height, health and history: nutritional status in the United Kingdom, 1750-1980* (Cambridge, 1990).

Fogel, R. W., 'Nutrition and the decline in mortality since 1700: some preliminary findings', in S. L. Engerman and R. Gallman, eds., *Studies in income and wealth*, 51: *long-term factors in American economic growth* (Chicago, 1987), pp. 439-555.

Jackson, R. V., 'Growth and deceleration in English agriculture, 1660-1790', *Econ. Hist. Rev.*, 2nd ser., XXXVIII (1985), pp. 333-52.

Komlos, J., 'Patterns of children's growth in east-central Europe', *Annals Hum. Biol.*, 13 (1986), pp. 33-48.

Komlos, J., 'The height and weight of West Point cadets: dietary change in antebellum America', *J. Econ. Hist.*, 47 (1987), pp. 897-927.

Komlos, J., 'A Malthusian episode revisited: the height of indentured servants in colonial America, 1720-1770', *Econ. Hist. Rev.*, XLVI (1993), forthcoming.

Komlos, J., *Nutrition and economic development in the eighteenth century Habsburg monarchy: an anthropometric history* (Princeton, N.J., 1989).

Komlos, J. and Kim, J. H., 'Estimating trends in historical heights', *Hist. Methods*, 23 (1990), pp. 116-20.

Mokyr, J., 'Is there still life in the pessimist case? Consumption during the industrial revolution, 1790-1850', *J. Econ. Hist.*, XLVIII (1988), pp. 69-92.

Mokyr, J. and Ó Gráda, C., 'Poor and getting poorer? Living standards in Ireland before the Famine', *Econ. Hist. Rev.*, 2nd ser., XLI (1988), pp. 209-35.

Nicholas, S. and Steckel, R. H., 'Heights and health of English workers during the early years of industrialization, 1770-1815', *J. Econ. Hist.*, 51 (1991), pp. 937-57.

Rose, R. B., 'Eighteenth-century price riots and public policy in England', *Internat. Rev. Soc. Hist.*, 6 (1961), pp. 277-92.

Sandberg, L. G. and Steckel, R. H., 'Heights and economic history: the Swedish case', *Annals Hum. Biol.*, 14 (1987), pp. 101-10.

Schwarz, L. D., 'Standard of living in the long run: London, 1700-1860', *Econ. Hist. Rev.*, 2nd ser., XXXVIII (1985), pp. 24-41.

Steckel, R., 'Growth, depression, and recovery: the remarkable case of American slaves', *Annals Hum. Biol.* 13 (1986), pp. 33-48.

Tanner, J. M., *A history of the study of human growth* (Cambridge, 1981).

Wachter, K. W., 'Graphical estimation of military heights', *Hist. Methods*, 14 (1981), pp. 31-42.

Wachter, K. W. and Trussell, J., 'Estimating historical heights', *J. Am. Stat. Assoc.*, 77 (1982), pp. 279-303.

Ward, W. P. and Ward, P. C., 'Infant birth weight and nutrition in industrializing Montreal', *Am. Hist. Rev.*, 89 (1984), pp. 324-45.

# VII

# *Further thoughts on the nutritional status of the British population*

I agree wholeheartedly with the assertion by Floud, Wachter, and Gregory that there are, indeed, a large number of issues upon which we do agree;[1] in retrospect, I feel I should have made this point even more clearly in my comment.[2] We do not even disagree completely on the nature of the trend in nutritional status in the late eighteenth and early nineteenth centuries: our divergent approaches yield quite similar estimates for the height of the soldiers born in the 1760s, 1770s, and 1780s.[3] We both found a definite downward trend in nutritional status during those three decades. As far as the military data were concerned, differences in our results were confined, in the main, to the decade prior to the 1760s, and to the few years between *c.*1807 and 1822. These were precisely the years, by the way, for which our analyses of the Marine Society boys diverged the most, too.[4]

I am not alone in questioning the nature of the cycles reported by Floud, Wachter, and Gregory.[5] Surely even more significant is the fact that seven different datasets examined recently unequivocally refuted the direction or intensity of the cycles posited by Floud et al.[6] Moreover, none of these studies have found an improvement in the biological standard of living among any segment of the population of the UK during the early years of the industrial revolution. Alluding specifically to the finding of Floud et al., Nicholas and Steckel, in fact, assert that their own evidence 'provides an alternative interpretation of living standards for English workers during the early (1770-1815) industrialization period' and then demonstrate definitively that the 'case of deteriorating living standards is . . . upheld for the pre-1820 period, particularly among the urban population'.[7] I take this to be an extremely important independent corroboration of my position.

[1] Floud, Wachter, and Gregory, 'Measuring historical heights'.

[2] Komlos, 'The secular trend'.

[3] Floud, Wachter, and Gregory, *Height, health and history*, p. 149.

[4] Compare fig. 4.7 in Floud, Wachter, and Gregory, *Height, health and history*, p. 166, and fig. 4 in Komlos, 'The secular trend', p. 129.

[5] 'My major complaint about *Height, health, and history* is the handling and interpretation of data on cycles in heights': Steckel, 'Heights, living standards, and history', p. 186.

[6] For the period prior to the 1760s see Komlos, 'A Malthusian crisis revisited' and Steegmann, 'Eighteenth-century British military stature', p. 85; for the 1780s and 1790s see J. Mokyr and C. Ó Gráda, 'The heights of the British and the Irish, *c.* 1800-1815: evidence from recruits to the East India Company's army', unpub. typescript, Dept. of Economics, Northwestern Univ.; for 1770-1815 see Nicholas and Steckel, 'Heights and living standards'; for the 1820s and 1830s see J. Riley, 'Height, nutrition and mortality risk reconsidered', unpub. typescript, Dept. of History, Univ. of Indiana. See also P. Riggs, 'The standard of living in Britain's Celtic fringe, *c.* 1800-1850: evidence from Glaswegian prisoners', unpub. typescript, Dept. of History, Univ. of Pittsburgh (1988), and S. Nicholas and R. H. Steckel, 'Tall but poor: nutrition, health and living standards in pre-Famine Ireland' (National Bureau of Economic Research, Historical Paper 39, 1992).

[7] Nicholas and Steckel, 'Heights and living standards', pp. 939, 949.

Because of this corroboration, it is superfluous to reiterate the statistical issues that continue to divide us. The discussion of those procedures by Floud and his co-authors seems useful in many respects, although not essential to the substance of my contention that the trend in nutritional status was negative in the late eighteenth and early nineteenth centuries.[8] Instead of focusing on technicalities, I prefer to confine the remainder of the few paragraphs allotted to me to two substantive issues: the need for the concept of the 'biological standard of living' and the notion of an incipient Malthusian crisis in the late eighteenth century.

In the early phases of the anthropometric research programme, historians hoped that physical stature would turn out to be a useful measure of the standard of living and a more representative proxy than, say, real wages. It soon became evident, however, that the relationship between real income and stature was more complex than had been anticipated.[9] While in some ranges of income the positive correlation between a population's income and stature might be good or even very good,[10] in other ranges, particularly during the early phases of acceleration in economic development, this turned out not to be the case.[11] We realized that we were not really measuring the standard of living as conventionally conceived in terms of the command over goods and services, and typically measured by real per caput or family income, or GNP per caput, or real wages; instead we had stumbled on an indicator of how well the human organism throve in its socio-economic and epidemiological environment.[12] Nutritional status, by itself, would not quite do to describe this indicator, because it does not connote satisfactorily a quality of life component used by social scientists. Hence, I saw the need to distinguish conceptually between the material standard of living and the notion of wellbeing from the biological viewpoint: the biological standard of living. The latter concept, however, encompasses more than does nutritional status; it is meant to capture such aspects of the quality of life as health in general, including the frequency and duration of sickness, the extent of exposure to diseases or pollutants, and age-specific life expectancy rates. To be sure, the physical stature of a population is a partial proxy for its nutritional status, but inasmuch as it is at all possible to infer changes

[8] I intended the results obtained by the Komlos-Kim method to have significance only as far as the direction of the trend was concerned, and not to draw inferences about the magnitude of the changes in height. I acknowledge that the Komlos-Kim method is not foolproof, and is as subject to the vagaries of sampling biases as any other statistical technique; Komlos and Kim, 'Estimating trends'.

I appreciate the clarification of my splicing technique; yet I believe that the discussion by Floud et al. of its effects on the results may have been exaggerated. Surely, none of the most important of the estimated trends, those of the English soldiers in tab. 6 is at risk of having been affected for the disputed years of the early nineteenth century, by having been spliced together in the 1770s; Komlos, 'The secular trend'. While it is true that a problem could exist if there were a kink in the curves at the point of splice, if the trends do not change direction, or if they are constant at the point of splice, the rank order obtained by the Komlos-Kim method remains valid. Thus, as far as the Marine Society boys are concerned, the changes at the point of splice are either in the right direction, or are so small that they can be considered constant to the first approximation, and therefore remain valid estimates of the trend in the height of these boys.

[9] Komlos, 'Anthropometric history'.

[10] Steckel, 'Height and per capita income'.

[11] Komlos, 'The height and weight of West Point cadets'.

[12] Komlos, *Nutrition and economic development*, pp. 29, 43, 183.

in the standard of living of a society from it, we ought to acknowledge that we are exploring one dimension of this multidimensional variable. That is why I thought it would be useful to introduce the idea of the biological standard of living as distinct from both nutritional status and the conventional material standard of living.[13]

The other issue worth elaborating is the rejection by Floud and his co-authors of my contention that an incipient Malthusian crisis threatened late eighteenth-century Britain. The occurrence of bread riots, and the decline in food consumption per caput, signalled, in my opinion, the onset of a temporary imbalance between population and the availability of food.[14] According to Crafts, 'it seems probable that per caput food consumption only regained its 1760 level at around 1840'.[15] This should not be surprising, inasmuch as during the course of the eighteenth century population increased by some 70 per cent,[16] and the urban share of the total population increased from 17 to 27 per cent.[17] Agricultural output and food imports did not keep pace with such a rate of growth in the number of mouths to feed. Consequently, towards the end of the century 'the bottom 3 per cent of . . . [the English] labor force lacked the energy for any work . . . [and] the balance of the bottom 20 per cent had enough energy for about six hours of light work (or 1.09 hours of heavy work) each day'.[18] Admittedly, nutritional status did not deteriorate sufficiently among a large enough segment of the population to increase aggregate mortality rates, but that is precisely why I think that the crisis was an incipient, rather than actual, one. The threat existed, but it did not materialize, as it had done earlier.[19] While death rates did not rise, 'poor relief increased at a real rate of 2.3 per cent per annum', that is, much faster than the population, and much faster even than the urban population.[20]

The study of physical stature is exciting, because it enables us to illuminate the interaction of complex demographic, economic, and biological processes, and to synthesize them into a more nuanced view of the industrial revolution and its effect on the lives of the men and women experiencing it, than was hitherto possible.[21] Without doubt, Floud, Wachter, and Gregory have made a pioneering and signal contribution to the economic history of this period. The fact that we continue to disagree over some details is a healthy sign in a relatively new area of the discipline, one full of statistical challenges,

[13] Others are also finding the concept useful; R. H. Steckel and J. C. Rose, 'Archaeological sites in the western hemisphere that are promising for integrative studies of the past', unpub. paper, Social Science History Association, New Orleans, Nov. 1991.

[14] Food consumption declined by as much as 20 per cent between 1740 and 1790; Jackson, 'Growth and deceleration'; Shammas, 'The eighteenth-century English diet', p. 263; see also Mokyr, 'Is there still life?'; and Stern, 'The bread crisis'.

[15] Crafts, *British economic growth*, p. 98.

[16] Wrigley and Schofield, *Population history*, p. 208.

[17] Wrigley, 'Urban growth', p. 54.

[18] Fogel, 'The conquest of high mortality', p. 46.

[19] Thomas, 'Escaping from constraints'.

[20] Fogel, 'The conquest of high mortality', p. 47.

[21] Komlos and Artzrouni, 'Mathematical investigations'.

sampling problems, and ambiguities in interpretation. The important thing, I would argue, is that progress is actually being made.

**Footnote references**

Crafts, N. F. R., *British economic growth during the Industrial Revolution* (Oxford, 1985).

Floud, R., Wachter, K. W., and Gregory, A., 'Measuring historical heights—short cuts or the long way round: a reply to Komlos', *Econ. Hist. Rev.*, XLVI (1993), pp. 145-54.

Floud, R., Wachter, K., and Gregory, A., *Height, health and history: nutritional status in the United Kingdom, 1750-1980* (Cambridge, 1990).

Fogel, R. W., 'The conquest of high mortality and hunger in Europe and America: timing and mechanisms', in P. Higonnet, D. S. Landes, and H. Rosovsky, eds., *Favorites of fortune: technology, growth, and economic development since the industrial revolution* (Cambridge, Mass., 1991), pp. 33-71.

Jackson, R. V., 'Growth and deceleration in English agriculture, 1660-1790', *Econ. Hist. Rev.*, 2nd ser., XXXVIII (1985), pp. 333-52.

Komlos, J., 'The height and weight of West Point cadets: dietary change in antebellum America', *J. Econ. Hist.*, 47 (1987), pp. 897-927.

Komlos, J., *Nutrition and economic development in the eighteenth-century Habsburg monarchy: an anthropometric history* (Princeton, N.J., 1989).

Komlos, J., 'Anthropometric history: what is it?', *J. Soc. & Biol. Structures*, 14 (1991), pp. 351-4.

Komlos, J., 'The secular trend in the nutritional status of the population of the United Kingdom, 1730-1860', *Econ. Hist. Rev.*, XLVI (1993), pp. 115-44.

Komlos, J., 'A Malthusian crisis revisited: the height of British and Irish servants in colonial America', *Econ. Hist. Rev.*, XLVI (1993), forthcoming.

Komlos, J. and Artzrouni, M., 'Mathematical investigations of the escape from the Malthusian trap', *Math. Pop. Stud.*, 2 (1990), pp. 269-87.

Komlos, J. and Kim, J. H., 'Estimating trends in historical heights', *Hist. Methods* 23 (1990), pp. 116-20.

Mokyr, J., 'Is there still life in the pessimist case?: consumption during the industrial revolution, 1790-1850', *J. Econ. Hist.*, 48 (1988), pp. 69-92.

Nicholas, S. and Steckel, R. H., 'Heights and living standards of English workers during the early years of industrialization, 1770-1815', *J. Econ. Hist.*, 51 (1991), pp. 937-57.

Shammas, C., 'The eighteenth-century English diet and economic change', *Expl. Econ. Hist.*, 21 (1984), pp. 254-69.

Steckel, R. H., 'Height and per capita income', *Hist. Methods*, 16 (1983), pp. 1-7.

Steckel, R. H., 'Heights, living standards, and history: a review essay', *Hist. Methods*, 24 (1991), pp. 183-7.

Steegmann, A. T., Jr., 'Eighteenth-century British military stature: growth cessation, selective recruiting, secular trends, nutrition at birth, cold and occupation', *Hum. Biol.*, 57 (1985), pp. 77-95.

Stern, W., 'The bread crisis in Britain, 1795-96', *Economica*, 31 (1964), pp. 168-87.

Thomas, B., 'Escaping from constraints: the industrial revolution in a Malthusian context', *J. Interdisc. Hist.*, 15 (1985), pp. 729-53.

Wrigley, E. A., 'Urban growth and agricultural change: England and the continent in the early modern period', in P. Borsay, ed., *The eighteenth-century town: a reader in English urban history, 1688-1820* (London, 1990), pp. 39-82.

Wrigley, E. A. and Schofield, R. S., *The population history of England, 1541-1871: a reconstruction* (Cambridge, Mass., 1981).

# VIII

## The growth of boys in the Stuttgart Carlschule, 1771–93

J. Komlos, J. M. Tanner, P. S. W. Davies and T. Cole

**Summary.** Height measurements taken in a mixed longitudinal manner on 1084 German-born boys aged 7 to 21+ at the Carlschule Academy in Stuttgart during the period 1771–93 have been examined. The boys can be divided into upper (aristocrat), middle and lower (artisan, servant) classes, nearly all housed and fed in this boarding school. Preece-Baines curves have been fitted to a subsample of 155 boys whose measurements cover at least the period 12–16 years at a density of two or three per year. In addition, the whole data, totalling 11,040 observations, have been examined as if purely cross-sectional; and the height-at-entry measurement for each of 670 students has been examined. The results of the longitudinal subsample and the cross-sectional analyses agree reasonably well. Social class differences existed both in tempo of growth as signified by age at peak height velocity, and in adult height. The longitudinal analysis gives adult differences of about 2 cm between upper and middle classes and a further 2 cm between middle and lower, even amongst these boys all resident in the same, very privileged, school. Tempo differences between upper and middle class were minor, amounting to only 0·3 year, but lower-class boys had their maximum growth increment about a year later than the others. Amongst middle classes a secular trend of about 2 cm averaged over all ages was found between those born before 1770 and those born later. This mainly represents a trend in tempo rather than in adult height. The heights of these boys are compared with those of contemporary Austrian upper and lower classes, English upper and lower classes, American Army cadets, and American slaves. The increase in German middle-class heights during the 18th century indicates that this group was improving its nutritional status and well-being, at a time when the heights of the remainder of the population were constant or declining. This is evidence in favour of the view that at the beginning of economic development the distribution of income tends to become more skewed.

### 1. Introduction

During the last quarter of the 18th century a unique educational institution flourished in Stuttgart. The creation of Carl Eugen (1728–99), Duke of Wurttemberg, it started as a residential school for training gardeners, plasterers, artists and musicians, all to be employed at the Duke's favourite hunting lodge. The educational level was so high, however, that officers, officials and even nobles began sending their sons there, and by 1775 the Carlschule had three divisions; a basic stage to which boys were admitted at 7 years of age, a secondary stage *Gymnasium*, and an upper section which was soon granted University status (Tanner 1981, p. 106). In general, fees were charged, but scholarships were also available. The graduates, not unlike the later *polytechniciens* of Napoleon, formed an elite that not only governed the Duchy, but spread far and wide beyond it. Amongst the graduates were a number of famous people; Friedrich Schiller, for example, and the zoologist Baron Cuvier.

Carl Eugen organized the school on military lines; consequently the pupils' heights were carefully measured, often more than twice a year. These records provide a unique archive of longitudinal data on boys' heights in the late 18th century, approximately contemporaneous with the famous measurements of De Montbeillard's son. The records were discovered in the Stadtarchive, Stuttgart, by Dr Robert Uhland, an archivist, and Wilhelm Theopold, Professor of Paediatrics. A preliminary analysis of a small sample was made by Hartmann (1970) and there is a general account in Theopold, Hövels, Hartmann and Uhland (1972). However, most of the data has remained unexplored, and in this paper we present both longitudinal and cross-sectional analyses based on a transcription of the entire available material. The fact that the school included noble, middle-class and artisan boys enables us to assess social class differences in the 18th century.

## 2. Subjects and methods

Data are available on 1084 German-born subjects, whose places of birth and social provenance are described in table 1. Though sons of the nobility and of commoners each had their own living quarters, and ate at separate tables, they shared the same food, and were taught in the same classes. Servants' portions may have been somewhat smaller. A few students lived in the town. There were in addition 196 subjects born outside Germany, but we have excluded these.

Table 1. Characteristics of the sample (number of subjects).

| Place of birth | | | |
|---|---|---|---|
| Wurttemberg | 358 | *Student's social status* | |
| Stuttgart[a] | 181 | Upper aristocrat[f] | 29† |
| Ludwigsburg[b] | 131 | Lower aristocrat[g] | 335 |
| Holstein[c] | 126 | Untitled[h] | 615 |
| | | Lower class[i] | 105 |
| | | *Occupation of father* | |
| Bavaria[d] | 100 | Upper class[j] | 237 |
| Rhineland[e] | 146 | Middle class[k] | 381 |
| Germany other | 42 | Lower class[l] | 197 |
| Total | 1084 | Military officer | 237 |
| | | Unclassified | 32 |

[a]Town in Wurttemberg. [b]Town near Stuttgart. [c]Includes Prussia, Schleswig, Saxony, Brandenburg, Hessen, Pomerania. [d]Includes Schwaben, Franken, Oberpfalz. [e]Includes Alsace, Lothringen, Speyer, Pflaz. [f]Prince, Graf (count). [g]Edelsmann (noble), Von used in name, chevalier (knight). [h]Presumed middle class. [i]Servant, musician, artist, dancer. [j]Landlord, mayor, noble. [k]Merchant, innkeeper, preacher, gentleman, hunter, teacher. [l]Soldier, artisan, labourer, artist, servant.

Source: Hauptstaatsarchiv, Stuttgart.

† Only 14 used for height calculations, being one from each family: see text.

Heights were measured in Wurttemberg feet, inches and lines (*strich*) and each time the previous as well as the current measurement was recorded, and the increment noted. There was no minimum height for entry, and height distributions were all approximately Normal. Longitudinal analysis (below) confirms in general the accuracy with which the measurements were taken. Some of the subjects were brothers, and no allowance has been made for this except that amongst the upper aristocrats we chose only one family member at random as this group was small, and the results might have been influenced by the particuliarities of certain families. Dates refer to the student's date of birth and not to the date of measurement. Age is as of the last birthday. There are in all 11,040 recorded observations, an average of 8·6 per student. However, since measurement continued whether or not the student had stopped growing (this being a military establishment) the number of measurements contributing useful information is somewhat lower than this.

## 2.1. *Longitudinal series*

Amongst these students there is a core who have measurements made over a sufficient period to warrant the fitting of Preece–Baines (1978) Model I curves to their heights. Only those series were fitted which began at or before the age of 12·0 years and ended at or after the age of 16·0 years. In a few subjects the fit was poor (usually due to a single, apparently erroneous, recording) and we eliminated those nine individuals whose residual standard deviations about their curves exceeded 0·75 cm. The remainder had an average residual SD of 0·31 cm, with a range of 0·03–0·74 cm.

Table 2. Characteristics of the Longitudinally-fitted Sub-Sample.

| Place of birth | | | |
|---|---|---|---|
| Wurttemberg | 51 | *Student's social status* | |
| Stuttgart | 32 | Aristocrat | 54 |
| Ludwigsburg | 27 | Middle class | 74 |
| Germany | | | |
| other | 45 | Lower class | 27 |
| Total | 155 | | |
| | | *Occupation of father* | |
| | | Upper class | 43 |
| | | Middle class | 38 |
| | | Lower class | 31 |
| | | Military officer | 43 |

This longitudinal subsample consisted of 155 individuals, some 14% of the whole. Seventy-one per cent of these subjects came from Wurttemberg, or the towns of Stuttgart or Ludwigsburg, compared to 62% of the main sample. The subsample is not greatly different from the whole sample in terms of social status and father's occupation (see table 2). Specifically, 35% of the subsample were of lower aristocratic status compared with 37% of the main sample, 48% middle class compared with 54% and 18% lower class compared with 9%. In terms of father's social rank, 28% of the subsample were aristocrats (excluding military officers) compared with 24%, 25% were middle class compared with 35%, 20% were lower class compared with 17%, and 28% were officers compared with 22%. Furthermore, comparison of the mean heights in this subsample with those of the whole sample shows no consistent difference from age to age (data not shown).

Values for adult height, heights at take-off of the pubertal spurt and at the point of peak height velocity, age at peak height velocity and at pubertal take-off, together with peak height velocity itself, were estimated in the usual way.

## 2.2. *Unweighted cross-sectional series*

A simple cross-sectional analysis has been done using all the data from the 1084 subjects as though each had been derived from a new person—that is, unweighted by individuals. Means and SDs for each year of age, according to social status and date of entry, are given, and within each year of age a multiple regression of height on social status, place of birth and birth date was carried out.

## 2.3. *Height-at-entry series*

An analysis has been made using only those first measurement made directly after entry to the Carlschule. The age range is 7+ to 21+ years, with a total number of 647 subjects (the remainder were not measured till some time had elapsed). Unlike section 2.2 above, this is a true cross-sectional analysis, with each subject entered only once. In this part of the analysis the secular trend was also investigated.

## 3. Results

### *3.1. Longitudinal Analysis*

Table 3 gives means and SDs of adult height, height at take-off and at PHV, age at take-off (TO) and at PHV, and PHV itself for the students grouped by social class and by father's occupation (the difference being that social class depended on title; occupation not necessarily so). The final height of the aristocrats is estimated as 171·0±0·77 cm, of middle class 168·7±0·68 cm, and of lower class 166·7±1·44 cm. Adult aristocrats were thus 2·3 cm taller than middle-class subjects and 4·3 cm taller than lower-class ones. The aristocrats were the earliest maturing with PHV at an average age of 15·2±0·13 years. The middle class averaged 15·5 years and the lower class 15·6 years. The aristocrats were 2·0 cm taller than middle class and 4·3 taller than lower class at the time each began their pubertal spurt, and by PHV the differences were 2·8 cm and 4·3 cm. Thus all the final height differences had already developed before the onset of puberty. Finally, the peak velocity reached by the aristocrats, 8·4±0·15 cm/year, was a little greater than that of the middle class, 8·1±0·16 cm/year or the lower class, 8·2±0·26 cm/year.

The results from classifying by father's occupation are very similar. The sons of officers were fractionally taller and earlier-maturing than the sons of aristocrats-not-officers. A subset of the lower class, consisting of 13 sons of labourers, servants and artisans, were the shortest of all, at 165·0±1·84 cm, and also the latest-maturing, with age at PHV 15·9±0·32 years.

Table 3. Fitted means for the longitudinal subsample, computed from individual Preece–Baines curves.

| | *n* | Adult height (mean and SD) | Height at TO (mean and SD) | Height at PHV (mean and SD) | Age at TO (mean and SD) | Age at PHV (mean and SD) | Velocity at PHV (mean and SD) |
|---|---|---|---|---|---|---|---|
| *Social class* | | | | | | | |
| Aristocrat† | 54 | 171·0 | 136·4 | 156·4 | 11·2 | 15·2 | 8·4 |
| | | 5·64 | 7·32 | 5·64 | 1·91 | 0·97 | 1·16 |
| Middle class | 74 | 168·7 | 134·4 | 153·6 | 11·3 | 15·5 | 8·1 |
| | | 5·85 | 7·68 | 5·33 | 2·00 | 1·23 | 1·18 |
| Lower class | 27 | 166·7 | 132·1 | 152·1 | 11·8 | 15·6 | 8·2 |
| | | 7·50 | 10·10 | 8·15 | 2·33 | 1·63 | 1·47 |
| Total | 155 | 169·1 | 134·7 | 154·3 | 11·3 | 15·4 | 8·2 |
| | | 6·15 | 8·12 | 6·18 | 2·03 | 1·23 | 1·16 |
| *Father's occupation*: | | | | | | | |
| Upper class | 43 | 169·9 | 135·6 | 155·0 | 10·9 | 15·4 | 8·0 |
| | | 5·98 | 8·46 | 6·22 | 2·51 | 1·20 | 1·84 |
| Middle class | 38 | 168·4 | 135·1 | 154·0 | 11·7 | 15·6 | 8·3 |
| | | 6·98 | 9·62 | 7·33 | 2·10 | 1·49 | 1·47 |
| Lower class | 31 | 167·6 | 132·1 | 152·2 | 11·5 | 15·6 | 8·3 |
| | | 5·93 | 6·91 | 5·04 | 1·42 | 1·03 | 1·09 |
| Officer | 43 | 170·2 | 135·2 | 155·4 | 11·3 | 15·2 | 8·4 |
| | | 5·59 | 6·86 | 5·37 | 1·68 | 1·10 | 1·19 |

†All but one lower aristocrats.

Table 4. Height by social status, (cm).

| | High aristocrat | | | Low aristocrat | | | | | | Middle class | | | | | | Low class | | |
|---|---|---|---|---|---|---|---|---|---|---|---|---|---|---|---|---|---|---|
| | | | | Before 1770 | | | After 1770 | | | Before 1770 | | | After 1770 | | | Before 1770 | | |
| Age (years) | *n* | Mean | SD | *n* | Mean | SD | *n* | Mean | SD | *n* | Mean | SD | *n* | Mean | SD | *n* | Mean | SD |
| 5+ | | | | | | | 6 | 102·5 | 2·7 | 3 | 98·1 | 4·0 | | | | | | |
| 6+ | 4 | 121·9 | 2·0 | 9 | 116·5 | 4·1 | 18 | 114·5 | 6·3 | 9 | 107·0 | 3·4 | 5 | 117·8 | 4·8 | 6 | 121·7 | 12·6 |
| 7+ | 7 | 124·4 | 3·1 | 15 | 117·1 | 4·1 | 19 | 116·4 | 6·3 | 38 | 114·5 | 4·8 | 17 | 114·6 | 5·0 | 9 | 121·5 | 13·1 |
| 8+ | 9 | 128·9 | 4·2 | 54 | 122·0 | 8·0 | 50 | 123·5 | 5·7 | 82 | 117·2 | 5·3 | 43 | 119·0 | 4·8 | 25 | 118·2 | 8·2 |
| 9+ | 11 | 133·9 | 6·2 | 99 | 125·5 | 8·0 | 116 | 128·2 | 6·2 | 115 | 122·1 | 5·8 | 62 | 124·6 | 5·5 | 36 | 122·0 | 7·2 |
| 10+ | 10 | 141·3 | 11·9 | 150 | 129·1 | 7·9 | 81 | 131·5 | 5·6 | 170 | 127·5 | 6·5 | 74 | 127·9 | 8·0 | 53 | 126·2 | 5·6 |
| 11+ | 23 | 144·0 | 7·1 | 186 | 133·4 | 7·7 | 79 | 135·2 | 6·2 | 250 | 130·8 | 6·6 | 77 | 132·4 | 5·5 | 87 | 128·8 | 7·1 |
| 12+ | 21 | 150·4 | 8·9 | 221 | 138·1 | 7·5 | 79 | 138·7 | 6·8 | 309 | 135·5 | 6·9 | 75 | 138·1 | 7·8 | 112 | 132·8 | 6·7 |
| 13+ | 12 | 154·9 | 4·8 | 265 | 143·3 | 8·1 | 55 | 143·3 | 6·7 | 380 | 140·1 | 7·3 | 60 | 143·2 | 8·0 | 141 | 136·6 | 7·3 |
| 14+ | 13 | 160·0 | 8·1 | 259 | 148·8 | 8·5 | 60 | 148·5 | 8·3 | 381 | 144·7 | 8·7 | 82 | 147·9 | 9·7 | 171 | 141·4 | 8·2 |
| 15+ | 9 | 166·4 | 5·5 | 282 | 154·6 | 9·2 | 46 | 153·3 | 9·7 | 384 | 150·0 | 9·5 | 82 | 155·4 | 9·8 | 179 | 146·1 | 8·9 |
| 16+ | 9 | 172·8 | 4·2 | 294 | 159·5 | 9·7 | 47 | 160·7 | 8·8 | 355 | 156·1 | 9·5 | 72 | 159·7 | 9·4 | 161 | 151·3 | 9·3 |
| 17+ | 8 | 173·5 | 4·9 | 243 | 163·7 | 8·3 | 18 | 164·6 | 6·9 | 360 | 161·3 | 8·3 | 58 | 164·8 | 6·4 | 149 | 157·7 | 9·2 |
| 18+ | 4 | 170·3 | 0·9 | 203 | 166·1 | 8·2 | 18 | 168·5 | 6·8 | 305 | 164·3 | 6·9 | 56 | 166·6 | 6·9 | 127 | 162·4 | 8·2 |
| 19+ | | | | 149 | 168·5 | 7·4 | 7 | 168·1 | 5·6 | 285 | 167·0 | 6·4 | 34 | 167·0 | 7·6 | 110 | 165·4 | 7·7 |
| 20+ | | | | 112 | 168·7 | 7·3 | 3 | 170·9 | 7·8 | 224 | 168·0 | 6·3 | 14 | 167·5 | 6·6 | 102 | 166·6 | 7·8 |
| 21+ | | | | | | | | | | 172 | 168·4 | 6·8 | 7 | 170·3 | 7·1 | 81 | 167·0 | 7·8 |

Table 5. Mean height increments (ΔH) of those boys present at two consecutive ages. Age is at centre of increment; *n* = number

| | | | Middle class | | | | | | | | | |
|---|---|---|---|---|---|---|---|---|---|---|---|---|
| | Lower aristocrat | | Before 1770 | | After 1770 | | All | | Lower class | | Whole sample | |
| Age (years) | *n* | ΔH | *n* | ΔH | *n* | ΔH | *n* | ΔH | *n* | ΔH | *n* | ΔH |
| 7 | 8 | 4·2 | 3 | 1·3 | 2 | 3·7 | 5 | 2·3 | 2 | 4·6 | 15 | 3·6 |
| 8 | 11 | 4·5 | 12 | 3·1 | 8 | 4·1 | 20 | 3·5 | 2 | 4·3 | 34 | 3·8 |
| 9 | 39 | 4·1 | 25 | 3·9 | 15 | 4·5 | 40 | 4·2 | 11 | 3·6 | 92 | 4·1 |
| 10 | 65 | 3·9 | 39 | 4·0 | 26 | 4·0 | 65 | 4·0 | 16 | 4·5 | 151 | 4·0 |
| 11 | 88 | 4·0 | 63 | 4·0 | 28 | 4·1 | 91 | 4·0 | 20 | 4·6 | 202 | 4·1 |
| 12 | 103 | 4·1 | 100 | 3·9 | 28 | 4·1 | 128 | 4·0 | 36 | 4·1 | 273 | 4·0 |
| 13 | 112 | 4·1 | 123 | 4·2 | 26 | 4·2 | 149 | 4·2 | 45 | 4·4 | 315 | 4·2 |
| 14 | 125 | 4·4 | 141 | 4·5 | 20 | 4·9 | 161 | 4·6 | 56 | 4·9 | 353 | 4·6 |
| 15 | 122 | 5·9 | 144 | 5·5 | 29 | 5·4 | 173 | 5·5 | 63 | 5·2 | 367 | 5·6 |
| 16 | 130 | 5·7 | 157 | 6·0 | 17 | 6·4 | 174 | 6·1 | 67 | 5·8 | 377 | 5·9 |
| 17 | 116 | 4·9 | 145 | 5·3 | 16 | 4·7 | 161 | 5·3 | 63 | 5·9 | 347 | 5·3 |
| 18 | 101 | 3·6 | 130 | 4·4 | 11 | 2·4 | 141 | 4·2 | 57 | 5·6 | 304 | 4·3 |
| 19 | 81 | 2·5 | 128 | 2·8 | 11 | 2·1 | 139 | 2·8 | 50 | 3·4 | 274 | 2·8 |
| 20 | 63 | 1·3 | 114 | 1·6 | 3 | 1·1 | 117 | 1·6 | 46 | 2·1 | 227 | 1·6 |
| 21 | 39 | 0·9 | 83 | 1·0 | 1 | 1·8 | 84 | 1·1 | 46 | 1·1 | 170 | 1·0 |
| 22 | 18 | 1·1 | 62 | 0·5 | | | 62 | 0·5 | 34 | 0·7 | 115 | 0·6 |

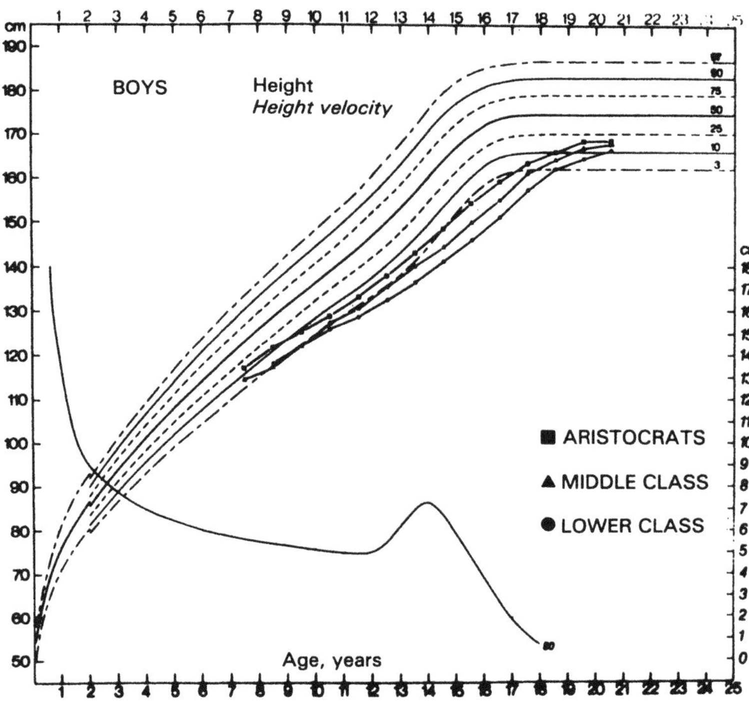


Figure 1. Mean heights of boys attending Stuttgart Carlschule 1871–93, mixed longitudinal data, unweighted cross-sectional means. Aristocrats (lower), middle-class and artisans all born before 1770, plotted on present-day British population charts.

*3.2. Unweighted cross-sectional series*

Table 4 gives the basic data on height by social status, with the two largest groups, lower aristocrats and middle class, subdivided into those born before 1770 and those born from 1770 onwards. In figure 1 the means of the lower aristocrats and middle class born prior to 1770 are plotted on modern British population charts. The difference between social classes is greatest at about 15 years, when aristocrats are 4·2 cm taller than middle class, and middle class 3·9 cm taller than lower class. Thereafter the differences decline, being about 2 cm each at age 19. The final height means are not well estimated in this analysis because sample sizes decline in the early 20s, and because those who stayed on at school past their late teens are probably not representative of the population as a whole.

In Table 5 the means of the actual increments of those individuals present in two consecutive years are given. (The mean of all values for each individual within the 1 year age bracket was subtracted from his previous mean value similarly calculated and the result annualized.) The age of maximum increment is at about 15·6 years for lower aristocrats (compared with 15·2 in the longitudinal sample), 15·8 in (all) middle class (compared with 15·5 in the longitudinal sample) and 16·7 in the lower class (compared with 15·6 in the longitudinal sample).

Table 4 also shows that the height of the middle-class boys increased over time, whereas that of the aristocrats did not. The secular trend is explored further in Table 6, where the regression coefficients of height on social status, place of birth and birth date before or after 1770 are given for each year of age from 10 to 21+. For the middle class the coefficients of the dummy variable for birth date are mostly positive and significant. The greatest difference is seen in the ages around puberty, and at ages 19 and 20 the

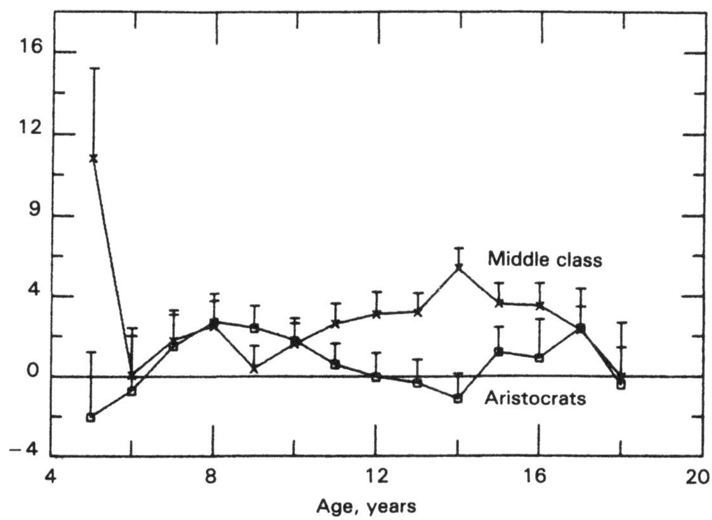


Figure 2. Secular change according to age between those born before and after 1770; middle class and aristocrats.

Table 6. Height by age profile.

| Age (years) | Social status | Constant | Birth date | Region of birth: Bavaria | Holstein | Ludwig | Rhine | Stuttgart | $n$ | $F$ | $R^2$ |
|---|---|---|---|---|---|---|---|---|---|---|---|
| 10+ | Low aristocrat | 129·8[a] | 0·3 | − 3·2[b] | 3·5[a] | 2·7[b] | −3·5[b] | −1·8 | 251 | 6·28[a] | .13 |
| | Middle class | 127·4[a] | 0·3 | 5·9[b] | 1·8 | 0·5 | 3·2[b] | −1·9[b] | 256 | 2·74[b] | ·06 |
| 11+ | Low aristocrat | 135·0[a] | −0·2 | − 4·5[a] | 1·5 | 0·4 | −4·1[a] | −2·6[b] | 291 | 4·91[a] | .09 |
| | Middle class | 131·0[a] | 1·1 | 2·7 | 1·9 | −0·3 | 0·6 | −0·6 | 349 | 1·44 | ·02 |
| 12+ | Low aristocrat | 138·4[a] | −0·4 | − 1·4 | 1·7 | 2·6[b] | −2·6[b] | −1·4 | 341 | 2·92[a] | ·05 |
| | Middle class | 134·8[a] | 2·3[a] | 1·6 | 6·4[a] | 1·5 | 2·3[b] | 0·5 | 424 | 4·10[a] | ·06 |
| 13+ | Low aristocrat | 144·2[a] | −0·9 | − 5·3[a] | 2·5[b] | 2·3[c] | −4·5[a] | −2·1 | 354 | 7·87[a] | ·12 |
| | Middle class | 139·4[a] | 2·7[a] | 2·1 | 6·1[a] | 0·7 | 1·7 | 0·8 | 477 | 3·22[a] | ·04 |
| 14+ | Low aristocrat | 150·1[a] | −1·7 | − 5·0[a] | 1·5 | 2·5 | −3·8[b] | −1·5 | 376 | 4·39[a] | ·07 |
| | Middle class | 144·8[a] | 2·5[b] | 0·6 | 6·5[b] | 0·2 | −0·2 | −0·2 | 509 | 2·32[b] | ·03 |
| 15+ | Low aristocrat | 155·8[a] | −2·2 | − 7·8[a] | 1·9 | −0·5 | −5·5[a] | 1·2 | 380 | 6·69[a] | ·10 |
| | Middle class | 150·0[a] | 4·6[a] | 2·6 | 5·1[b] | 1·9 | −1·7 | 1·0 | 508 | 4·96[a] | ·06 |

| | | | | | | | | | | | |
|---|---|---|---|---|---|---|---|---|---|---|---|
| 16+ | Low aristocrat | 160·2[a] | 0·8 | −7·7[a] | 1·5 | 1·8 | −2·6 | 3·5[b] | 398 | 6·51[a] | ·09 |
| | Middle class | 155·8[a] | 2·9[b] | 1·9 | 4·0 | 2·2 | 1·3 | 0·0 | 460 | 2·15[b] | ·03 |
| 17+ | Low aristocrat | 164·8[a] | 0·4 | −6·1[a] | −0·4 | 0·4 | −3·5[b] | 2·9[b] | 309 | 4·12[b] | ·08 |
| | Middle class | 160·6[a] | 2·9[a] | 1·2 | 0·7 | 4·0[a] | 2·0 | 0·0 | 450 | 3·47[a] | ·04 |
| 18+ | Low aristocrat | 166·1[a] | 1·4 | − 4·1[b] | 1·1 | −2·1 | −0·4 | 3·2[b] | 245 | 2·26[b] | ·05 |
| | Middle class | 163·5[a] | 2·6[a] | 2·4 | 1·4 | 0·5 | 1·9 | 0·9 | 389 | 1·99 | ·03 |
| 19+ | Low aristocrat | 168·2[a] | −2·2 | − 6·0[a] | 1·3 | −1·5 | 3·1 | 3·3[b] | 182 | 2·77[b] | ·09 |
| | Middle class | 166·1[a] | −0·2 | 1·9 | 3·2 | 2·6[b] | 0·1 | 1·3 | 329 | 1·20 | ·02 |
| 20+ | Low aristocrat | 168·4[a] | 0·6 | − 3·9[b] | 1·7 | −2·7 | 4·7 | 2·8 | 125 | 1·62 | ·08 |
| | Middle class | 167·6[a] | −0·7 | 4·1[b] | 0·0 | 2·0 | −0·1 | −0·4 | 243 | 1·36 | ·03 |
| 21+ | Low aristocrat | 170·5[a] | 3·1 | −13·4[a] | −2·5 | 0·2 | 6·7[b] | 0·1 | 124 | 4·19[a] | ·18 |
| | Middle class | 167·0[a] | 4·5[b] | 6·1[a] | 2·4 | 4·2[a] | 1·5 | 2·0[b] | 426 | 5·06[a] | ·07 |

Dependent variable is height in centimetres. The constant represents the height of a boy born in Wurttemberg after 1770, of the age indicated.

[a]1% level of significance; [b]5% level of significance.

coefficients are not significant (figure 2). None of the coefficients for aristocrats are significantly different from zero. It seems that the secular trend shown by the middle class is largely or wholly one of tempo, not adult size.

Place of birth does not have a consistent association with stature, except that boys from Holstein are a little taller and earlier maturing, both amongst the lower aristocrats and the middle class.

We have to bear in mind in considering these results, however, the bias introduced by the mixed longitudinal nature of the data, with some boys over-represented and others under-represented. There may well be a height-related bias also in time of leaving the school.

### 3.3. *Height-at-entry series*

To get away from a possible mixed-longitudinal bias, height at first measurement at school has also been analysed. Average age at entry to the school was 12·3 years. Table 7 shows the means for the 670 boys, and the difference due to secular trend. The secular trend was analysed as a whole by fitting a multiple regression for height with pre-1770/post-1770 as a dummy variable, year of age as dummy variables and social status and place of birth (five categories) as dummies also. The results in general bear out those of the analysis above in relation to social class and place of birth and estimate a secular trend, averaged over all ages, of 2·3 cm, significant at 5%.

One further value in Table 7 needs emphasis: the overall estimate of height difference between middle class and lower class at entry to the school is 4·9cm, about twice as much as the difference after some time at the school.

## 4. **Discussion**

The three questions that examination of the data set out to answer are:

(1) What was the height of boys in the late 18th-century Europe, particularly Germany; are the cases of De Montbeillard's son and the children of the Berlin Orphanage reported by Jampert (see Tanner 1989) exceptional? How do the heights compare with present-day ones?

Table 7. Analysis of heights at first entry to school. Regression on social status, place of birth and birth date. Height means refer to middle-class boys born in Wurttemberg before 1771.

| | Regression coefficient | Age (years) | Height (cm) |
|---|---|---|---|
| Social status | | | |
| Upper aristocrat | 5·3* | 7+ | 112·8 |
| Lower aristocrat | 1·3** | 8+ | 118·1 |
| Lower class | −4·9* | 9+ | 122·4 |
| Birth place | | 10+ | 128·5 |
| Stuttgart | 0·5 | 11+ | 130·1 |
| Ludwigsburg | 0·6 | 12+ | 136·5 |
| Bavaria | 0·8 | 13+ | 140·4 |
| Holstein | 2·5* | 14+ | 144·6 |
| Rhineland | −1·8* | 15+ | 151·0 |
| Birth date | 2·3* | 16+ | 155·6 |
| $R^2$ 0·8 | | 17+ | 163·5 |
| $F$ 100·5* | | 18+ | 164·9 |
| | | 19+ | 171·1 |
| | | 20+ | 169·0 |
| | | 21+ | 171·5 |

* Significant at 5% level; ** significant at 10% level.

(2) What social class differences in height existed at that time, in tempo and size?
(3) Are secular trends demonstrable over the rather short period of some 20 years?

To answer these questions we examined the data in three ways. First we abstracted from the records of the 1084 German-born boys those whose longitudinal run of measurements began before age 12 and ended after age 16. On average, these boys had 20 measurements (range seven to 30) between their initial examination and age 21. After nine out of this subgroup has been eliminated for poor fits due to errors in recorded measurements 155 subjects remained, who were fitted by Preece–Baines curves with residual errors very similar to those of modern data. These 155 boys, though only 12% of the whole, seemed nevertheless to be a fairly good subsample in terms of social class and fathers' occupation, though they entered the school somewhat earlier than the majority of students and therefore were subjected for longer than the others to the nutritional and hygienic discipline of the school, which perhaps was superior to what most of them would have experienced at home. Certainly this is true of the lower-class subjects; and indeed they caught up in height while at the school from being some 5 cm below the middle class at entry to 2·7 cm later (tables 4 and 7).

The second way we analysed the data was by ignoring its mixed longitudinal nature by simply supposing that each measurement was of a separate child. This introduces biases which are hard to evaluate, but makes use of the whole data set. Thirdly, we analysed only the measurement-at-entry data for each boy, so the longitudinal bias was eliminated. However, bias may well occur in relation to age at entry. Those relatively few who entered for the first time at age 19 and over are considerably taller than those who entered before 19, so this is not a suitable method for estimating final height.

*Social class differences*

Both longitudinal and cross-sectional analyses reveal social class differences in tempo and in final height. The longitudinal subsample estimates final differences as about 2 cm between upper and middle class and a further 2 cm between middle and lower class (a result not greatly different from some modern European data, for instance Poland; Bielicki 1986). The longitudinal subsample estimates tempo differences as quite small: age at PHV 15·2 years for the upper class, 15·5 for the middle and 15·6 for the lower. Actual-increment analysis of the whole sample gives a wider range of PHV means; *viz*: 15·6 years, 15·8 years and 16·7 years. The differences in final height are established before puberty begins. A similar pattern is seen in the difference between Indian and English boys (Hauspie, Das, Preece and Tanner, 1980) and in differences between tall and short children in a well-nourished population of today (see Gasser, Kneip, Binding, Largo, Prader and Molinari, 1989).

*Comparison with other contemporary data*

We have used the unweighted cross-sectional means to compare with other data in tables 8 and 9. The German aristocrats were some 3 cm taller than Austrians of comparable status at ages 14–16. They were considerably shorter than English boys, derived from similar landed classes, attending Sandhurst Military Academy and shorter, too, than cadets at the US West Point Military Academy (table 8). In contrast, the lower-class students of the Carlschule were about the same height as the sons of German or Austrian soldiers serving in the Habsburg army (table 9). The lower-class boys were some 5 cm taller than the destitute London boys recruited to the Marine Society (Floud and Wachter 1982, Tanner 1981), but shorter than American slaves by an

Table 8. Stature of upper-class youth in international comparison (cm)

| Age (years) | German (A), 1758–69 n | Height (cm) | Austrian (B), 1745–55 n | Height (cm) | English (C), 1800 Height (cm) | American (D), 1820s n | Height (cm) |
|---|---|---|---|---|---|---|---|
| 7 | 21 | 118·6 | | | | | |
| 8 | 64 | 122·5 | | | | | |
| 9 | 108 | 125·8 | | | | | |
| 10 | 162 | 129·4 | | | | | |
| 11 | 198 | 134·0 | | | | | |
| 12 | 246 | 138·7 | 18 | 140·4 | | | |
| 13 | 289 | 143·9 | 51 | 144·2 | | | |
| 14 | 272 | 149·1 | 68 | 145·7 | 157·0 | | |
| 15 | 297 | 154·9 | 43 | 152·4 | 163·0 | | |
| 16 | 305 | 159·7 | 22 | 156·2 | | 35 | 167·9 |
| 17 | 252 | 164·1 | | | | 45 | 169·4 |
| 18 | 211 | 166·4 | | | | 35 | 172·0 |
| 19 | 153 | 168·5 | | | | 36 | 173·0 |
| 20 | 112 | 168·7 | | | | | |

*n* = Number of observations; age is as at last birthday.
A = Aristocrat, Carlschule; B = Aristocrat at the Austrian Maria Theresianische Militarakademie; C = English gentry attending the Sandhurst Military Academy; D = Cadets at the United States West Point Military Academy.
Sources: Komlos (1986, 1987); Floud, Wachter and Gregory, (1990).

Table 9. Stature of lower-class youth in international comparison (cm)

| Age[a] (years) | German (A), 1758–69 | German (B), 1790s | Austrian (C), 1800s | English (D), 1800s | American (E), 1800s |
|---|---|---|---|---|---|
| 8 | 118·2 | 116·3 | 116·4 | | 114·6 |
| 9 | 122·0 | 119·4 | 119·4 | | 121·0 |
| 10 | 126·2 | 123·9 | 124·1 | | 125·1 |
| 11 | 128·8 | 127·6 | 127·8 | | 131·9 |
| 12 | 132·8 | 131·5 | 131·9 | | 135·2 |
| 13 | 136·6 | 136·4 | 137·0 | 132·0 | 141·0 |
| 14 | 141·4 | 140·6 | 140·9 | 135·0 | 146·6 |
| 15 | 146·1 | 146·8 | 147·7 | 142·0 | 153·0 |
| 16 | 151·3 | 151·1 | 151·4 | 147·0 | 158·1 |
| 17 | 157·7 | | | | 163·2 |
| 18 | 162·4 | | | | 165·7 |
| 19 | 165·4 | | | | 167·7 |
| 20 | 166·6 | | | | 168·5 |
| 21 | 167·0 | | | | 170·5 |

[a]Age at last birthday.
A = At the Carlschule; see table 1 for definition; B = sons of German soldiers serving in the Habsburg army; C = sons of soldiers serving in the Habsburg army; D = London destitute; E = slaves.
Sources: Komlos (1986); Floud and Wachter (1982); Margo and Steckel (1982).

amount reaching 7 cm at ages 15 and 16 and reducing to about 2·5 at ages 19–21 (Steckel 1987).

*Secular trend*

There seems to have been an increase in height in the cohort of middle-class boys born after 1770 compared with those born before. This was greatest at pubertal ages, reaching 4 cm then, and considerably less at maturity. The aristocrats showed no such increase; for the lower class we have no data, as measurements of lower-class students were discontinued halfway through the existence of the Carlschule. The middle-class increase in height, or, more accurately, growth tempo, indicates that their nutritional status was improving at a time when other evidence shows that that of the lower classes was declining all over Europe (Komlos 1989). In contrast the aristocrats may have reached a stable position: data on elite boys in the Habsburg Monarchy show that their stature was already increasing earlier in the century, with a difference of 5·6 cm in the teenage years between cohorts born in the 1730s and the 1750s (Komlos 1986). Evidently at this time improvements in diet and in well-being generally were quite possible amongst the upper echelons of society. This is consistent with the fact that the life expectancy of the British peerage improved considerably during the second half of the 18th century in comparison to the nationwide average because of a lowered infant mortality, and because nutritional status is one of the determinants of life expectancy (Hollingsworth 1964, Wrigley and Schofield 1981).

The increase in the nutritional status of the middle-class students is in sharp contrast to the pattern evident among European lower classes, whose age-by-height profile was consistently declining during the closing decades of the 18th century. The Stuttgart students, whose socioeconomic status was far above average, remained unaffected by the substantial increase in food prices which occurred at this time. This is an indication that their privileged socioeconomic position enabled them to remain isolated from the widespread Malthusian nutritional crises threatening Europe (see Komlos 1989).

**Acknowledgements**

J. Komlos gratefully acknowledges the financial support of the US National Science Foundation.

**References**

BIELICKI, T., 1986, Physical growth as a measure of the economic well-being of populations: the 20th century. In *Human Growth: A Comprehensive Treatise*, edited by F. Falkner and J. M. Tanner (New York: Plenum), 2nd edition, Vol. III, pp. 283–306.

FLOUD, R., and WACHTER, K., 1982, Poverty and physical stature, evidence on the standard of living of London boys 1770–1870. *Social Science History*, **6**, 422–452.

FLOUD, R., WACHTER, K., and GREGORY, A., 1990, *Height, Health and History* (Cambridge: Cambridge University Press).

GASSER, Th., KNEIP, A., BINDING, A., LARGO, R., PRADER, A., and MOLINARI, L., 1989, Flexible methods for nonparametric fitting of individual and sample growth curves. In *Auxology 88: Perspectives in the Science of Growth and Development*, edited by J. M. Tanner (London: Smith-Gordon), pp. 23–30.

HARTMAN, W., 1970, *Beobachtungen zur Akzeleration des Längenwachstums der zweiten Hälfte des 18. Jahrhunderts* (unpublished dissertation) (Frankfurt: Goethe University).

HAUSPIE, R. C., DAS, S. R., PREECE, M. A., and TANNER, J. M., 1980, A longitudinal study of the growth in height of boys and girls of West Bengal (India) aged 6 months to 20 years. *Annals of Human Biology*, **7**, 429–441.

HOLLINGSWORTH, T., 1964, *The Demography of the British Peerage*. Supplement to *Population Studies*, **18**, 1–108.

KOMLOS, J., 1986, Patterns of children's growth in East-Central Europe in the 18th century. *Annals of Human Biology*, **13**, 33–48.

KOMLOS, J., 1987, The height and weight of West Point cadets: dietary change in antebellum America. *Journal of Economic History*, **47**, 897–928.

KOMLOS, J., 1989, *Nutrition and Economic Development in the Habsburg Monarchy: an Anthropometric History* (Princeton: Princeton University Press).

MARGO, R. A., and STECKEL, R. H., 1982, The heights of American slaves: new evidence on slave nutrition and health. *Social Science History*, **6**, 516–538.

PREECE, M. A., and BAINES, M. J., 1978, A new family of mathematical models describing the human growth curve. *Annals of Human Biology*, **5**, 1–24.

STADTARCHIVE STUTTGART, GERMANY, *Documents of the Hohe Carlschule, A 272.*

STECKEL, R. H., 1987, Growth depression and recovery: the remarkable case of American slaves. *Annals of Human Biology*, **14**, 111–132.

Tanner, J. M., 1981, *A History of the Study of Human Growth* (Cambridge: Cambridge University Press).

Tanner, J. M., 1989, The first study of human growth: Christian Friedrich Jampert. *International Journal of Anthropology*, **4**, 19–26.

Theopold, W., Hövels, O., Hartmann, W., and Uhland, R., 1972, Beobachtung über das Längenwachstum in der zweite Hälfte des 18 Jahrhunderts. *Deutsches Arzteblatt: Arztliche Mitteilungen*, **11**, 611–617.

Wrigley, A., and Schofield, R., 1981, *The Population History of England 1541–1871: A Reconstruction* (Cambridge, Mass: Harvard University Press).

**Zusammenfassung.** Es werden Körperhöhendaten von 1084 deutschen Jungen im Alter von 7 bis 21+ Jahren ausgewertet, die in den Jahren 1771 bis 1793 an der Stuttgarter Karlsschule mit einer Methodik erhoben worden waren, die einem gemischtenlongitudinalen Ansatz entspricht. Die Jungen, die nahezu alle im Internat untergebracht waren und dort ernährt wurden, können in Angehörige der oberen (aristokratischen) sowie der mittleren und unteren (Handwerker und Diener) sozialen Schicht unterteilt werden. Den Daten einer ausgewählten Stichprobe von 155 Jungen, für die mindestens über die Altersspanne von 12 biz 16 Jahren 2 bis 3 Messungen pro Jahr vorlagen, wurden Preece-Baines Kurven angepaßt. Darüber hinaus wurde das insgesamt 11 040 Beobachtungen umfassende Datenmaterial so ausgewertet, als ob es sich ausschließlich um Querschnittsdaten handeln würde. Darüber hinaus wurde für 670 Schüler auch die Körperhöhe bei Eintritt in die Karlsschule untersucht.

Die Ergebnisse für die longitudinale Teilstichprobe stimmen ziemlich gut mit denen der Querschnittsanalyse überein. Soziale Klassenunterschiede existierten sowohl im Wachstumstempo, wie durch das Alter beim Auftreten der maximalen Wachstumsgeschwindigkeit angedeutet, als auch in der Erwachsenenhöhe. Die longitudinale Analyse läßt Körperhöhendifferenzen von 2 cm zwischen der oberen und der mittleren sozialen Schicht erkennen und weitere 2 cm zwischen der mittleren und der unteren—sogar bei diesen Jungen, die alle in derselben priviligierten Schule lebten. Während die Unterschiede im Wachstumstempo zwischen der oberen und der mittleren sozialen Schicht nur geringfügig waren, etwa 0,3 Jahre, erreichten die Jungen aus der unteren sozialen Schicht ihre maximale Körperhöhenzuwachsrate jedoch etwa ein Jahr später als die aus den anderen sozialen Schichten. In der mittleren sozialen Schicht ließ sich ein über alle Altersklassen gemittelter säkularer Trend von 2 cm zwischen Jungen, die vor 1770 geboren wurden und denen, die nach 1770 geboren wurden, beobachten. Dies belegt im wesentlichen einen Trend im Wachstumstempo, weniger in der Erwachsenenhöhe.

Die Körperhöhen dieser Jungen werden mit zeitgenössischen Daten von österreichischen Jungen aus oberen und unteren sozialen Schichten, englischen Jungen aus oberen und unteren sozialen Schichten, amerikanischen Armeekadetten und amerikanischen Sklaven verglichen. Der Anstieg der Körperhöhe in der deutschen mittleren sozialen Schicht im 18. Jahrhundert zeigt, daß diese Gruppe ihren Ernährungszustand und ihren Lebensstandard verbesserte und dies zu einer Zeit, in der die Körperhöhe der restlichen Population konstant blieb oder abnahm. Dies is ein deutlicher Beleg dafür, daß zu Beginn der wirtschaftlichen Entwicklung die Einkommensverteilung eine Tendenz in Richtung zunehmender Schiefe erkennen ließ.

**Résumé.** On a examiné des mesures de taille prises semi-longitudinalement sur 1084 garçons allemands de 7 à 21 ans, à l'Académie de Carlschule à Stuttgart de 1771 à 1793. Les garçons, pratiquement tous internes dans l'école, peuvent être subdivisés en classe supérieure (aristocrates), moyenne et inférieure (artisans, domestiques). Les courbes de Preece-Baines ont été ajustées à un sous-échantillon de 155 garçons dont les mensurations recouvrent au moins la période de 12 à 16 ans, à raison de 2–3 par an. De surcroît, les données d'ensemble totalisant 11040 observations, ont été examinées sous forme purement transversale, ainsi que la mesure de "la stature à l'entrée" de 670 étudiants. Les résuetats des analyses semi-longitudinales et transversales convergent raisonnablement. Des différences de classes sociales se marquent aussi bien dans le tempo de la croissance, comme l'exprime l'âge au pic de croissance, que dans la stature adulte. L'analyse longitudinale donne des différences adultes de près de 2 cm entre classes supérieure et moyenne et encore de 2 cm entre classe moyenne et inférieure, même parmi ces garçons tous pensionnaires de cette école très privilégiée. Les différences de tempo entre classes supérieure et moyenne sont mineures, de 0,3 an seulement, mais les garçons de classe inférieure présentent leur accroissement maximum environ un an après les autres. Parmi les classes moyennes, une tendance séculaire moyenne sur tous les âges d'environ 2 cm, a été trouvée entre ceux nés avant 1770 et ceux nés plus tard. Ceci représente plus une tendance évolutive du tempo que de la taille adulte. Les statures de ces garçons sont comparées avec celles d'autrichiens contemporains de classes supérieure et inférieure, d'anglais de classes supérieure et inférieure, de cadets de l'armée américaine et d'esclaves américains. L'augmentation des statures de la classe moyenne allemande au cours du 18ème siècle, indique que ce groupe améliorait son bien-être et son statut nutritionnel à une époque où les statures du reste de la population étaient constantes ou en déclin. Ceci prêche en faveur de l'idée que la distribution des revenus tend à devenir plus biaisée lors du début du développement économique.

# IX

# The Standard of Living of Jews in Austria-Hungary: The Anthropometric Evidence, 1860-1920

Until fairly recently, the income and wealth of Jews in the Habsburg Empire have not been the subject of serious analysis. Generally, Jews are thought to have prospered during the economic expansion that lasted until the outbreak of the First World War.[1] Impressionistic evidence on the business elite is sometimes used to support the notion that the Jews fared well relative to their Christian compatriots. Yet, evidence from the extreme right tail of the income-distribution curve can hardly be indicative of the relative socio-economic position of a group whose diversity must have been similar to that of the other elements in the society. The occupational structure of the Jews no doubt diverged in significant ways from that of the rest of the population, but this difference need not have been translated into a financial advantage. Nevertheless, many commentators on the subject suggest, on the basis of these two kinds of evidence, that the economic mobility of the Jews was above average. Thus, according to Arthur J. May, Hungarian Jews "obtained a controlling influence in industry, banking, commerce and the press."[2]

In short, the impression is often generated that the Jews controlled practically all of Austro-Hungarian economic life. By implication, they must have enjoyed a higher standard of living than the rest of the population, because their "control" of the economy could not have been for altruistic purposes. Such generalizations should be regarded with some skepticism, however. While emphasizing the concentration of Jews in certain occupations, scholars have tended to overlook their lack of representation in others. There may have been

1 David F. Good, *The Economic Rise of the Habsburg Empire 1750-1914* (Berkeley, 1984) and John Komlos, *The Habsburg Monarchy as a Customs Union: Economic Development in Austria-Hungary in the Nineteenth Century* (Princeton, 1983).

2 Arthur J. May, *The Habsburg Monarchy 1867-1914* (New York, Norton, 1968), p. 243 (first published in 1951.) See also Andrew C. Janos, *The Politics of Backwardness in Hungary 1825-1945* (Princeton, 1982), p. 113 and Istvan Deak, "The Convert," *New York Review of Books* 34 (1987), no. 4, p. 41.

many Jewish lawyers, for instance, but there were few Jewish law professors at the universities. Moreover, it is dangerous to make inferences from occupational structure to the distribution of income and wealth, because the relationship between the two has not been adequately explored. The Jewish population, too, had its underclass, whose omission from the analysis prevents us from acquiring a complete picture of the standard of living. This notion is vividly reinforced by the visual evidence of Roman Vishniac's photographs,[3] which reveal the existence of a broad stratum of Jews living near the very bottom of the income distribution. In short, we do not know how well the Jews fared financially relative to their Christian compatriots, because this question has eluded rigorous examination. From a theoretical point of view, the problem is hardly intractable. It cannot, however, be tackled unless one has a broad random sample of the population at large, representing the full range of incomes. Then one could calculate the mean per-capita annual income of the Jews and compare it with the societal average. One could use probate records to make a similar comparison for assets.

In this study I attempt to illuminate these issues by using anthropometric evidence to develop a set of proxy measures for income. Economic historians have recently begun to explore the ways in which biological, economic, and demographic variables interacted in the eighteenth and nineteenth centuries, using nutritional status as a crucial link among these processes.[4] Human stature plays an important role in this research because data are plentiful and because their use allows one to quantify nutritional status, something hitherto unfeasible. A population's nutritional status—that is, the quality and quantity of nutrients it consumes, less the nutrients lost to work and disease—has a major influence on stature at a particular age, on the annual growth rate, on the timing and intensity of the adolescent growth spurt, and on the age at which final height is reached.[5] The nutritional status of a population depends mainly on its food intake, which, in turn, depends on a wide variety of factors, such as the relative price of food, family income and therefore the occupation of the head of household, and the distribution of wealth. More formally, the demand for food F is a function of

3 Roman Vishniac, *The Vanished World* (New York, 1983).

4 Robert W. Fogel, "Nutrition and the Decline in Mortality since 1700: Some Preliminary Findings," in Stanley Engerman and Robert Gallman, eds., *Long-Term Factors in American Economic Growth* [National Bureau of Economic Research, Studies in Income and Wealth, vol. 51] (Chicago, 1987); John Komlos, *Nutrition and Economic Development in the Eighteenth-Century Habsburg Monarchy: An Anthropometric History* (Princeton, 1989) and John Komlos, "The Height and Weight of West Point Cadets: Dietary Change in Antebellum America, 1820-1880," *Journal of Economic History* (December, 1987), 897-927.

5 J. M. Tanner, *A History of the Study of Human Growth* (Cambridge, 1981).

real income Y, the price index of a typical basket of food products $P_F$ relative to all other prices $P_{AOG}$ and tastes T. Let $P_F/P_{AOG}$ be designated by Z. Then

$$F_t = (f_1, f_2, \ldots, f_n) = q(Y_t, Z_t, T) \quad (1)$$

where $f_i$ is the quantity of the $i^{th}$ foodstuff. The vector $F_t$ is determined by utility maximization. The $i^{th}$ foodstuff has a caloric content $c_i$ and price per unit $p_i$. The total caloric intake C is given by:

$$C = j(f_1, f_2, \ldots, f_n) = j(F_t) = f_1c_1 + f_2c_2 + \ldots + f_nc_n \quad (2)$$

The consensus among biologists is that the terminal height of a population, H, must be between the genetically determined minimum $H_{min}$ and maximum $H_{max}$. The degree to which H exceeds $H_{min}$ is determined by the caloric intake C above the requirements of basic metabolism, the work effort, and the nutrients claimed by diseases. Let these claims on the nutrients be designated W. Then

$$H = H_{min} + \int_{-1}^{30} [h(C_t - W_t, t)]\, dt \leq H_{max} \quad (3)$$

where h is the instantaneous growth rate and t designates time. The integral is from -1 to 30 in order to signify, on the one hand, that to some extent nutritional status is hereditary, i.e., that intrauterine nutrition also influences nutritional status, and, on the other hand, that growth cannot continue past age 30. From equations (1), (2), and (3) we obtain:

$$H = H_{min} + \int_{-1}^{30} [g\,(Y_t, Z_t, T, W_t, t)]\, dt \leq H_{max} \quad (4)$$

If Z, T, and W remain constant, there is a positive correlation between Y and H. Moreover, if one studies $\delta H$, the change in stature over time, the genetic component of stature, $H_{min}$, drops out of equation (4). The correlation between $\delta H$ and $\delta Y$ has been amply documented.[6] In France, draftees who could afford to pay for a substitute to take their place in the army averaged one centimeter taller than those who replaced them.[7]

Research has also supported the notion of a strong correlation between social status and stature, presumably because family income correlates positively with

6 Richard H. Steckel, "Height and Per Capita Income," *Historical Methods* 16 (1983), 1–7 and John Komlos, "Stature and Nutrition in the Habsburg Monarchy: The Standard of Living and Economic Development in the Eighteenth Century," *American Historical Review* 90 (1985), 1149–1161.

7 Emmanuel Le Roy Ladurie and N. Bernageau "Étude sur un contingent militaire (1868): Mobilité géographique, delinquance et stature, mises en rapport avec d'autres aspects de la situation des conscrits," *Annales de démographie historique* (1971), 311–37.

social status. Thus in Germany the sons of titled nobles were on average 4 centimeters taller than the sons of the lower aristocracy in the 1760s. The latter, in turn, were about 3.5 centimeters taller than the sons of middle-class families, who were 4.5 centimeters taller than lower-class boys.[8] Thus there was a 12-centimeter gap between the height of the upper and lower classes in eighteenth-century Germany. In the 1790s, boys in the London slums were so malnourished that they were shorter than boys living in practically any society of today, and a full 20 centimeter shorter than their gentlemanly contemporaries attending the Sandhurst Military Academy.[9] Class distinctions in America also translated into differences in stature, although they were not as large as in Europe. American slaves were only about 2.5 centimeters shorter than their masters.[10] Cadets at the West Point Military Academy whose fathers had blue collar occupations were 1.1 centimeters shorter than those whose fathers were professionals.[11] Thus it has been amply demonstrated that there is a strong and persistent correlation in historical populations between human stature and socio-economic variables such as status and income. This is particularly the case if the groups that are compared face the same relative prices for foodstuffs. Whenever income has been found not to correlate positively with stature it is because food prices varied across space or time.[12] This does not pose a problem in this study of Viennese Jews, because they faced the same food prices as their fellow countrymen.

I therefore compare the stature of Viennese Jews with that of the rest of the population in order to ascertain whether their standard of living in the late nineteenth and early twentieth centuries was above average. The data were obtained from the Viennese *Kriegsarchiv* (*Stellungslisten* 1873-1914).[13] Only those Jews who were born in Vienna were sampled for the purposes of this study. Given the large number of lower-class Jews who migrated to Vienna from Galicia, it is clear that the average income of all Jews was less than that of

8 John Komlos, "Height and Social Status in Eighteenth-Century Germany," *Journal of Interdisciplinary History* 20 (1990).

9 Roderick Floud and Kenneth Wachter, "Poverty and Physical Stature, Evidence on the Standard of Living of London Boys 1770-1870," *Social Science History* 6 (1982), 422-452.

10 Richard Steckel, "Slave Height Profiles from Coastwise Manifests," *Explorations in Economic History* 16 (1979), 363-380; Robert Margo and Richard Steckel, "Heights of Native-Born Northern Whites during the Antebellum Period," *Journal of Economic History* 43 (1983), 167-74; Richard Steckel, "A Peculiar Population: The Nutrition, Health and Mortality of American Slaves from Childhood to Maturity," *Journal of Economic History* 46 (1986), 721-742.

11 Komlos, "The Height and Weight of West Point Cadets."

12 Komlos, "Stature and Nutrition in the Habsburg Monarchy."

13 Vienna, Kriegsarchiv. *Stellungslisten*, Niederösterreich. 1873-1914.

those in the sample. Consequently, the standard of living of all Jews who resided in Vienna will be overestimated.

Table 1: Height of 21- and 22-Year-Old Recruits in the Austro-Hungarian Army (cm)

| Place of Birth | Vienna | | | | Lower Austria | |
|---|---|---|---|---|---|---|
| | Jews | | Non-Jews | | Non-Jews | |
| Decade of Birth | N | X | N | X | N | X |
| 1860 | 41 | 167.6 | 345 | 167.5 | 276 | 166.1 |
| 1870 | 1055 | 166.8 | 1589 | 167.4 | 521 | 165.4 |
| 1880 | 1154 | 167.1 | 1525 | 167.6 | 544 | 165.9 |
| 1890 | 727 | 167.8 | 1192 | 168.7 | 229 | 166.1 |

N=Number of observations X=mean value

Source: *Stellungslisten*, 1873-1914

Jewish recruits born in Vienna were consistently shorter at age 21 and 22 than their non-Jewish Viennese counterparts (Table 1). The birth cohorts of the 1870s were 0.6 centimeters shorter; by the 1890s the difference was 0.9 centimeters. Viennese Jews were between 1.4 and 1.7 centimeters taller, however, than the recruits born in the rural districts of Lower Austria. A similar pattern emerges among 19- and 20-year-old birth cohorts of the 1890s. The 19- and 20-year-old Jews (N=351) were 1.6 centimeters shorter than non-Jews born in Vienna in the same decade (N=654), but they were 1.2 centimeters taller than 19- and 20-year-old Christian men born in the rural areas of Lower Austria. This result is confirmed by examining the change in stature between the birth cohorts of the 1870s and 1890s. During these two decades the height of the 21- and 22-year-old Jewish recruits increased by 1.0 centimeter, while the height of the Christian recruits born in Vienna increased by 1.3 centimeters. In contrast, recruits born in Lower Austria grew only 0.7 centimeters in this period. Hence the economic position of the Viennese Jews improved by more than that of the Lower Austrians, but by less than that of the Viennese Christians.

Because stature is a cumulative measure of nutritional status, it reflects the individual's well-being prior to reaching terminal height. Hence the height of the recruits is actually a proxy for family income. On the basis of this evidence it appears that Jewish family income was not extraordinarily high by any stretch of the imagination. In fact, in Vienna they appear to have enjoyed a lower standard of living at the end of the nineteenth and beginning of the twentieth centuries than the rest of the urban population, even leaving out of consideration those poor Jews who had recently migrated into the city from Galicia.

Table 2: Distribution of Recruits' Occupation by Birth Decade

| Occupation | Jews 1860s* | Jews 1870s | Jews 1880s | Jews 1890s | Non-Jews 1890s |
|---|---|---|---|---|---|
| A Student | 2.8 | 11.9 | 18.7 | 43.3 | 17.3 |
| B Bureaucrat | 3.3 | 10.1 | 14.2 | 12.5 | 9.2 |
| C Commercial employee | 34.3 | 18.9 | 17.2 | 10.6 | 3.6 |
| D Clerk | 17.4 | 24.7 | 18.0 | 9.2 | 11.4 |
| E Professional | 8.9 | 7.5 | 4.8 | 6.2 | 5.8 |
| F Blue Collar/Unskilled | 1.4 | 10.4 | 9.2 | 5.7 | 13.3 |
| G Technician | 2.8 | 4.2 | 6.6 | 5.2 | 5.9 |
| H Artisan | 13.6 | 4.7 | 2.8 | 2.1 | 13.1 |
| I Salesman | 5.6 | 2.1 | 2.0 | 1.9 | 0.6 |
| J Performing Artist | 0.5 | 2.1 | 1.6 | 1.4 | 2.4 |
| K None | 0.0 | 1.8 | 1.9 | 1.2 | 1.5 |
| L Blue Collar | 9.4 | 1.5 | 2.6 | 1.1 | 15.9 |
| Number of observations | 213 | 1101 | 1053 | 1677 | 667 |
| 1 Student | 2.8 | 11.9 | 18.7 | 43.3 | 17.3 |
| 2 Professional | 12.2 | 17.6 | 19.0 | 18.7 | 15.0 |
| 3 Third Sector/Other | 57.8 | 47.8 | 39.0 | 23.1 | 18.0 |
| 4 Unskilled Worker | 1.4 | 10.4 | 9.2 | 5.7 | 13.3 |
| 5 Skilled Worker | 25.8 | 10.4 | 13.0 | 8.4 | 34.9 |

*Includes the birth cohorts of the 1850s
Source: *Stellungslisten*, 1873-1914

Note: Classification includes the following occupations:
A Hochschüler, Gymnasiast, Handelsakademiker
B Beamter, Privatbeamter
C Kommis, Handelsangestellte
D Kontorist, Korrespondent, Praktikant, Buchhalter
E Mediziner, Jurist, Ingenieur, Chemiker, Apotheker, Lehrer
F Gehilfe, Hilfsarbeiter, Geschäftsdiener, Fabriksarbeiter, Manipulant, Lehrling, Taglöhner
G Techniker
H Bäcker, Bildhauer, Photograf, Weber, Schneider, Taschner, Glaser, Schlosser, Installateur, Uhrmacher, Goldschmied, Porzellanerzeuger, Buchbinder, Bürstenbinder, Maler, Schuhmacher, Fleischhauer, Friseur
I Kaufmann, Reisender, Vertreter, Agent, Händler
J Schauspieler, Künstler, Musiker, Sänger, Tänzer
K Privat, Ohne Beschäftigung
L Goldarbeiter, Elektromonteur, Mechaniker, Steinschleifer, Metalldreher

1 = A; 2 = B; 3 = C, D, J; 4 = F; 5 = G, H, L.

The military data can also be used to explore the occupational distribution of the recruits and the dynamics of economic change. During the approximately half a century under consideration, there was a large increase in the share of students and bureaucrats among the Jewish recruits, but a significant decline in the number of artisans, lower-level clerks, and blue-collar workers. The evidence also confirms the hypothesis of a segmentation of the labor market (Table 2). Among the birth cohorts of the 1890s, that is, those who reached young adulthood just prior to the outbreak of World War I, many Christians, but few Jews, entered blue-collar occupations (e.g., mechanic) or became artisans. Whereas almost half the Christians were blue-collar workers, only 14 percent of the Jews were. Another major difference is that a large number of Jewish youth decided to postpone their entrance into the labor force. While 43 percent of the Jewish birth cohorts of the 1890s were students in their early twenties, only 17 percent of the Christians followed the same life-cycle pattern. While this is an indication of differences in both labor-force participation rates and of labor-market segmentation, these differences apparently did not translate into higher average nutritional standards for the Jews.

Differences in occupational distribution are also evident among women. This is clear from the protocols of one of the public welfare clinics in Vienna where expectant mothers who were unable to pay for a midwife came to deliver at government expense. The women were generally lower class and unmarried. While half of the Jewish women had semi-skilled occupations, such as seamstress, almost all of the Christian lower-class women were unskilled, primarily domestics (Table 3). The nutritional status of the women can be inferred from the age at which they reached menarche, a biological variable influenced by nutritional intake during childhood and adolescence. Among Vienna-born women, Jews reached menarche at age 15.4 (N=41), while Christians reached menarche at age 15.1 (N=453), that is, 3.6 months earlier. By contrast, Christians born in Lower Austria reached menarche at age 15.8 (N=531). These data indicate that the nutritional status of Viennese lower-class Jewish women during childhood was similar to that of Jewish men: inferior to that of Christians born in Vienna, but better than that of those born in Lower Austria.

In conclusion, this preliminary study on the standard of living of Viennese Jews indicates that the issue is much more complex than the stereotypical view would lead one to believe. Certainly much more research needs to be done before we can more fully understand this question. One should be careful not to confuse the situation of a few business leaders at the top of the social pyramid with that of the Jewish masses. The data on recruits into the Austro-Hungarian army and on lower-class women giving birth at the Viennese *Allgemeines Krankenhaus* confirm the hypothesis of a segmentation of the labor market while showing that on average this segmentation did not provide an advantage to the Jewish members of the society. While the standard of living of Jewish men and women born in Vienna was below that of the Christians born in the city, it was above that of the Lower Austrian population. Even from this preliminary research, it is quite clear that it is dangerous to draw inferences about the distribution of income and wealth from occupational distributions.

Table 3: Occupational Distribution of Lower-Class Viennese Women, 1907-1908

| Occupation | Jewish | Non-Jewish |
|---|---|---|
| A Domestic | 7.2 | 7.6 |
| B Maid | 24.6 | 63.7 |
| C Seamstress | 39.1 | 4.7 |
| D Blue-collar unskilled | 15.9 | 24.1 |
| E Commercial | 10.1 | 0.0 |
| F Student | 1.4 | 0.0 |
| G Prostitute | 1.4 | 0.0 |
| Number of Observations | 69 | 278 |
| 1 Servants | 31.6 | 71.3 |
| 2 Unskilled worker | 17.3 | 24.1 |
| 3 Skilled worker | 49.2 | 4.7 |
| 4 Student | 1.4 | 0.0 |

Source: *Protokollbücher*, 1907-08.

Note: Classification includes following occupations:
A Bedienerin
B Magd
C Schneiderin, Näherin, Strickerin
D Büglerin, Gehilfin, Tageslöhnerin, Arbeiterin, Lehrling
E Kontoristin, Verkäuferin, Hausiererin, Floristin
F Schülerin
G Prostituterte

1=A,B; 2=D,G; 3=C,E; 4=F

# X

# The Age at Menarche in Vienna: The Relationship between Nutrition and Fertility

The purpose of this article is to join the debate on the effect of nutrition on fertility rates in chronically malnourished populations. Without doubting the existence of such a relationship under famine conditions, some scholars have questioned the relevance of nutritional status above a critical threshold level to population growth because of the absence of a plausible link between these two biological variables. For instance, the fact that the age of menarche is sensitive to nutritional intake is well known, and that conceivably could be one of the biological control mechanisms of the crude birth rate. Yet in Europe, a long hiatus generally separated the onset of menses from the onset of reproductive life. Thus, the thesis of the existence of a connection between nutritional status, as reflected in the age at which females reach menarche, and the rate of fertility appears illogical on the surface. Nonetheless, the appearance of implausibility should not deter one from conducting an empirical investigation of the extent to which the onset of childbearing, an important determinant of total fertility in a precontraceptive society, was a function of the age at menarche, that is, of nutritional status. 158

Another important reason demographers argue that fertility (the number of live births per one thousand females of childbearing age) is independent of nutritional status is that in the contemporary Third World the relationship is often insignificant. In areas such as rural Bangladesh, the crude birth rate is high even among poor, apparently undernourished women. Because of the substantial improvements in the epidemiological environment in the twentieth century, however, the relationship between nutritional intake and fertility rates need not be the same now as it was in European malnourished populations of the past. Because nutritional status is a measure of the food intake minus the

Historical Methods 22, 4, pp. 158–163, 1989. Reprinted with permission of the Helen Dwight Reid Educational Foundation. Published by Heldref Publications, 1319 18th Street, N.W., Washington, DC 20036-1802. 

nutrients claimed by diseases, the nutritional status of women in the underdeveloped countries of today could be higher than that of women in preindustrial Europe. This is possible even if food consumption itself might not have increased, because improvements in medical technology eliminated or greatly reduced the incidence of many diseases. Consequently, the occasional failure to find significant correlations between nutritional status and fertility rates among contemporary women should not be used as evidence that such patterns were lacking among historical populations; their situation is not analogous.

Modern biological research has established beyond doubt that nutritional status is an important determinant of the age at which sexual maturity is reached.[1] The age at menarche has declined by about three years during the course of the last century primarily because of improvements in nutrition. The average age at menarche in mid-nineteenth-century Europe was about sixteen, but it is under thirteen in developed countries today.[2] The age at menarche has also been found to vary negatively with socioeconomic status: middle-class girls experience menarche earlier than lower-class ones.[3] In America, for instance, well-nourished girls reach menarche two years earlier than undernourished ones.[4] Historians, too, have shown interest in this biological variable in order to obtain quantitative indicators of secular changes in female nutritional status and to explore cross-sectional differences in dietary intake within a society.[5]

While a consensus has emerged on the relationship between the age at menarche and nutritional status, the relationship between it and vital demographic rates remains controversial. One side has argued that the relatively late age of menarche was associated with submaximum fecundity in historical populations: "The fact that undernourished human beings . . . are less fecund than well-nourished populations can be regarded as an ecological adaptation to reduced food supplies of the environment."[6]

In contrast, others have argued that chronic malnutrition probably did not affect fecundity (the probability of conception per menstrual cycle without contraception) in demographically significant ways. According to one demographer, "Age at menarche signals the beginning of potential childbearing, but actual reproduction starts at marriage. The mean age at marriage is almost always higher than the mean age at menarche. . . . In populations where the mean age at marriage is near 20 or higher, one can hardly expect a fertility effect from nutritional variations in age at menarche. . . ."[7] Because the mean age at first marriage of women in preindustrial Western Europe was in the mid-twenties, while menses began on average at about age sixteen, the long interval separating the two events led historical demographers to conclude that the two variables must have been unrelated, given that most births occurred within marriage.[8] In a recent review of the literature, this position was reiterated succinctly: "With marriage so late, . . . [the] age of menarche . . . would have had little effect on fertility . . . Only in populations in which marriage is closely tied to menarche is it conceivable that nutrition could significantly influence overall fertility through a fall in age at menarche."[9] Consequently, even if improvements in nutrition had induced a fall in the age at menarche in historical populations, the age at first birth would not have been affected by the rise in nutritional status because marriage was so delayed. The consensus has thus emerged that chronic undernutrition did not influence birth rates: "Little support is provided for the existence of a significant link between food intake and childbearing in situations of chronic or endemic malnutrition. . . ."[10] Of course, this argument does not apply to famines, during which, all agree, fertility is invariably sharply curtailed.[11]

Because of the absence of evidence, the relationship between the age at menarche and the age at first birth in chronically malnourished historical populations has not been explored. The issue has, however, been of some interest to sociologists and demographers interested in contemporary adolescent behavior. Some of these studies have, in fact, found significant positive correlations between the age at menarche and such fertility-determining factors as thinking about sex, the onset of dating and coitus, or the age at first conception.[12] This can be the case for social as well as for biological reasons because "pubertal hormones lead to early development of secondary sexual characteristics which are attractive to males, providing early opportunity for intercourse" and because "peers provide encouragement for early intercourse . . . to the woman with early menarche." In addition, "increased release of sex hormones leads to increased libido, and consequently to early intercourse. . . ." Last, "Those women with early puberty are more fecund than women with later puberty, leading to early births for a given exposure to the risk of pregnancy.''[13] Girls who reach menarche prior to age twelve experience ovulatory menstrual cycles 50 percent of the time within one year thereafter, while girls who reach it above age thirteen do not experience ovulatory menstrual cycles 50 percent of the time until four and a half years after menarche.[14]

In historical populations, data limitations prohibit the testing of an explicit combined sociobiological model with complex interactions linking hormone levels with sexuality, as can be done for contemporary adolescents.[15] Hence, the goal of this study could hardly have been to unravel the interaction of the social and biological mechanisms at work. Yet if one could show that, in spite of the long delay between the age at menarche and age at first birth, there was a positive relationship between them, then the case for the nutrition sensitivity of conceptions could still be made. Even if one grants the plausibility of the argument that the long hiatus between the two events diminishes the likelihood that menarche was related to first birh, the hypothesis should still be tested.

## DATA

In the early-twentieth century, one of the public women's clinics in Vienna collected information on the age at menarche of women who gave birth there. (The information was based on recollection. Although the women were not well educated, it is said that the onset of menses is a sufficiently important event for girls to be able to recall their age quite accurately. Random error in recollection would not jeopardize the findings of a study based on this data, as would systematic error. If, for instance, those who gave birth at younger ages tended to have poorer memories and systematically believed that they experienced menses at younger ages [an unlikely occurrence], findings would be biased. Because the question was asked in the course of a medical examination, the women would not have had any reason to deceive willfully.)

The women were generally lower class and unmarried. The overwhelming share of the sample, fully two-thirds, was made up of domestic servants. The rest was composed of blue-collar working-class women such as seamstresses. In other words, the subpopulation studied was made up of the ''respectable'' part of those urban lower-class women who bore illegitimate children at

public expense. Absent were such groups as the criminal element, the prostitutes, and the homeless. This peculiarity of the sample hardly mattered for the purposes of the study. No claim is made that the sample was representative of all women or even of the lower-class women of Vienna. It is important only that these were women whose socioeconomic status meant that they probably experienced an inferior diet during the course of their childhood and adolescence. That they were undernourished by modern standards is indicated also by their small average adult stature (152 cm, or 5′0″) and the low birth weight of their babies (3,244 grams for boys [7 lb. 2 oz.] and 3,114 grams for girls [6 lb. 14 oz.]).[16] Their terminal height was that of twelve-year-old American girls today, and their newborns weighed 200 grams less than those born in the Philadelphia almshouse in the 160
1870s.[17] Because the age at which they reached menarche was comparable to that of preindustrial European women, their nutritional status, too, must have been quite similar.

Other available data include age at the time of delivery (as of the last birthday), place of birth, occupation, number of prior pregnancies, weight of the newborn, and duration of stay in the clinic. Age at prior pregnancies is not available. Of the circa 1,700 observations extracted from the protocols for the year 1907, 806 were for women delivering for the first time and 373 for women having their second delivery.[18]

## RESULTS

The data indicated an obviously positive nonlinear relationship between the age at menarche and the age at first birth in this sample of lower-class women (table 1 and figure 1). The time elapsed between menarche and first birth fell consistently from about ten years for early maturers to seven years for those who reached menarche at age eighteen. For the thirty-two super-late maturers in the sample, who reached menarche at age nineteen or twenty, the elapsed time was about eight years (table 1). Whether this idiosyncratic pattern was merely the result of the small number of women in the upper-age category or was caused by the sociobiological attributes of these women needs further study. The standard deviations of the age at first birth fell (by .7 years) from those who reached sexual maturity at age eleven to those who reached it at age eighteen, only to be reversed for the nineteen- and twenty-year-olds. The variance of the age

**TABLE 1**
**The Relationship between Age at Menarche and Mother's Age at First Birth**

| Age at menarche[a] | *N* | Age at first birth | *SD* | Difference[b] |
|---|---|---|---|---|
| 11 | 10 | 20.9 | 4.6 | 9.9 |
| 12 | 41 | 21.8 | 4.3 | 9.8 |
| 13 | 102 | 22.1 | 4.8 | 9.1 |
| 14 | 160 | 22.8 | 4.3 | 8.8 |
| 15 | 170 | 23.3 | 4.2 | 8.3 |
| 16 | 118 | 23.3 | 3.7 | 7.3 |
| 17 | 87 | 24.4 | 3.9 | 7.4 |
| 18 | 55 | 24.9 | 3.9 | 6.9 |
| 19 | 18 | 26.7 | 5.8 | 7.7 |
| 20 | 14 | 28.2 | 5.1 | 8.2 |

*Source:* Vienna, Austria, Stadtarchiv. Protokollbücher 1907; Allgemeines Krankenhaus, Gebärklinik III.
*Note:* *N* is the number of observations; the sample was restricted to Christian women (the difference between the total sample of this table of 775 and the sample of 806 of column 1 in table 2 is made up of 31 Jewish women).
[a]As of last birthday.

**FIGURE 1**
**Age at Menarche and Age at First Birth in Vienna, 1907**

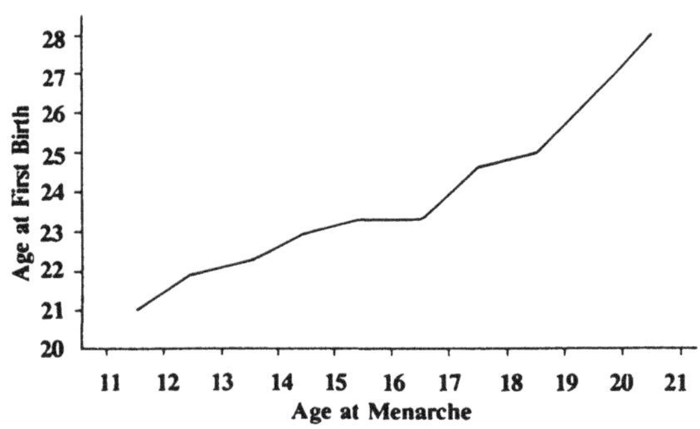

at which the upper-age group of women gave first birth was also an outlier.

Multivariate regression analysis controlled for factors that could have influenced the age at first birth (table 2). Occupation was not entered in the regression because most of the women were domestic or had other similarly low-status positions. There was, in fact, very little socioeconomic variation in the sample. The little variation that existed was captured by the religion regressor because more Jewish women tended to have semiskilled occupations, such as that of commercial employee. Because of the small number of Jewish women in the sample, the fear of multicollinearity inhibited the inclusion of an occupational variable in the regression.

The multivariate analysis confirmed the positive relationship between the age at menarche and the age at first birth. The trend was statistically significant at the 1 percent level; the strength of the relationship, however, varied with the age menarche was reached. For instance, a one-year delay in menarche, from age sixteen to seventeen, delayed first birth , on average, by thirteen months[19] (table 2). Yet between ages fifteen and sixteen the effect was minuscule. The regression equation explained only 20 percent of the variation in the mother's age at first birth. This is not surprising, insofar as one would expect a considerable random variation in the age at first birth. After all, it depended on a large number of chance factors, such as the identification of a suitable partner, the transition to sexual intercourse, and the frequency of intercourse.

The variation of the age at menarche with place of birth indicates that those who were born in the economically most-backward areas of the Habsburg Monarchy reached menarche later than those who were born in the more-developed provinces[20] (table 3). Women born in Vienna, which had the highest per capita income in the monarchy, began childbearing about a year earlier than women born in the provinces, with a given age at menarche (tables 2 and 3). They were also better nourished because they reached menarche earlier than the other women. The regression controlled for this regional variation, and consequently the trend could not have been caused by it.



Unmarried women gave birth earlier than married women, with the age at menarche held constant. This result is plausible, for it must have been more difficult for these lower-class women to find a suitable husband than a sexual partner. They also needed to save some money for marriage in order to make themselves more attractive as mates. Hence, women whose transition to coitus preceded marriage could be considerably younger at first conception than women who were not at risk of pregnancy until after marriage. Prenuptial pregnancies seldom led to marriage among this social group. Instead, illegitimate children were frequently given to orphanages.

## DISCUSSION

Although the age at menarche and the age at first birth were separated from one another by as long as seven years in this sample of chronically malnourished Viennese women, the two events were nonetheless positively correlated. The strength of the relationship between these biological variables depended on the age at menarche. A reduction of one year in the age at menarche, from age seventeen to sixteen, reduced the age at first birth by thirteen months, whereas a similar reduction from age eighteen to seventeen would have reduced the age at first birth by only five months, on average. A reduction from sixteen to fifteen would have had no effect (table 2). Nonetheless, these preliminary results indicate that the relationship does exist, although it is not linear.

There is no reason the correlation between the age at menarche and age at first birth found in this sample would not have held for low-status, chronically malnourished women of other societies in the past. It is true that in contrast to preindustrial peasant societies, conception among the Viennese women seldom led to marriage, but that hardly affects the argument. The primary task of this study was to establish that a positive relationship between the age at menarche and the age at first birth in a historical population is within the realm of the possible even though the two events were separated from one another by a fairly long time span. Whether the subsequent birth occurred in or out of wedlock is immaterial to the demographic consequences of this relationship.

The fact that a decline in the average age at menarche of one year can account for as much as an equal decline in the age at first birth establishes the possibility of a link between female nutritional status and fertility under conditions of chronic malnutrition. This is quite important in historical demography because the decline in the age at marriage and at first birth of less than a year sufficed to account for the acceleration of the English population growth in the first half of the eigh-

161

**TABLE 2**
**Determinants of the Mother's Age at Birth**

| | (1) | (2) |
|---|---|---|
| Constant | 28.6* | 28.4* |
| Unmarried | −5.8* | −3.4* |
| Jewish | 0.1 | 3.5* |
| Age at menarche | | |
| 11–12 | −2.0* | −3.4* |
| 13 | −2.6* | −2.0* |
| 14 | −1.6* | −2.0* |
| 15 | −1.2* | −1.5* |
| 16 | −1.1* | −1.6 |
| 18 | 0.4 | −0.2 |
| 19–20 | 2.8* | 0.7 |
| Birthplace | | |
| Galicia | 0.7 | −1.6 |
| L. Austria | 0.7* | 2.1* |
| Bohemia[a] | 1.2* | 2.1* |
| Hungary | 1.4* | 2.2* |
| $R^2$ | 0.2 | 0.2 |
| $N$ | 806 | 373 |
| $F$ | 18.8* | 8.1* |

*Notes:* Dependent variable was mother's age at (1) first birth, (2) second birth. Constant represents the age of a married Christian woman, born in Vienna, whose age at menarche was seventeen.
[a]Includes Moravia and Silesia.
*Significant at the 1 percent level; all independent variables were dummy variables.

162

**TABLE 3**
**Age at Menarche by Place of Birth, 1907**

| Place of birth | $N$ | | Age[a] |
|---|---|---|---|
| Vienna | 453 | | 15.1 |
| L. Austria | 531 | | 15.8 |
| U. Austria | 42 | | 15.9 |
| Silesia | 35 | | 15.9 |
| Hungary | 88 | | 16.0 |
| Moravia | 162 | | 16.1 |
| Bohemia | 244 | | 16.3 |
| Galicia | 21 | | 16.5 |
| Carinthia[b] | 18 | | 16.8 |
| Total | 1,594 | Average | 16.0 |

*Source:* Vienna, Austria, Stadtarchiv. Protokollbücher, 1907; Allgemeines Krankenhaus, Gebärklinik III.
[a]Half a year was added because ages were recorded as of last birthday.
[b]Including Carniola.

teenth century.[21] These facts underscore the plausibility of the view that the English demographic upswing of the 1730s, for instance, was caused by a rise in nutritional status brought about by the extremely favorable harvest conditions of the time.[22] A rise in fertility was more important than a fall in mortality in bringing about this acceleration in population growth. The rise in fertility, in turn, was brought about by the fall in celibacy and by the fall in the mean age of first marriage of less than one year. Hence, there is reason to believe that the increased fertility rate could have been caused by a fall in the age at menarche resulting from the rise in nutritional status.

A positive theory of marriage is difficult to test in preindustrial Europe. Data are not available to ascertain when and how the decision was made to begin looking for a mate, what costs were associated with such a search, how long it took to find a mate, and what factors determined the onset and frequency of intercourse.[23] The negative correlation between grain prices and marriage rates has been amply documented, and the argument has been put forth that during such good times the propensity to form families rose. Yet one would think that such a decision would depend on permanent income rather 162
than on a transitory event such as a good harvest.

Another possible route of influence on the age at first conception and at marriage, however, has been overlooked. The fall in grain prices undoubtedly led to a rise in average nutritional status, which, in turn, could have led to a decline in the age at menarche. If the positive correlation between the age at menarche and at first conception found in this sample of women held for other societies of the past, then the fall in grain prices could have affected marriage rates indirectly through changes in biological processes.[24] First marriages depended to a considerable degree on conceptions because the reproductive capability of the couples was frequently tested prior to marriage. Hence, conception very often preceded marriage. Consequently, the decision to marry also could have been a function of conceptions, and therefore of nutritional intake, and not only of permanent income. In this way, improvements in the diet might have influenced family formation. The importance of this factor is shown by the fact that during the eighteenth century, prenuptial pregnancies as well as illegitimate births increased greatly in England. At the end of the seventeenth century, 15 percent of the brides were pregnant; a century later, 35 percent were.[25] Thus, there was a large increase in marriages consummated after conception. This pattern suggests that the age at marriage declined in eighteenth-century England partly because conceptions rose; therefore, the rise in the crude birth rate conceivably could have been obtained through a nutrition-related fall in the age at menarche. In other words, the decline in the age at marriage was not necessarily exogenous.

In conclusion, a statistically significant positive relationship was found between the age at menarche and the age at first birth among chronically malnourished, lower-class Viennese women born in the late-nineteenth century. The relationship is demographically significant. The fall in the age at marriage of less than a year sufficed to induce the acceleration of English population growth, which became the harbinger of the demographic explosion.[26] Because the age at menarche depends on nutritional status, the above relationship indicates that nutritional status could have correlated positively with fertility rates, and therefore with population growth, in a noncontraceptive society. In the population under study, conceptions did not lead to marriage, as they did in preindustrial peasant societies. Nonetheless, those societies in which conceptions did lead to marriage, changes in the age at menarche could have exerted an influence on the age at marriage by regulating conceptions. Thus, if menarche is sensitive to nutrition and if it could influence the onset and frequency of intercourse and thereby the number of conceptions, then the hypothesis that the age at first marriage, the age at first birth, and the crude birth rate could all be functions of nutritional status in a chronically malnourished population cannot be rejected out of hand. The extent to which it held in other societies and at other times needs further study. In any event, the findings of this study refute the oft-cited argument that the relatively long hiatus separating the onset of menses and first birth by itself logically eliminates the possibility of a relationship between them.

## NOTES

I am indebted to Professor W. Peter Ward for bringing this data set to my attention, and to Markus Hein and Erich Foltyn for helping with the data processing.

1. Frisch, R. 1978. Population, food intake and fertility. *Science*, 199:22–30; *Idem*. 1975. Demographic implications of the biological determinants of female fecundity. *Social Biology*, 22:17–22. Bongaarts, J. 1980. Does malnutrition affect fecundity? A summary of evidence. *Science*. 208:564–69.

2. Frisch, Population, food intake and fertility, 24. Yet in eighteenth-century France, malnourished women could reach menarche as late as age eighteen, nineteen, or twenty. Hufton, O. H. 1974. *The poor of eighteenth-century France 1750–1789*. Oxford: Clarendon Press, 45.
3. Bongaarts, Does malnutrition affect fecundity?, 566.
4. Frisch, R. 1972. *Pediatrics*, 50. As cited in Bongaarts, Does malnutrition affect fecundity?, 566.
5. Laslett, P. 1985. Age at menarche in Europe since the eighteenth century. *Journal of Interdisciplinary History*, 16:221–36. Brudevoll, J. E., K. Liestol, and L. Walloe. 1979. Menarcheal age in Oslo during the last 140 years. *Annals of Human Biology*, 6:407–16. Shorter, E. 1981. L'age des premières règles en France, 1850–1950. *Annales: Economies, Sociétés, Civilisations*, 36. Tanner, J. M. 1973. Trend towards earlier menarche in London, Oslo, Copenhagen, the Netherlands, and Hungary. *Nature* (London), 243:95–96. Ducros, A., and P. Pasquet. 1978. Evolution de l'âge d'apparition des premières règles (ménarche) en France. *Biométrie Humaine*, 13:35–43. Manniche, E. 1983. Age at menarche: Nicolai Edvard Ravn's data on 3,385 women in mid-nineteenth century Denmark. *Annals of Human Biology*, 10:79–82. Peter Stearns emphasized the importance of this topic in his statement: "When the history of menarche is widely recognized as equal in importance to the history of monarchy, we will have arrived." In Stearns, P. 1976. Coming of age. *Journal of Social History*, 10: 250. I am indebted to my colleague, Reid Andrews, for this citation.



6. Frisch, Population, food intake and fertility, 29. She suggests that the cause of this relationship between malnutrition and subfecundity might be the dependence of estrogen levels on the consumption of fats. For a counterargument, see Trussell, J. 1978. Menarche and fatness: Reexamination of the critical body composition hypothesis. *Science*, 200:1506–09. Trussell has not claimed to "invalidate [Frisch's] hypothesis, but [to] establish clearly that more careful testing is needed."
7. Bongaarts, Does malnutrition affect fecundity?, 566. This statement is, however, quite misleading. Reproduction does not generally start at marriage. Rather, it starts either with the first intercourse or with the first conception.
8. Hajnal, J. 1965. European marriage patterns in perspective. In *Population in history: Essays in historical demography*, edited by D. V. Glass and D. E. C. Eversley. London: E. Arnold. Wrigley, E. A., and R. S. Schofield. 1983. English population history from family reconstitution: Summary results, 1600–1799. *Population Studies*, 37:157–84.
9. Menken, J., J. Trussell, and S. Watkins. 1981. The nutrition fertility link: An evaluation of the evidence. *Journal of Interdisciplinary History*, 11:425–41.
10. Ibid., 425.
11. Le Roy Ladurie, E. 1969. L'aménorrhée de famine (XVII$^{e}$–XX$^{e}$ siècles). *Annales S. E. C.*, 24:1589–1601.
12. Sandler, D. P., A. J. Wilcox, and L. F. Horney. 1984. Age at menarche and subsequent reproductive events. *American Journal of Epidemiology*, 119:765–74. Gottschalk, L. A. et al. 1964. Psychosocial factors associated with pregnancy in adolescent girls: A preliminary report, *Journal of Nervous and Mental Disorders*, 138:524–34. Ryder N. B., and C. F. Westoff. 1971. *Reproduction in the United States: 1965*. Princeton: Princeton University Press. Presser, H. B. 1978. Age at menarche, socio-sexual behavior, and fertility. *Social Biology*, 25:94–101. Presser found "some evidence that in the absence of contraceptive use—at least at the time of first sexual experience—early age at menarche may be related to early age at first birth."

13. Udry, J. R., and R. L. Cliquet. 1982. A cross-cultural examination of the relationship between ages at menarche, marriage, and first birth. *Demography*, 53-63.
14. Apter, D., and V. Reijo. 1983. Early menarche, a risk factor for breast cancer, indicates early onset of ovulator cycles. *Journal of Clinical Endocrinology and Metabolism*, 57:82-6. (This may have been the reason the Viennese women who reported regular menstural cycles had an average age at menarche of 15.6 years, whereas those who reported irregular cycles had an average menarcheal age of 16.1 years.) According to another study, even four years after menarche only 43 percent of the cycles were ovulatory. Metcalf, M. G. et al. 1983. Incidence of ovulation in the years after the menarche. *Journal of Endocrinology*, 97:213-19. In a Norwegian sample, "younger age at menarche was found to be associated with a [somewhat] shorter duration of postpartum amenorrhoea." Liestol, K., M. Rosenberg, and L. Walloe. 1988. Lactation and post-partum amenorrhoea: A study based on data from three Norwegian cities, 1860–1964. *Journal of Biosocial Science*, 20:423-34,
15. For a sophisticated combined sociobiological model, see: Udry, J. R. 1988. Biological predispositions and social control in adolescent sexual behavior. *American Sociological Review*, 53(5):709-22. Female sexuality is quite responsive to the level of testosterone. For other studies linking female sexual behavior to hormone (androgens) levels, see Persky, H. L. et al. 1982. The relation of plasma androgen levels to sexual behaviors and attitudes of women. *Psychosomatic Medicine*, 44:305-19; Bancroft, J. et al. 1980. Androgens and sexual behavior in women using oral contraceptives. *Clinical Endocrinology*, 12:327-40.
16. The Viennese women were six cm shorter than female slaves in antebellum America. Steckel, R. H. 1979. Slave height profiles from coastwise manifests. *Explorations in Economic History*, 16:363-80.
17. U.S. Department of Health, Education and Welfare, National Center for Health Statistics, NCHS Growth Charts, 1976. *Monthly Vital Statistics Report*, 25, No. 3 supplement (June 22, 1976) (HRA 76-1120). Personal communication from Robert Margo and Claudia Goldin.
18. For the sake of clarity, it should be pointed out that the ages at first birth are unavailable for these 373 women. The remainder of the sample pertained to women delivering for the third time or more. Vienna, Austria, Stadtarchiv. Protokollbücher, Allgemeines Krankenhaus, Gebärklinik III.
19. The birth intervals, measured cross sectionally, were also affected slightly by the age at menarche. The average time elapsed between the first and second births of those women who reached menarche at or before age 14 was 2.4 years; for the women who were older at menarche, the comparable interval was 2.8 years. The calculation is not presented in tabular form. The average age at first birth of women who reached menarche at a given age was subtracted from the average age at second birth of women who also reached menarche at the same age. Clearly, they were not the same women. That is why these are called cross-sectional rather than longitudinal estimates.
20. Good, D. 1984. *The economic rise of the Hapsburg empire 1750–1914*. Berkeley: California University Press, Table C6.
21. Wrigley, E. A., and R. S. Schofield. 1981. *The population history of England, 1541–1871: A reconstruction*. Cambridge: Harvard University Press.
22. Jones, E. L. 1965. Agriculture and economic growth in England, 1660–1750: Agricultural change. *Journal of Economic History*, 25:1-18.

23. That few teenage girls worked as prostitutes might be indicative of the general delay in the transition to intercourse. The average age of prostitutes in Paris in the eighteenth century was twenty-six. Benabou, E. M. 1987. *La prostitution et la police des moeurs au XVIIe siècle*. Paris: Librairie Acadèmique Perrin, 267. Yet there were young girls who became prostitutes even before they menstruated. Hufton, *The poor of eighteenth-century France*, 316.
24. In contemporary America, for instance, there is a positive correlation between the age at menarche and age at first intercourse. Udry, J. R. 1979. Age at menarche, at first intercourse, and at first pregnancy. *Journal of Biosocial Science*, 11:433–41.
25. Wrigley, E. A., and R. S. Schofield. 1983. English population history from family reconstruction: Summary results, 1600–1799. *Population Studies*, 37:157–84.
26. Simulations done by Marc Artzrouni of the Department of Mathematics, Loyola University of New Orleans, indicate that with vital rates similar to those of England of the 1730s, a fall in the age at marriage of one year could have brought about a doubling of population growth from .3 to .6 percent per annum.

NB. This article has been indexed under its original pagination, as indicated by the page references in the margin.

# XI

# Patterns of children's growth in East-central Europe in the eighteenth century

**Summary.** Records in Vienna of the heights of (*a*) Military Academy youths born between 1730 and 1760, (*b*) orphanage children born between 1760 and 1780 and (*c*) military boarding-school children born between 1775 and 1815 have been retrieved and analysed. This constitutes the earliest extant set of measurements of the heights of a group of individuals. Stature was increasing in the late 1740s and decreasing after the 1770s. This evidence indicates a rise and subsequent fall in nutritional status and is consistent with the known pattern of European agricultural conditions in the eighteenth century. Shifts in the age of maximum increment support the notion of the secular changes in the nutritional status of these boys. The stature of the Habsburg boys was greater than the poorest boys of contemporary London but compared unfavourably with the height of the English gentry and American cadets of the nineteenth century and, of course, with the height of today's populations.

## 1. Introduction

The purpose of this study is to explore the nature of human growth in the eighteenth and early-nineteenth centuries in order to discover its salient changes up to the present, and in the process to ascertain the effects of the environment on biological processes.

Until recently, only two data sets were known concerning the growth of children and adolescents in the eighteenth century, and both pertain to the later decades. The first set comes from the Carlschule in Stuttgart, Germany, and has only been partially analysed (Hartman 1970, Tanner 1981, p. 110). It appears to show that youth of aristocratic descent were taller than those of middle-class origins by 3–7 cm throughout adolescence. Aristocrats reached their final height earlier than did middle-class youth; consequently, after age 20 the height difference diminished to 1 cm. Final heights were considerably below modern standards. Timing of the adolescent growth spurt was not unlike that of modern youth, though its magnitude seemed to have been several centimetres less.

The second eighteenth-century data set pertains to poor London boys recruited into the Marine Society, a charitable foundation (Tanner 1981, p. 112, Floud and Wachter 1982). These data, too, suffer from several limitations. Since minimum height requirements were enforced, and at times flexibly, population height is difficult to estimate from sample height. In addition, the data are limited to four age-groups between the ages of 13 ± and 16 ± . In spite of these limitations, this data set reveals quite well the decline in stature of this socio-economic group during the last decades of the eighteenth century, a trend not reversed until the cohorts born in the 1820s.

These data leave many questions unanswered about patterns of growth in the past. As a consequence, a recently discovered data set in the Viennese Military Archive promises to be a valuable source of information on this subject.

## 2. Subjects and methods

Measurements of the height of aristocratic youth entering the Military Academy (founded in the 1750s by Maria Theresia to train officers) are the earliest known records

Table 1. Height of boys in the Habsburg Monarchy (all nationalities).

| | Age at last birthday | | | | | | | | | | | | | | | | | |
|---|---|---|---|---|---|---|---|---|---|---|---|---|---|---|---|---|---|---|
| | 6 yr | | | 7 yr | | | 8 yr | | | 9 yr | | | 10 yr | | | 11 yr | | |
| | *N* | *H* | σ | *N* | *H* | σ | *N* | *H* | σ | *N* | *H* | σ | *N* | *H* | σ | *N* | *H* | σ |
| 1735/55† | | n.a. | | | n.a. | | | n.a. | | | n.a. | | | n.a. | | 30 | 131·5 | 6·4 |
| 1765/69‡ | | n.a. | | | n.a. | | | n.a. | | | n.a. | | 50 | 127·6 | 5·3 | 27 | 134·4 | 7·5 |
| 1770/74‡ | 20 | 108·7 | 5·4 | 29 | 114·2 | 3·2 | 38 | 117·3 | 4·3 | 44 | 121·9 | 6·1 | | n.a. | | | n.a. | |
| 1775/79§ | | n.a. | | | n.a. | | | n.a. | | | n.a. | | | n.a. | | | n.a. | |
| 1780/84§ | | n.a. | | | n.a. | | | n.a. | | 11 | 119·5 | 6·3 | 18 | 123·7 | 6·8 | 15 | 130·8 | 5·5 |
| 1785/89§ | | n.a. | | | n.a. | | | n.a. | | 17 | 122·3 | 4·6 | 12 | 123·9 | 4·5 | 72 | 129·2 | 5·4 |
| 1790/94§ | 19 | 107·8 | 5·1 | 108 | 112·1 | 4·8 | 217 | 117·0 | 4·7 | 263 | 120·6 | 5·1 | 316 | 124·0 | 5·7 | 312 | 128·7 | 6·0 |
| 1795/99§ | 94 | 106·9 | 4·9 | 234 | 112·1 | 5·1 | 263 | 116·1 | 5·1 | 496 | 119·8 | 5·3 | 422 | 124·0 | 5·9 | 634 | 128·2 | 6·3 |
| 1800/04§ | 164 | 107·8 | 4·6 | 401 | 111·8 | 5·2 | 700 | 115·9 | 5·4 | 659 | 119·4 | 5·7 | 649 | 123·6 | 6·1 | 387 | 127·8 | 6·3 |
| 1805/09§ | 99 | 109·1 | 4·7 | 274 | 112·1 | 4·5 | 436 | 116·2 | 5·3 | 699 | 119·6 | 5·5 | 986 | 123·8 | 5·9 | 1031 | 128·0 | 6·2 |
| 1810/14§ | 118 | 109·2 | 4·4 | 311 | 112·9 | 4·4 | 344 | 117·1 | 4·6 | 196 | 121·1 | 4·9 | 27 | 125·5 | 4·5 | | — | |

| | 12 yr | | | 13 yr | | | 14 yr | | | 15 yr | | | 16 yr | | | 17 yr | | | 18 yr | | |
|---|---|---|---|---|---|---|---|---|---|---|---|---|---|---|---|---|---|---|---|---|---|
| | *N* | *H* | σ | *N* | *H* | σ | *N* | *H* | σ | *N* | *H* | σ | *N* | *H* | σ | *N* | *H* | σ | *N* | *H* | σ |
| 1735/44† | 55 | 134·2 | 7·2 | 114 | 138·4 | 7·2 | 88 | 142·7 | 7·8 | 49 | 148·9 | 8·1 | | n.a.‖ | | | n.a. | | | n.a. | |
| 1745/55† | 18 | 140·4 | 5·2 | 51 | 144·2 | 7·7 | 68 | 145·7 | 6·4 | 43 | 152·4 | 8·6 | 22 | 156·2 | 3·8 | | n.a. | | | n.a. | |
| 1760/64‡ | | n.a. | | | n.a. | | | n.a. | | 20 | 154·0 | 6·5 | 16 | 156·2 | 7·2 | | n.a. | | 9 | 163·7 | 5·5 |
| 1765/69‡ | 41 | 137·0 | 6·7 | 40 | 141·4 | 6·7 | 32 | 146·2 | 6·3 | | n.a. | | | n.a. | | | n.a. | | | n.a. | |
| 1770/74‡ | | n.a. | | | n.a. | | | n.a. | | | n.a. | | | n.a. | | 9 | 157·4 | 6·4 | | n.a. | |
| 1775/79§ | | n.a. | | | n.a. | | | n.a. | | 27 | 149·9 | 11·7 | 14 | 153·8 | 10·2 | 9 | 161·2 | 4·5 | | n.a. | |
| 1780/84§ | | n.a. | | | n.a. | | 22 | 144·8 | 7·3 | 30 | 148·6 | 9·4 | 110 | 154·8 | 7·5 | 207 | 160·8 | 7·2 | 49 | 161·7 | 5·2 |
| 1785/89§ | 188 | 132·5 | 6·5 | 238 | 137·3 | 6·3 | 242 | 141·1 | 6·8 | 383 | 147·0 | 8·4 | 365 | 154·3 | 8·0 | 208 | 160·2 | 6·1 | 48 | 163·3 | 4·0 |
| 1790/94§ | 353 | 132·2 | 5·9 | 366 | 137·0 | 6·8 | 478 | 141·5 | 7·2 | 442 | 146·7 | 8·3 | 553 | 151·6 | 8·3 | 430 | 157·3 | 7·6 | 115 | 159·1 | 5·9 |
| 1795/99§ | 720 | 132·0 | 6·6 | 855 | 136·4 | 7·1 | 661 | 140·6 | 7·5 | 573 | 146·3 | 8·5 | 336 | 152·0 | 7·9 | 270 | 158·4 | 7·4 | 61 | 160·0 | 6·1 |
| 1800/04§ | 460 | 132·0 | 6·2 | 535 | 136·5 | 6·6 | 735 | 140·9 | 7·4 | 814 | 145·9 | 8·2 | 648 | 152·1 | 8·2 | 447 | 158·3 | 7·8 | 39 | 161·5 | 6·1 |
| 1805/09§ | 732 | 132·0 | 6·5 | 434 | 136·2 | 6·4 | 223 | 141·3 | 7·3 | 21 | 147·2 | 8·4 | | n.a. | | | n.a. | | | n.a. | |

*N*, Number of observations.
*H*, Mean height in cm.
σ, Standard deviation of the height in cm.

Source: Kriegsarchiv, Vienna, Austria, Standestabellen Erziehungshäuser Faszikel 3925, 3926, 3927, Josephinische Waisenhaus, Faszikel 3922, and Theresianische Militärakademie, Faszikel 434.

† Aristocrats. ‡ Orphans. § Students in military schools. ‖ Datum discarded because it was an outlier due to sampling or measurement error.

on stature (Kriegsarchiv, Vienna, Theresianische Militärakademie, Faszikel 434). Five hundred and sixty observations on aristocratic adolescents born in the 1730s and 1740s have survived. In addition, 366 observations have been found among records of an orphanage, the Josephinische Waisenhaus, founded by Maria Theresia's son Joseph II. This data set contains information on children of various ages born in the 1760s and 1770s (Kriegsarchiv, Vienna, Josephinische Waisenhaus, Faszikel 3922). Data of the Military Academy and orphanage consist only of a single observation on an individual.

The third, and major source of data for this study is from schools run by the military for the sons of ordinary soldiers (Kriegsarchiv, Vienna, Erziehungshäuser Musterlisten, Faszikel 3925, 3926, 3927). Exactly 25 071 valid data have been put on tape from this source for children born between 1775 and 1815. Here there are multiple measurements on each individual; however, the data have been treated as if cross-sectional. Also, the height, age at last birthday and birthplace of the boys were recorded. Stature was measured in Austrian feet and inches. One Austrian foot is 31·6 cm divided into 12 inches, which, in turn, is subdivided into four parts, the strich. Measurements are recorded in feet, inches and strich.

## 3. Results

The height data are displayed in table 1. Examination reveals that the youth in the officer academy (hence aristocratic) born between 1735 and 1744 were shorter than orphans born during the next decade. The aristocratic youth, born in the late 1740s and early 1750s, were taller than those born earlier, by as much as 3–6 cm, depending on their ages.

In contrast, the orphans as well as the students in military schools, born between 1760 and 1800, declined in stature consistently at all ages (figure 1, table 1). The difference between the heights attained in the 1760s or 1770s and those of the 1780s ranged from 0·3 to 4·5 cm at various ages. Even though the number of observations at the beginning of the period is small, the pattern is consistent at all ages. The stature of boys of all ages continued to decline until we reach the cohorts born in the early-nineteenth century, although after the 1790s the decline was no longer substantial, with the exception of boys of 15 years or older, whose average height diminished by 3 cm or more (table 1). The stature of boys increased after the end of the Napoleonic Wars (figure 1).

Figure 2 shows the mean height increments; the largest increment of the aristocrats (1735–1755) and the orphans (1760s) was reached between the ages of 14·5 and 15·5 (figure 2, table 2). The size of the mean growth increment of the orphans was greater than that of the aristocrats. The average growth increment of the cohorts born in the 1760s at its maximum was almost 8 cm.

During the last decades of the century, the adolescent growth spurt of boys attending the military boarding schools was delayed on average (figure 3). Compared with the orphans born in the 1760s who experienced their greatest growth increment on average between the ages of 14·5 and 15·5, the boys born in the 1780s experienced it between the ages of 15·5 and 16·5, and those born in the 1790s between the ages of 16·5 and 17·5 (figure 4). Thereafter, the trend was reversed (figure 5). The largest growth increment in the 1800s was reached between the ages of 15·5 and 16·5. The mean adolescent growth spurt diminished over time.

Preece–Baines curves were fitted (by J. M. Tanner) to the data in table 1. Although the fit was not good for the 1770's cohort, the mean age of maximum increment for children born in the 1780s was 16·1 years, the 1790s 15·7 years and the 1800s 16·0 years.

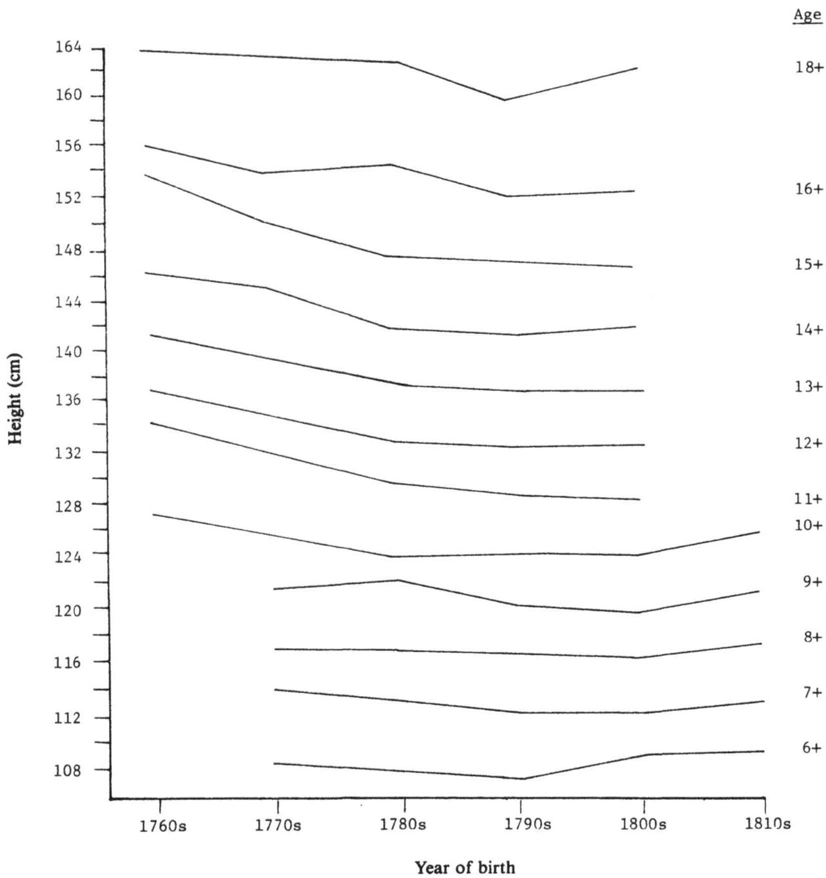


Figure 1. Height of boys in the Habsburg Monarchy by year of birth, 1760–1815. n.b., Boys born in the 1760s and 1770s up to the age of nine plus were orphans, others were sons of ordinary soldiers enrolled as students in military boarding schools.

Final adult stature was estimated to have been between 163 and 166 cm, with perhaps a larger estimate for the 1800's cohorts.

*Regional variation*

Regression analysis of the aristocratic youth by place of birth reveals that their height did not, in general, depend on their place of birth. Dummy variables have been entered in the regression for the ages, and year of birth, as well as the place of birth of the boys. Those born in Bohemia, alone, appear to have been slightly taller than average (by 2·7 cm) (table 3, equation (1)).

Among the orphans (1760s), Galician boys were significantly shorter than average (table 3, equation (3)). Regression analysis of data-base spanning the period 1760–1815 (table 4) and including boys from both the military schools and the orphanage reveals

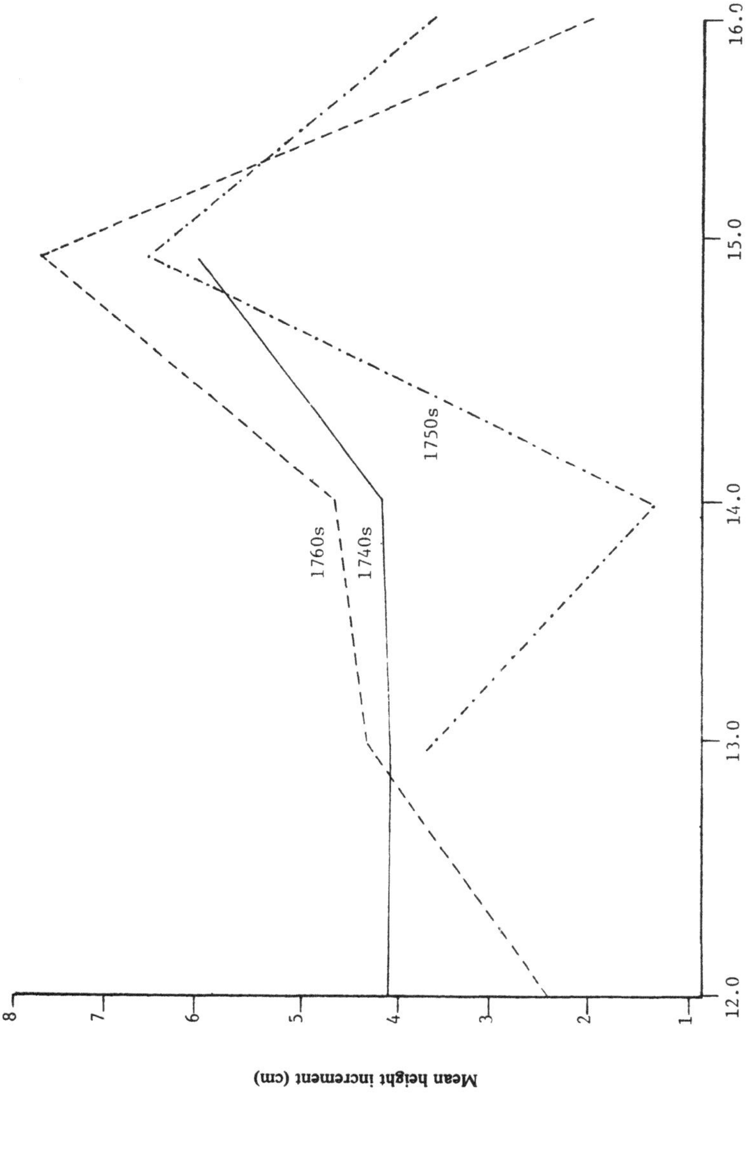


**Figure 2.** Mean growth increment of boys in the Habsburg Monarchy, 1735–1775. ___, Aristocrats born between 1735 and 1744; _._., aristocrats born between 1745 and 1755; _ _ _, orphans born between 1760 and 1775.

Table 2. Mean height increment of boys in the Hapsburg Monarchy (cm).

| | Age at centre of increment period (yr) | | | | | | | | | | | |
|---|---|---|---|---|---|---|---|---|---|---|---|---|
| | 7 | 8 | 9 | 10 | 11 | 12 | 13 | 14 | 15 | 16 | 17 | 18 |
| 1735/44† | n.a. | n.a. | n.a. | n.a. | n.a. | 4·2‖ | 4·2 | 4·3 | 6·2 | n.a. | n.a. | n.a. |
| 1745/55† | n.a. | n.a. | n.a. | n.a. | n.a. | n.a. | 3·8 | 1·5 | 6·7 | 3·8 | n.a. | n.a. |
| 1760/74‡ | 5·5 | 3·1 | 4·6 | n.a. | 6·8 | 2·6 | 4·4 | 4·8 | 7·8 | 2·2 | 3·8¶ | 3·8 |
| 1775/79§ | n.a. | n.a. | n.a. | n.a. | n.a. | n.a. | n.a. | n.a. | n.a. | 3·9 | 5·5†† | n.a. |
| 1780/84§ | n.a. | n.a. | n.a. | 4·2 | 7·1 | n.a. | n.a. | n.a. | 3·8 | 6·2 | 6·0 | 0·9 |
| 1785/89§ | n.a. | n.a. | n.a. | 1·6 | 5·3 | 3·3 | 4·8 | 3·8 | 5·9 | 7·3 | 5·9 | 3·1 |
| 1790/94§ | 4·3 | 4·9 | 3·6 | 3·4 | 4·7 | 3·5 | 4·8 | 4·5 | 5·2 | 4·9 | 5·7 | 1·8 |
| 1795/99§ | 5·2 | 4·0 | 3·7 | 4·2 | 4·2 | 3·8 | 4·4 | 4·2 | 5·7 | 5·7 | 6·4 | 1·6 |
| 1800/04§ | 4·0 | 4·1 | 3·5 | 4·2 | 4·2 | 4·2 | 4·5 | 4·4 | 5·0 | 6·2 | 6·2 | 3·2 |
| 1805/09§ | 3·0 | 4·1 | 3·4 | 4·2 | 4·2 | 4·0 | 4·2 | 5·1 | 5·9 | n.a. | n.a. | n.a. |
| 1810/14§ | 3·7 | 4·2 | 4·0 | 4·4 | n.a. | n.a. | n.a. | n.a. | n.a. | n.a. | n.a. | n.a. |

† Aristocrats. ‡ Orphans. § Boys in military boarding schools. ‖ 1735/55.
¶ Estimated difference of mean height. †† Average of 1770/79.

Table 3. Equations (1)–(3) (dependent variable: height in cm).

| | Equations† | | |
|---|---|---|---|
| | (1)‡ | (2)‡ | (3)§ |
| Intercept | 142·6* | 143·1* | 106·0* |
| Birth year | | | |
| 1740/44 | −1·4 | −1·3 | |
| 1745/49 | 3·2* | 3·3* | |
| 1750/59 | 5·6* | 5·9* | |
| Birthplace | | | |
| Moravia | 1·1 | | −0·7 |
| Bohemia | 2·7* | | |
| Silesia | | | 2·3 |
| Galicia | −2·2 | | −6·6* |
| Styria | | | −0·8 |
| Hungary | 0·3 | | 0·7 |
| Upper Austria | | | −5·2 |
| Lower Austria | −0·4 | | −4·3 |
| Italy | −0·3 | | |
| Other | | | −0·6 |
| Age | | | |
| 7+ | | | 8·2* |
| 8+ | | | 11·3* |
| 9+ | | | 16·4* |
| 10+ | | | 22·6* |
| 11+ | −10·4* | −10·4* | 28·4* |
| 12+ | −7·2* | −7·4* | 30·4* |
| 13+ | −3·0* | −3·1* | 34·7* |
| 14+ | | | 40·2* |
| 15+ | 5·6* | 5·7* | 46·6* |
| 16+ | 9·5* | 9·9* | 50·0* |
| 17+ | | | 51·6* |
| 18+ | | | 58·0* |
| *F* | 23·8* | 36·7* | 120·7 |
| *P* | | | ·0001 |
| $\bar{R}^2$ | 0·40 | 0·39 | 0·83 |
| *N* | 556 | 556 | 489 |

n.b., All independent variables are dummy variables.
(1) Intercept indicates the height of a 14-year-old boy born outside of the six provinces before 1740;
(2) intercept indicates the height of a 14-year-old boy born before 1740 without regard to place of birth;
(3) intercept indicates the height of a 6-year-old boy born in Bohemia.
‡ Habsburg Aristocrats, 1735–1755. § Orphans, 1760–1779. * Significant at the 1% level.

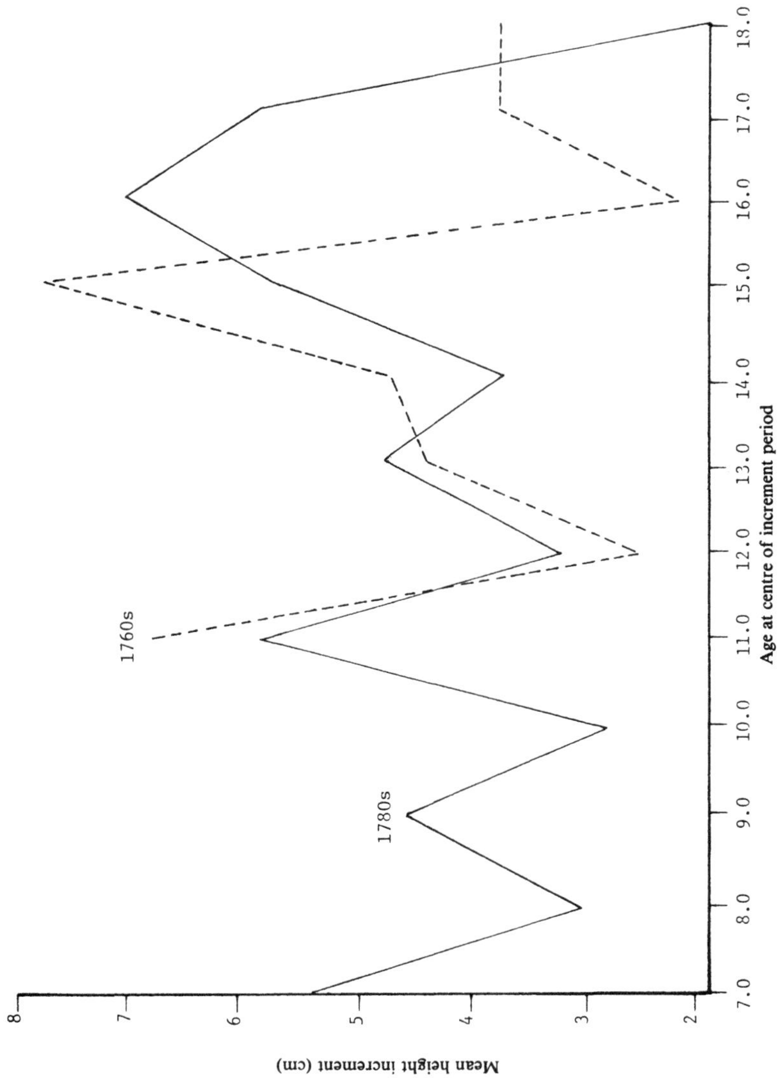


Figure 3. Mean growth increment of boys in the Habsburg Monarchy, 1760–1789. _ _ _, Orphans born between 1760 and 1775; ___ students in military boarding schools born between 1780 and 1789.

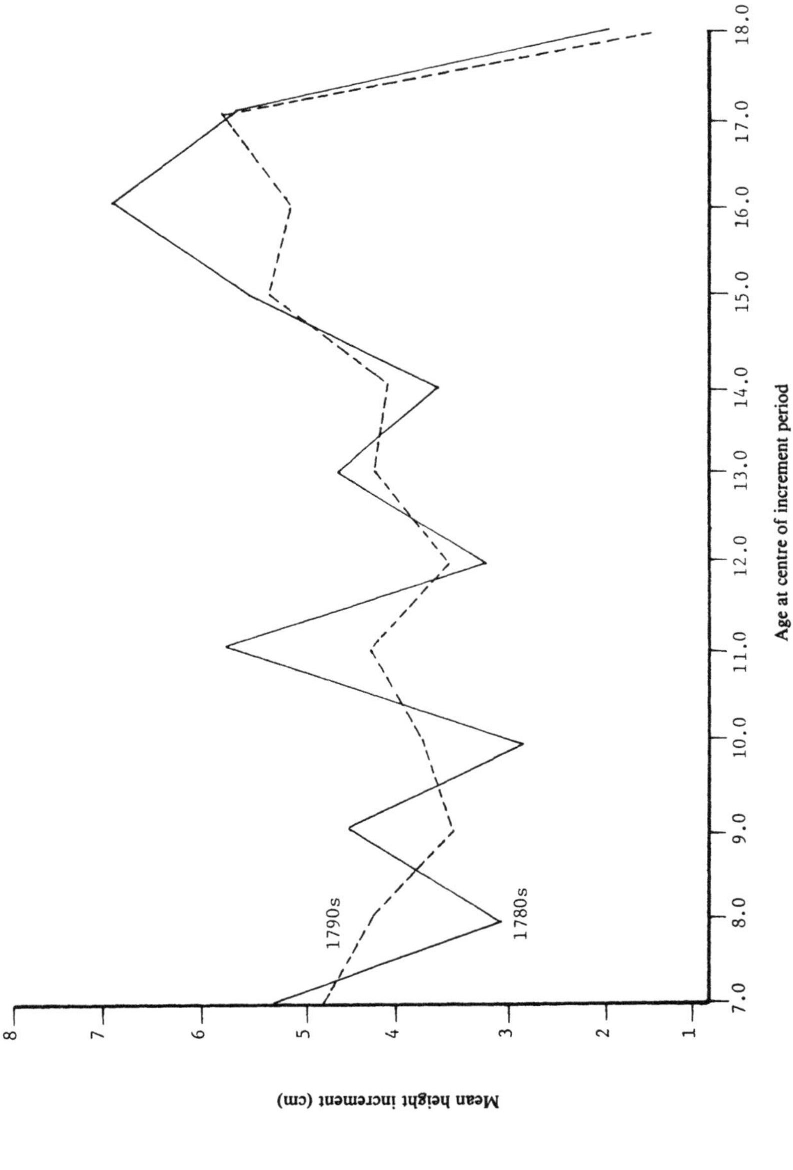


Figure 4. Mean growth increment of boys in the Habsburg Monarchy, 1780–1799. ——, Students in military boarding schools born in the 1780s. – – –, students in miliary boarding schools born in the 1790s.

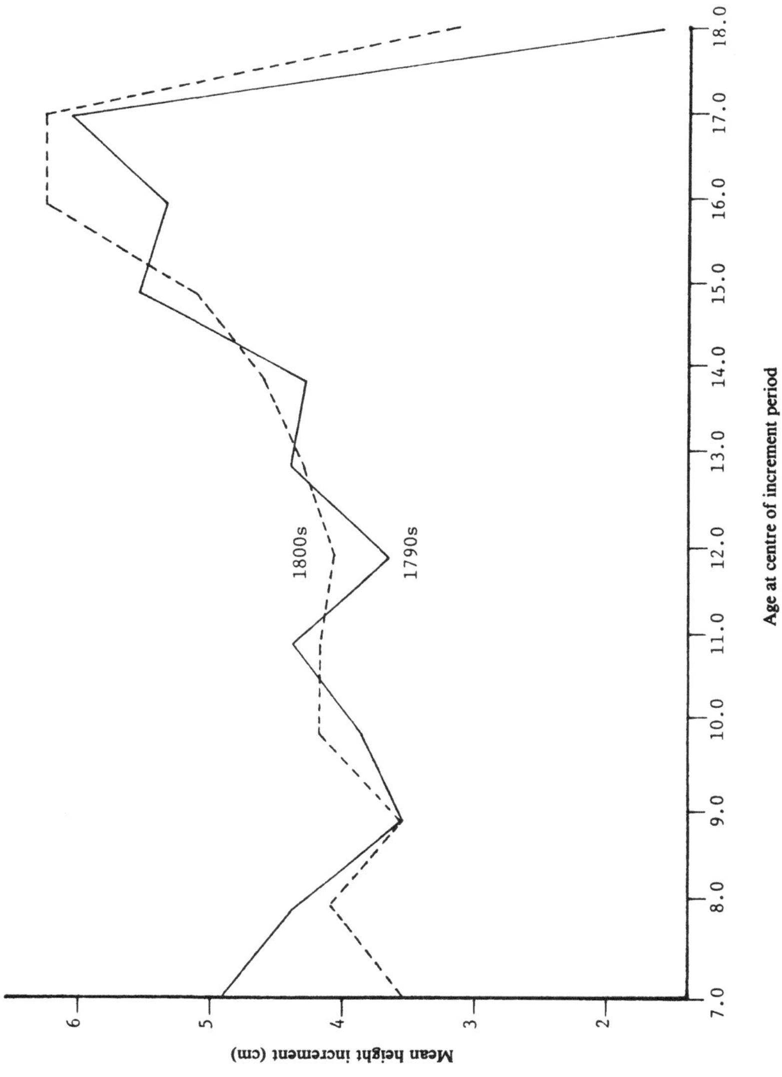


**Figure 5.** Mean growth increment of boys in the Habsburg Monarchy, 1790–1815. ——, Students enrolled in military boarding schools born in the 1790s; – – –, students enrolled in military boarding schools born between 1800 and 1815.

Table 4. Regression equations (dependent variable: height in cm).

| | Age (yr) | | | | | | | | | | | | |
|---|---|---|---|---|---|---|---|---|---|---|---|---|---|
| | 6·0 | 7·0 | 8·0 | 9·0 | 10·0 | 11·0 | 12·0 | 13·0 | 14·0 | 15·0 | 16·0 | 17·0 | 18·0 |
| Intercept | 106·1** | 111·0** | 115·6** | 118·5** | 122·5** | 125·8** | 130·0** | 133·9** | 139·0** | 144·2** | 150·2** | 156·0** | 160·1** |
| Month | 0·36** | 0·30** | 0·29** | 0·31** | 0·28** | 0·41** | 0·33** | 0·37** | 0·35** | 0·47** | 0·36** | 0·39** | −0·12 |
| Birth year | | | | | | | | | | | | | |
| 1760s | | | | | 4·57** | 5·78** | 4·60** | 4·75** | 6·02** | 8·09** | | | |
| 1770s | 3·51** | 4·35** | 0·59 | 1·72* | | | | | | 3·52** | | | |
| 1780s | | | | −0·28 | −0·64 | 1·74** | 0·60 | 1·14** | 0·68 | 0·70 | 3·21** | 3·14** | 2·04 |
| 1800s | 0·29 | 0·36 | −0·49 | −0·38 | −0·27 | −0·36 | 0·12 | 0·16 | 0·19 | −0·18 | 0·71 | 1·21** | 2·83 |
| 1805 | 2·47** | 0·58 | −0·45 | −0·32 | 0·02 | −0·11 | 0·10 | −0·14 | 0·51 | −0·08 | | | |
| 1810s | 1·69** | 1·15** | 0·19 | 0·83* | 1·63 | | | | | | | | |
| Birthplace | | | | | | | | | | | | | |
| Moravia | −1·27* | −1·05** | −0·64* | −0·15 | −0·47 | −0·10 | −0·31 | 0·07 | −1·12** | −1·70** | −0·17 | −0·76 | |
| Silesia | | | | | −0·84 | −1·57** | −1·58* | −1·59* | −1·71* | −2·26** | | | |
| Galicia | | −2·16** | −1·32** | −1·52** | −1·19** | −0·64 | −0·65 | −0·26 | −1·07 | −1·25 | −0·89 | −0·79 | |
| Styria | | −1·94** | 0·17 | −1·08* | −0·54 | −1·85** | −0·74 | −1·13 | 1·15 | −1·47 | −1·39 | | |
| Hungary | −2·42** | −1·57** | −0·75* | −0·15 | −0·09 | 0·74 | 0·80* | −1·80** | 1·20** | 0·25 | 0·34 | 0·25 | |
| Upper Austria | | | −0·56 | −0·54 | −0·47 | −0·68 | −0·64 | −0·71 | −1·75* | −2·34** | −0·70 | −0·81 | |
| Lower Austria | | −0·14 | −1·0** | −0·08 | 0·70 | 0·69 | 0·73 | 1·06* | −0·22 | 0·39 | 0·19 | −0·40 | |
| Other | −1·03 | −1·64** | −0·67 | 0·44 | 0·51 | 0·42 | 0·49 | 0·57 | 0·73 | 0·43 | 0·37 | −0·01 | |
| Carinthia | | | −1·55** | −0·39 | 0·32 | 1·13 | 0·08 | 1·77** | 2·20** | 2·0* | | | |
| Italy | | | | | 0·37 | 0·85 | 0·69 | 1·37* | | −0·52 | | | |
| Germany | | 0·14 | −0·75 | −0·18 | −0·36 | −0·38 | −0·25 | 0·20 | −0·47 | 0·39 | −0·06 | 0·11 | |
| Tyrol | | | | | | −0·46 | −0·02 | 0·48 | −1·40 | −1·01 | | | |
| Transylvania | | | 0·34 | 1·04 | −0·08 | 0·98 | 1·52** | 0·28 | 0·28 | 0·85 | | | |
| Croatia | | | | | | 3·24** | 4·07** | 4·70* | 2·11** | 2·49** | | | |
| $F$ | 6·6 | 7·8 | 5·7 | 7·2 | 6·1 | 11 | 7·5 | 9·2 | 7·8 | 7·7 | 7·7 | 8·3 | 1·3 |
| $\bar{R}^2$ | 0·12 | 0·07 | 0·04 | 0·04 | 0·04 | 0·07 | 0·05 | 0·06 | 0·05 | 0·06 | 0·04 | 0·05 | ·01 |
| $N$ | 320 | 1128 | 1860 | 2250 | 2411 | 2480 | 2497 | 2469 | 2337 | 2300 | 1793 | 1328 | 82 |

* Coefficient significant at the 10% level.
** Coefficient significant at the 5% level.
n.b., Intercept indicates the height of a Bohemian Boy born in the 1790s.

the following pattern of height by place of birth: boys born in Croatia were regularly the tallest; boys from Carinthia, Transylvania and Italy were often taller than Bohemian boys, whose height is estimated by the intercept; Lower Austrian and German boys were about as tall as Bohemian boys; and Hungarian boys were shorter before 10 years, but taller than Bohemian boys thereafter; Moravian, Silesian, Galician, Styrian and Upper Austrian boys were consistently shorter than Bohemian boys (table 5).

**4. Discussion**

The average height of the aristocratic youth attending the Viennese officer academy increased between about 1755 and 1765. This result is not likely to have been caused by sampling error; even though the number of observations is small at each age, the pattern is consistent at all ages and for all nationalities. All of the increases in stature were attained before entering the military academy, and hence cannot be attributed to changes in the diet they received at the institution. The increases in stature can be explained by better nutritional status of the cadets which was due to better harvest conditions in Europe (Jones 1974). Similar increases in stature are seen in adult Hungarian peasant recruits at the same time (J. Komlos, unpublished data). One might have thought that the improved agricultural conditions would not have affected the stature of the aristocratic youth the same way as it affected the stature of the peasantry. One should note, however, that the cadets were chosen by Maria Theresia from the lower aristocracy. In fact, only offspring of those aristocrats were eligible who did not possess sufficient financial means to educate their sons themselves, although those whose father served in government administration for 20 years or who were sons of 'brave' officers were also admitted. Maria Theresia usually chose the poorest candidates (Leitner von Lietnertreu 1852). As a consequence of this selection bias, it is indeed plausible that not only the peasantry's but also the lower aristocracy's real income increased in conjunction with the fall in cereal prices.

In spite of the increases in their stature, the aristocratic youth in the Habsburg Monarchy remained relatively short. Those aged 14 and above were not even taller than boys in the orphanage born two decades later (table 1). In addition, they were shorter than the German aristocrats of the Carlschule (table 1), who, in turn, were shorter than the English gentry of the same age attending Sandhurst military academy (R. Floud, personal communication, unpublished table). This pattern supports the notion that aristocrats in different countries enjoyed a different diet, perhaps because of different real incomes.

To be sure, height is a function of net nutrition—net, that is, after the claims of work and of the disease environment have been satisfied. Nonetheless, it is not likely that either of these factors would have varied considerably either over the geographical range or over the time-span covered by this study. As a consequence, a change in stature is interpreted in this study as having been caused by a change in nutritional intake.

The diet of all the aristocrats was a combination of institutional food and food consumed in their own households. The importance of the latter is shown by the Carlschule aristocrats having been taller than middle-class boys of the same age in spite of their consuming the same institutional food after they were eight years old (table 1). Since the data on the German and Habsburg aristocratic youth are separated in time by a generation, a direct comparison is not fully warranted during a period when nutritional status was changing rapidly in Europe. Nonetheless, one might note that Habsburg

Table 5. Height of boys in the Habsburg Monarchy by place of birth in cm.

| Place of birth | 7 yr old 1790s | 7 yr old 1800s | 7 yr old 1810s | 8 yr old 1790s | 8 yr old 1800s | 8 yr old 1810s | 9 yr old 1790s | 9 yr old 1800s | 10 yr old 1790s | 10 yr old 1800s | 11 yr old 1790s | 11 yr old 1800s | 12 yr old 1790s | 12 yr old 1800s |
|---|---|---|---|---|---|---|---|---|---|---|---|---|---|---|
| Moravia | | 111·98 | | | 115·96 | 117·20 | 119·54 | 119·30 | | 123·01 | 129·55 | 127·61 | 132·10 | 131·70 |
| Bohemia | 112·53 | 113·40 | 113·40 | 117·23 | 116·50 | 118·20 | 120·40 | 119·70 | 124·04 | 123·85 | 127·77 | 127·99 | 131·70 | 131·85 |
| Silesia | | | | | | | | | | 123·66 | | 126·59 | | |
| Galicia | | 110·90 | | | 115·20 | | | 118·00 | | 122·29 | | 126·51 | 130·22 | 131·54 |
| Other | 110·66 | 112·05 | | 116·10 | 116·00 | | | 120·78 | 124·22 | 124·80 | 128·24 | 129·16 | 132·30 | 132·90 |
| Steyermark | | | | | | | | 118·95 | | 123·50 | | 127·38 | | 132·26 |
| Hungary | | 110·39 | 113·00 | | 115·80 | 116·00 | 120·55 | 119·10 | 124·65 | 123·80 | 128·00 | 128·88 | 132·52 | 133·09 |
| L. Austria | | | | | 116·40 | | 120·58 | 119·40 | 126·00 | 124·08 | 130·20 | 127·80 | 133·30 | 131·86 |
| Germany | | | | 116·28 | | | 119·40 | | 123·89 | | 127·60 | | 131·46 | |
| Transylvania | | | | | | | | 119·86 | | 123·40 | | 128·36 | | 133·20 |

| Place of birth | 13 yr old 1790s | 13 yr old 1800s | 14 yr old 1790s | 14 yr old 1800s | 15 yr old 1780s | 15 yr old 1790s | 15 yr old 1800s | 16 yr old 1780s | 16 yr old 1790s | 16 yr old 1800s | 17 yr old 1780s | 17 yr old 1790s | 17 yr old 1800s |
|---|---|---|---|---|---|---|---|---|---|---|---|---|---|
| Moravia | 136·51 | 135·90 | 139·96 | 139·97 | 145·00 | 145·56 | 144·50 | 154·23 | 151·60 | 152·00 | 159·20 | 157·90 | 157·90 |
| Bohemia | 135·71 | 136·30 | 140·70 | 141·40 | 148·10 | 146·30 | 146·10 | 155·04 | 151·80 | 152·50 | 161·60 | 157·60 | |
| Silesia | | | | | | | | | | | | | |
| Galicia | | 136·52 | 140·06 | 140·20 | | 144·81 | 145·90 | | 150·40 | | | | |
| Other | 136·20 | | 141·44 | 142·00 | | 146·50 | 148·30 | | 152·28 | 153·00 | | 157·24 | 158·90 |
| Steyermark | | | | | | | | | | | | | |
| Hungary | 138·45 | 136·90 | 142·00 | 142·25 | | 147·50 | 146·20 | 156·53 | 152·37 | 151·20 | | 158·71 | |
| L. Austria | 137·50 | 137·00 | 140·30 | 140·85 | | 147·69 | 145·80 | | 151·36 | | | | |
| Germany | 136·40 | | 140·61 | | | 146·79 | | | 151·09 | | | | |
| Transylvania | | 135·10 | | | | | | | | | | | |

aristocrats born between 1745 and 1755 were, in fact, about the same height as the German aristocratic youth born between 1760 and 1780 until the age of 13. Thereafter, German aristocratic youth experienced an earlier and greater adolescent growth spurt than the Habsburg aristocratic youth, whose height became closer to that of German middle-class boys between the ages of 14 and 18.

That the Habsburg aristocratic youth were not taller than the boys in the Austrian orphanage is surprising, because one would have expected that the diet of orphans was nutritionally not as sound as that of members of the lower aristocracy. The two data-sets, however, are separated from one another in time by two decades. Hence it is conceivable that the height of aristocrats continued to increase and that by the 1760s they would have been taller than the boys in the orphanage.

The height of the children by place of birth varied greatly from decade to decade (table 5), and differences in stature by place of birth were generally very small.

The consistent decline in the height of the children at all ages between the 1760s and the turn of the century is extraordinary (figure 1, table 4). In fact, one would have expected the boys in the oprhanage, on which the sample of the 1760s and 1770s is based, to have been less well nourished than those in the military boarding schools, and therefore to have been shorter.

The diminution of height throughout the late-eighteenth century is in accordance with what is known about the economic circumstances of the period, particularly about agricultural conditions (Abel 1966, 1974). The rise in grain prices and the concomitant fall in real wages imply a decline in the standard of living; the data on the trend in stature substantiates this notion with direct evidence for the first time. Adult male heights in the Habsburg monarchy also diminished during the latter half of the century (Komlos 1985). Data on Swedish conscripts evinced a similar pattern, with the decline particularly noticeable among 19-, 20- and 21-year-old soldiers, but less so with those over 25 (Sandberg and Steckel 1980, R. Steckel, personal communication). American slaves as well as slum boys in London also suffered during the last decades of the eighteenth century. Both groups recovered, however, shortly thereafter. The London poor, in fact, experienced a phenomenal increase in stature of as much as 12 cm after the end of the war in 1815 (Floud and Wachter 1982, Fogel *et al.* 1983). From this evidence one sees that the nutritional crisis was not confined to East-central Europe.

The adolescent growth spurt was sensitive to the boys' nutritional status. The delay in the age at which most boys experienced their greatest growth increment after the 1760s is yet another indication that their nutritional status was deteriorating.

The sudden upturn in stature after the end of the Napoleonic Wars is unexpected because unusually bad harvests prevailed throughout Europe in 1816–1817, and the immediate end of a war has not been associated with an increase in the standard of living, or with increased economic activity (Post 1977, Komlos 1983).

The nutritional status of the boys in the military boarding schools compares favourably with the height of poor London boys (table 6). The boys in our sample were 4–6 cm taller than the poor boys of London. Of course, the gentry boys attending Sandhurst Military Academy in England were much taller, as were the cadets attending the West Point Military Academy in the United States. At age 18, the Sandhurst boys were 173 cm (R. Floud, personal communication) and those of West Point were 171 cm (table 6); this was 4–6 cm taller than the eighteenth-century German aristocrats, who in turn were taller than the Habsburg aristocrats (table 7).

Compared with early-twentieth-century standards, the boys of the Habsburg monarchy were small in stature. At age 14, for instance, they lagged 10 cm behind the

Table 6. Comparison of the height (cm) of boys of selected ages in various countries 1800–1981.

| | Age of boys (yr) | | | | | | |
|---|---|---|---|---|---|---|---|
| Countries† | 6 | 10 | 13 | 14 | 15 | 16 | 18 |
| 1. 1800 Habsburg Monarchy | 108 | 124 | 137 | 141 | 146 | 152 | 162 |
| 2. 1760/80 German middle class | | 128 | 140 | 144 | 150 | 155 | 163 |
| 3. 1760/80 German aristocrat | | 131 | 144 | 151 | 156 | 161 | 167 |
| 4. 1800 London poor | | | 132 | 135 | 142 | 147 | |
| 5. 1800 British gentry | | | 148 | 156 | 161 | | |
| 6. 1800 France | | | | | | 167 | 170 |
| 7. United States slave | | 127 | 142 | 146 | 152 | 159 | 166 |
| 8. 1840 United States | | | | | | 167 | 171 |
| 9. 1900 Russian poor Jews | | | | | 147 | 153 | 162 |
| 10. 1920 Vienna | | | | | 152 | 155 | 160 |
| 11. 1923 Vienna | | | | | 151 | 158 | 164 |
| 12. 1981 Austria | 124 | 141 | | 166 | | | 180 |

† Sources: 1, table 2; 2 and 3, Hartmann (1970); 4 and 5, (R. Floud personal communication); 6, Sutter, Izac and Ngoc Toan (1958); 7, Steckel (1979); 8, United States Military Academy (1850); $N = 517$, stature at ages other than reported above was 17 yrs 169·2 cm, 19 yrs 171·6 cm, 20 yrs 172·7cm; 9, 10 and 11, Nobel (1924); 10, Österreichischen Statistischen Zentralamt (Forthcoming).

Table 7. The height of boys born in Germany and in the Habsburg Monarchy, 1745–1799.

| | Height of boys (cm) | | | |
|---|---|---|---|---|
| Age | A | B | C | D |
| 8 | 122·5 | 119·9 | | 116·3 |
| 9 | 128·3 | 121·7 | | 119·4 |
| 10 | 130·7 | 126·2 | | 123·9 |
| 11 | 134·6 | 132·0 | | 127·6 |
| 12 | 139·3 | 135·8 | 140·4 | 131·5 |
| 13 | 144·4 | 139·6 | 144·2 | 136·4 |
| 14 | 150·6 | 144·4 | 145·7 | 140·6 |
| 15 | 156·4 | 149·5 | 152·4 | 146·8 |
| 16 | 161·1 | 155·4 | 156·2 | 151·1 |
| 17 | 164·3 | 159·9 | | |

A, German aristocrat 1760–1780; B, German middle class 1760–1780; C, Habsburg aristocrats 1745–1755; D, German peasant 1790s.

norm in Western Europe established around 1900. At the turn of the twentieth century there were, however, still some European populations to whom the Habsburg boys were comparable. One such group was the poverty-stricken Jews of Russia (table 6). The children of Vienna who lived through the First World War were only slightly taller in 1920. Shortly after the war, however, their nutritional status began to improve (Nobel 1924).

Compared with contemporary Austrian children, the boys in the Habsburg military schools appear to have been tiny (table 6): they were between 16 and 18 cm shorter—a difference of about four years' worth of growth behind contemporary children. Although the means differed greatly, the height distribution which prevailed in the 1800s is almost identical to that obtained in today's youth in Austria (Österreichischen Statistischen Zentralamt, unpublished observations).

The seven-year-old Habsburg boys of the turn of the nineteenth century had heights in the neighbourhood of the fifth centile of modern European boys. Thereafter, their stature declined relatively as their average growth velocity was 4 cm per year from age 7 to about 14, which is about 1 cm less than the growth velocity of modern boys. Their peak height velocity was smaller, and after 1775 was reached later than that of modern boys. In fact there is not a single European population or population in North or South America of European ancestry with whom the Habsburg boys of the eighteenth or early-nineteeth century are comparable. The children of Taipei, Hong Kong, Japan, Alaska and South Korea are all taller. One can find comparable heights only among such truly underdeveloped nations as Bangladesh, Thailand, India, rural Burma, some areas of Mexico, or among the Maya in Guatemala, the Hutu in Rwanda, the Bush negroes of Surinam and the children of mixed ancestry of Bolivia (Eveleth and Tanner 1976).

In conclusion, this study, based on the earliest data hitherto discovered, shows that the stature of the Habsburg boys increased in the late 1740s, and decreased after the 1760s. This evidence confirms directly for the first time the notion often found in anecdotal accounts that the nutritional status of the population was deteriorating during the latter decades of the eighteenth century (Abel 1974).

**Acknowledgements**

The support of an N.I.C.H.D. National Research Service Award No. 2 T32 HD07168 from the Center for Population Research is hereby gratefully acknowledged. Computation was done at the computing centre of the Wirtschaftsuniversität, Wien. I appreciate the hospitality of the Rector of the University, Professor Herbert Matis, and the help of Walter Story, former Director of the Centre, in bringing this study to fruition. I gratefully acknowledge, furthermore, the assistance of Klaus Erhard, Erich Foltyn, Paul Mageli, Asha Narang and Nona Schlegel in extracting the data from the archives and rendering them machine-readable. Markus Hein helped diligently with the computation. I benefitted greatly from Professor J. M. Tanner's comments.

**References**

ABEL, W., 1966, *Agrarkrisen und Agrarkonjunktur*, 2nd edn. (Berlin, Hamburg: Parey), p. 241.

ABEL, W., 1974, *Massenarmut und Hungerkrisen im vorindustriellen Europa* (Berlin, Hamburg: Parey), p. 14.

EVELETH, P. B., and TANNER, J. M., 1976, *Worldwide Variation in Human Growth* (Cambridge: Cambridge University Press), pp. 277, 307, 331, 388.

FLOUD, R., and WACHTER, K., 1982, Poverty and physical stature, evidence on the standard of living of London boys, 1770-1870. *Social Science History*, **6**, 432-433.

FOGEL, R., ENGERMAN, S., FLOUD, R., FRIEDMAN, G., MARGO, R., SOKOLOFF, K., STECKEL, R., TRUSSELL, J., VILLAFLOR, G., and WACHTER, K., 1983, Secular changes in American and British stature and nutrition. *Journal of Interdisciplinary History*, **14**, 445-481.

HARTMAN, W., 1970, Beobachtungen zur Akzeleration des Längenwachstums in der zweiten Hälfte des 18. Jahrhunderts. Doctoral Dissertation, Goethe University, Frankfurt.

JONES, E. L., 1974, *Agriculture and the Industrial Revolution* (New York: Wiley).

KOMLOS, J., 1983, *The Habsburg Monarchy as a Customs Union: Economic Development in Austria-Hungary in the Nineteenth Century* (Princeton, NJ: Princeton University Press). p. 99.

KOMLOS, J., 1985, Stature and nutrition in the Habsburg Monarchy: the standard of living and economic development in the eighteenth century. *American Historical Review*,**90**, 1149-1161.

KRIEGSARCHIV, VIENNA, AUSTRIA, Standestabellen Erziehungshäuser Faszikel 3925, 3926, 3927; Josephinische Waisenhaus, Faszikel 3922; Theresianische Militärakademia, Faszikel 434.

LEITNER VON LEITNERTREU, Th., 1852, *Geschichte der Wiener Neustädter Militärakademie*, Hermannstadt.

NOBEL, E., 1924, Anthropometrische Untersuchungen an Jugendlichen in Wien. *Zeitschrift für Kinderheilkunde*, **36**, 13-16.

PIRQUET, C., 1913, Eine einfache Tafel zur Bestimmung von Wachstum und Ernährungszustand bei Kindern. *Zeitschrift für Kinderheilkunde*, **6**, 255.

POST, J., 1977, *The Last Great Subsistence Crisis in the Western World* (Baltimore: The Johns Hopkins University Press).

SANDBERG, L. G., and STECKEL, R., 1980, Soldier, soldier, what made you grow so tall? *Economy and History*, **23**, 91–105.
STECKEL, R., 1979, Slave height from coastwise manifests. *Explorations in Economic History*, **16**, 363–380.
SUTTER, J., IZAC, R., and TRAN NGOC TOAN, 1958, L'evolution de la taille des polytechniciens. *Population*, **13**, 373–406.
TANNER, J., 1978, *Fetus into Man; Physical Growth from Conception to Maturity* (Cambridge, MA: Harvard University Press), p. 7.
TANNER, J., 1981, *A History of the Study of Human Growth* (Cambridge: Cambridge University Press).
UNITED STATES MILITARY ACADEMY, 1859, West Point, Archive, List of Candidates for Admission, etc.

Address correspondence to: Dr J. Komlos, Carolina Population Center, University Square 300A, University of North Carolina at Chapel Hill, NC 27514, USA.

**Zusammenfassung.** Wiener Aufzeichnungen der Körperhöhe von (*a*) Jugendlichen der Militärakademie, geboren 1730–1760, (*b*) Waisenhauskindern, geboren 1760–1780, und (*c*) Militärinternatsschülern, geboren 1775–1815, worden ausgezogen und analysiert. Dies stellt den frühesten noch vorhandenen Satz von Maßen der Köperhöhe einer Gruppe von Individuen dar. Die Körperhöhe nahm in den späten 1740er Jahren zu und nahm nach den 1770ern ab. Dieser Befund weist auf eine Zunahme und eine folgende Abnahme des Ernährungsstatus hin und stimmt überein mit dem bekannten Ablauf des europäischen landwirtschaftlichen Zustands im 18. Jahrundert. Veränderungen im Alter des größten Zuwachses unterstützt die Vorstellung des säkularen Wandels im Ernährungszustand dieser Knaben. Die Körperhöhe der Habsburger Knaben war größer als bei den ärmsten Knaben des zeitgenössischen London, hielt aber nicht dem Vergleich stand mit der Körperhöhe des englischen Landadels und amerikanischer Kadetten des 19. Jahrhunderts und natürlich mit der Körperhöhe der heutigen Bevölkerungen.

**Résumé.** Les enregistrements à Vienne de la taille de (*a*) les jeunes gens de l'Académie Militaire nés entre 1730 et 1760, (*b*) les enfants d'orphelinat nés entre 1760 et 1780 et (*c*) les enfants de pensionnats militaires nés entre 1775 et 1815 ont été réunis et analysés. Ceci constitue l'ensemble existant le plus ancien de mensurations des tailles d'un groupe d'individus. La stature a augmenté dans les dernières années 1740 et diminué après les années 1770. Ce phénomène indique une hausse puis une baisse de l'état de nutrition et est cohérent avec le tableau connu des conditions de l'agriculture européenne au 18e siècle. Des glissements de l'âge d'accroissement maximum confortent la notion des changements séculaires de l'état de nutrition de ces garçons. La stature des garçons de Habsburg était plus haute que celle des garçons les plus pauvres du Londres contemporain mais se comparait défavorablement à la taille de l'aristocratie anglaise et des cadets américains du 19e siècle et, évidemment, à la taille des populations d'aujourd'hui.

# XII

# Stature and Nutrition in the Habsburg Monarchy: The Standard of Living and Economic Development in the Eighteenth Century

THE RELATIONSHIP BETWEEN INDUSTRIALIZATION and the well-being of a population has been controversial since the classical economists discoursed about the issue. The debate has been particularly heated in Great Britain: the "optimists" have argued forcefully that the Industrial Revolution caused a rise in real income and an increase in the standard of living; the "pessimists," while conceding the long-term benefits of industrialization, have argued with equal fervor that at least for one or two generations it lowered workers' standard of living.[1] So conceived, the controversy is difficult to resolve,[2] not only because of the shortage of vital economic indicators but also because of myriad problems posed by the psycho-

[1] This investigation was supported by an NICHD National Research Service Award #2 T32 Hd071 from the Center for Population Research and a grant from the American Council of Learned Societies. I am indebted to Robert Fogel for awakening in me an interest in the history of human stature. I am grateful to him, Roderick Floud, and Richard Steckel for their comments on another version of this paper. I thank Roderick Floud for providing me with a computer program to estimate population height from sample height. Paul Mageli, Peter Waugh, Nona Valeriano Schlegel, Asha Narang, Erich Foltyn, Markus Hein, Klaus Erhard, Gregory Deklava, and Barbara Enger contributed invaluable service in helping to collect and process the data. I have also benefited from comments made during workshops at the following universities, institutes, and conferences: Chicago, Vienna, Linz, Salzburg, Vanderbilt, Harvard, and Northwestern; the Institute of Advanced Studies, Vienna; the 1984 Cliometrics conference at Miami University of Ohio; the Historical Institute of the Hungarian Academy of Sciences, Budapest; the Carolina Population Center, Chapel Hill, N.C.; and the Triangle Economic History Workshop in North Carolina. This project could not have been undertaken without research in Vienna, which was facilitated by the support of Gerhard Botz, Eduard März, Herbert Matis, and Erich Streisser. Computation was done at the EDV Zentrum of the Wirschaftsuniversität Wien. I thank its former director, Walter Story, for his valuable services. I am also indebted to the staff of the Kriegsarchiv, Vienna, and the staff of the Carolina Population Center for the invaluable services they rendered me and my assistants.

[1] For a negative view of the effect of the Industrial Revolution on the standard of living in Britain, see, for example, J. L. Hammond, "The Industrial Revolution and Discontent," *Economic History Review* (hereafter, *EHR*), 1st ser., 2 (1930): 215–28. For a positive view, see R. M. Hartwell, "The Rising Standard of Living in England, 1800–1850," *EHR*, 2d ser., 13 (1961): 397–416. Also see E. J. Hobsbawm, "The Standard of Living during the Industrial Revolution: A Discussion," *EHR*, 2d ser., 16 (1963): 119–34; and R. M. Hartwell, "The Standard of Living," *EHR*, 2d ser., 16 (1963): 135–46.

[2] Gertrude Himmelfarb, *The Idea of Poverty* (New York, 1983), 136.

logical disamenities that accompany industrialization. Psychological factors could have offset the additional income that might have been generated as a result of industrialization.[3] A recent effort to address this issue uses infant mortality rates as a proxy measure of disamenities associated with urban life.[4] Even if, for the sake of simplicity, one were to disregard the psychological component of the industrializing process, there are at least two distinct questions yet to be resolved, as G. N. von Tunzelmann recently pointed out: "What actually happened to the standard of living . . . [and] what might have happened" in the absence of industrialization?[5] This study analyzes the first of these questions; the results obtained lead me to speculate on the second.

THE PESSIMISTIC VIEW GREW out of an intellectual tradition dating from Malthus and Ricardo, reinforced by the young Engels's indictment of the English middle class, which had an important role in disseminating the idea of the misery of one segment of the population to future generations.[6] An important aspect of Engels's argument was that "the poverty of the proletariat was quantitatively different from the poverty of the old poor—the preindustrial poor were less impoverished, less hardworking, less miserable."[7] This view was later shared by Marx and his

[3] P. R. G. Layard and A. A. Walters, *Microeconomic Theory* (New York, 1975), 146; and Alan Milward and S. B. Saul, *The Economic Development of Continental Europe, 1780–1870* (London, 1973), 21. For a recent controversy over various indexes of well-being, see M. W. Flinn, "English Workers' Living Standards during the Industrial Revolution: A Comment," *EHR*, 2d ser., 37 (1984): 88–92. Also see Peter Lindert and Jeffrey Williamson, "English Workers' Living Standards during the Industrial Revolution," *EHR*, 2d ser., 36 (1983): 1–25; N. F. R. Crafts, "English Workers' Real Wages during the Industrial Revolution: Some Remaining Problems," *Journal of Economic History* (hereafter, *JEH*), 45 (1985): 139–44; and L. D. Schwarz, "The Standard of Living in the Long Run: London, 1700–1860," *EHR*, 2d ser., 38 (1985): 24–41.

[4] Jeffrey G. Williamson, "Was the Industrial Revolution Worth It? Disamenities and Death in Nineteenth-Century British Towns," *Explorations in Economic History* (hereafter, *EEH*), 19 (1982): 221–45, and "Urban Disamenities, Dark Satanic Mills, and the British Standard of Living Debate," *JEH*, 44 (1981): 75–84. (The weakness in Williamson's approach lies in the assumption that the urban marginal worker was fully compensated for the greater incidence of infant mortality in cities than in rural areas.) One might note that the American experience has been less controversial than the European one. A recent regional study has shown that real hourly wages stagnated during the first two decades of the nineteenth century but then increased by two-thirds by 1860. Annual earnings increased even faster. Donald Adams, "The Standard of Living during American Industrialization: Evidence from the Brandywine Region, 1800–1860," *JEH*, 42 (1982): 903–17. Historians are in agreement that real wages fell in France during the half-century prior to the revolution. O. H. Hufton, *The Poor of Eighteenth Century France, 1750–1789* (Oxford, 1974), 16.

[5] Von Tunzelmann, "The Standard of Living Debate and Optimal Economic Growth," in J. Mokyr, ed., *The Economics of the Industrial Revolution* (New York, 1985).

[6] Friedrich Engels, *The Condition of the Working Class in England*, trans. W. O. Henderson and W. H. Chaloner (New York, 1958). See Henderson and Chaloner's introduction to the 1958 Oxford edition of Engels's book.

[7] Himmelfarb, *Idea of Poverty*, 285. Engels's eyewitness account of the misery of a segment of British workers should not be considered fully accurate or a representative sample of the total working population. He documented the suffering of the very poor and attributed their plight to the rise of the factory system only because he was ignorant of their condition before the onset of the Industrial Revolution. Children in the preindustrial era, he thought, "grew up in idyllic simplicity." Weavers "earned enough to live on" and "enjoyed a comfortable and peaceful existence." The culprit, he made clear, was "the expansion of trade" that "led to severe competition, which had an adverse effect on the standard of living of the workers." "Industry alone," he asserted time and again, "has been responsible for . . . their wretched conditions. . . . Every new machine brings with it unemployment, want, and

followers, who, in the words of an eminent historian, "saw the pauperization of the workers as the baptismal act of capitalism."[8]

The German historian Wilhelm Abel pointed out that the Marxian contention is unacceptable because it ignores the prevalence of preindustrial rural poverty.[9] Abel devoted a whole book to describing the process of immiseration, which, with long-term fluctuations, stretched into the nineteenth century.[10] Consequently, the poverty of the working class in England and elsewhere can be better understood in a comparative perspective rather than analyzed in isolation. T. S. Ashton put it this way: "Some writers have compared the semi-skilled operatives in the new factories with the small farmers and craftsmen of an earlier generation. If comparison is to be made at all, it must be with the squatters of the countryside and the paupers of the towns, from whose ragged ranks the factory workers were largely drawn."[11] What, in other words, would have happened to the landless segment of the population and their descendants in the nineteenth century without industrialization? Bruno Hildebrand, for instance, argued that the poverty of the working class in Germany during the first half of the nineteenth century was a result of the low level of industrialization.[12] Emmanuel Le Roy Ladurie summed up his assessment of European immiseration by pointing out that it "seems to have been above all the distinctive mark of a blocked society, one incapable of raising productivity and hence the living standards of the lower classes."[13]

In order to be convinced of the intensity of rural poverty before the spread of industrialization, one might consider the case of Bohemia during the subsistence crisis of 1771. An eyewitness description of it is appalling. Sickness was prevalent because of hunger; malnutrition was endemic; the quality of clothing was poor. Children started strenuous work early in life. Peasants were uprooted, since they often had to look for work in areas where the harvest came before their own. People were dirty. Adults slept on straw, and children lay naked near the hearth. People ate grass and bread made of weed and flour dust. Some peasants worked

---

suffering." Engels admitted that some workers were better off than the ones he documented, but he did not attempt to write about their lives or try to ascertain or even think about income distribution among workers. Consequently, his analysis is one-sided; however, Marx accepted it and probably incorporated it into his immiseration doctrine. Engels did mention several times that industrialization increased wages, but he apparently meant nominal, not real, wages. For he insisted that the preindustrial artisan's "standard of life was much better than that of the factory worker of today [1845]." He also accepted that urban workers were better off than the rural proletariat whose lot had deteriorated even more since the beginning of the Industrial Revolution. Engels, *Condition of the Working Class in England*, xxv, xxx–4, 9–11, 24, 29, 90–92, 295. It follows from this, since in the absence of industrialization the factory workers would have joined the ranks of the rural proletariat, that the living standards of factory workers would have declined to the level of rural laborers had industrialization not proceeded as it did.

[8] Emmanuel Le Roy Ladurie, *The Territory of the Historian* (Chicago, 1979), 13.

[9] Abel, *Agrarkrisen und Agrarkonjunktur* (1935; 2d edn., Hamburg, 1966), 241.

[10] Abel, *Massenarmut und Hungerkrisen im vorindustriellen Europa* (Berlin, 1974), 14.

[11] Ashton, *An Economic History of England: The Eighteenth Century* (1955; reprint edn., London, 1972), 234.

[12] He cited the case of Oberhessen and the unindustrialized areas of Prussia; Bruno Hildebrand, *Die Nationalökonomie der Gegenwart und Zukunft* (Jena, 1922), as cited by Abel, *Agrarkrisen und Agrarkonjunktur*, 227.

[13] Le Roy Ladurie, *Territory of the Historian*, 13.

as long as factory workers, spinning and weaving flax day and night.[14] This should not be construed to suggest that this description is any more representative of the condition of the rural population than are the sketches we have from Engels's pen of the condition of the industrial worker. Yet it does put the situation of the nineteenth-century working class in a different light: its preindustrial counterparts often did not fare better.

Scholars have concluded from the rise in cereal prices in the second half of the eighteenth century that the standard of living was falling. "We see how great a slice had been taken out of wages, and how much the purchasing power of the classes lowest in the social scale had dwindled. It was a process of impoverishment, against which they stood helpless."[15] Of course this makes sense. It is difficult to imagine that the majority of the population would have been able to protect itself against violent and unanticipated price rises. Contemporaries, too, noted that wages did not keep pace with prices.[16] Even so, the case for a deteriorating standard of living is not proved. It is difficult to draw conclusions from data on the price of grain because wage data are less reliable and scarcer than price data and daily wages are not easily converted into annual income. In addition, wage data are typically limited to a few occupations and often remained constant over decades, sometimes centuries, even in the face of great fluctuations in commodity prices. It is necessary to know if length of the workday, intensity of effort, and incidence of underemployment, unemployment, or compensations in kind varied sufficiently to offset the fluctuations in real wages obtained by deflating the wage series with the price of grain. These economic indicators are also suspect because there was considerable geographic variation in prices and rents, since some regions maintained their isolation from the world market into the nineteenth century. Initially, grain prices did not rise uniformly.[17] Furthermore, a large share of the population of the Habsburg monarchy, like those in other parts of Europe, was engaged in subsistence agriculture at the beginning of industrialization and as a result was isolated to some extent from market forces reflected in cereal prices. Finally, annual harvest data became available only during the latter third of the nineteenth century, and, consequently, estimates of the grain supply before that time are possible for only a limited number of years.[18] We can draw generalizations about the standard of living from the rise of cereal prices, but only with many caveats.

There is much evidence of the disappearance of meat from, or at least its greater scarcity in, the diet of the lower classes at the beginning of the nineteenth century.[19]

[14] Franz Mayer, "Die volkswirtschaftliche Zustände Böhmens um den Jahren 1770," *Mittheilungen des Vereins für Geschichte der Deutschen in Böhmen*, 1876, 125–49. The eyewitness pointed out the inadequate nourishment in Prachiner County by stating that the people were short; *ibid.*, 131. One might also consider the children in preindustrial Tyrol who journeyed at an early age to Germany in search of work. Otto Uhlig, *Die Schwabenkinder aus Tirol und Vorarlberg* (Innsbruck, 1978). I am indebted to Roman Sandgruber for bringing this work to my attention.

[15] B. H. Slicher Van Bath, *The Agrarian History of Western Europe, A.D. 500–1850* (London, 1963), 226.

[16] See Abel, *Agrarkrisen und Agrarkonjunktur*, 188.

[17] On the estate of Teschen in Bohemia, for instance, the price of wheat did not begin to rise steadily until 1800; Arthur Salz, *Geschichte der böhmischen Industrie in der Neuzeit* (Munich, 1913), 539.

[18] Roman Sandgruber, *Österreichische Agrarstatistik 1750–1918* (Vienna, 1978).

[19] Johann Heinrich Von Thünen theorized that, during the initial phase of economic development,

We can quantify probable per capita meat consumption, since some data on animal stock exist, but from these estimates it is difficult to construct an index, since in the short run the rate of slaughter and animal weight could have varied dramatically and the census of the animal stock is not reliable enough for such an exercise. We can be fairly certain, however, that per capita meat consumption in Austria declined considerably, perhaps by as much as 50 percent between 1750 and 1850. The decline in meat consumption in Vienna may have started later but reached its nadir, as it did in the cities of Linz and Graz, in 1850.[20]

In sum, the problems associated with the construction of real wage series on either regional or national levels seem insurmountable despite recent noteworthy efforts.[21] This difficulty is partly a result of the poor quality or absence of data, the ambiguity of weights attached to the various component series, and the many assumptions that must be made to overcome the limitations in our knowledge. How well different segments of the population fared during these changing economic circumstances is another complicating factor. It appears that research in this area has been pushed to the limit without resolving the controversy.[22] Even if the most recent estimate of real wages in England appears plausible,[23] its independent verification by an entirely different method should be welcomed.

A new approach uses an indirect method to investigate the trend in nutrition in the eighteenth and nineteenth centuries.[24] Food clearly accounted for a large, though slowly declining, share of the European population's total expenditure. Typically, close to three-fourths of personal income was spent on alimentation in the eighteenth century.[25] Consequently, nutrition in this era is a fairly good, perhaps the best possible, indicator of the material well-being of a population. Until recently, efforts to measure systematically variations in food consumption have been frustrated by the almost complete absence of suitable data on the amount, variety, and nutritional adequacy of the food intake of populations in the past. Economic historians have therefore turned to a proxy—height—to measure nutrition.[26]

---

meat consumption of the poorer segments of the population would decline. See Abel, *Agrarkrisen und Agrarkonjunktur*, 238–39.

[20] Roman Sandgruber, *Die Anfänge der Konsumgesellschaft* (Vienna, 1982), 158.

[21] Jeffrey Williamson, *Did British Capitalism Breed Inequality?* (Boston, 1985).

[22] One recent article refers to this debate as having become "confused . . . arcane . . . of ever greater statistical complexity"; Roderick Floud and Kenneth Wachter, "Poverty and Physical Stature, Evidence on the Standard of Living of London Boys, 1770–1870," *Social Science History* (hereafter, *SSH*), 6 (1982): 422.

[23] Lindert and Williamson, "English Workers' Living Standards during the Industrial Revolution."

[24] Roderick Floud, "The Heights of Europeans since 1750: A New Source for European Economic History" (typescript, Department of History, University of London, 1983), 2.

[25] Fernand Braudel, *Civilization and Capitalism, 15th–18th Century*, volume 1: *The Structures of Everyday Life: The Limits of the Possible*, trans. Siân Reynolds (New York, 1981), 138. Grain alone made up approximately one-half of a man's daily food intake. In 1600, 78 percent of a bricklayer's family income was spent on food. In 1800, the share was 73 percent; Abel, *Agrarkrisen und Agrarkonjunktur*, 138.

[26] Heights were among the first data studied by social scientists. See Robert Fogel *et al.*, "Exploring the Uses of Data on Height: The Analysis of Long-Term Trends in Nutrition, Labor Welfare, and Labor Productivity," *SSH*, 6 (1982): 403, 417. An 1829 publication, for instance, stated that "human height becomes greater and growth takes place more rapidly, other things being equal, in proportion as the country is richer [and] nourishment better"; L. R. Villerme, "Mémoire sur la taille de l'homme en France," *Annales d'hygiène publique*, 1 (1829): 551–59, as cited in Floud and Wachter, "Poverty and

During the eighteenth century, when industrialization once again accelerated, data were kept on the height of soldiers and children in various institutions. From a correlation of income, nourishment, and height, the well-being of a population can be inferred, although the relationship between the height-by-age profile of a population and the nutritional status is complex. This is not the place to go into that problem in detail. It suffices to say that in this study emphasis is placed on measuring changes in the height of a population, which genetic factors do not affect.[27]

Through natural selection, populations can adapt to different levels of nutrition over millennia, but, within the European population, nutrition has been found to be more important than genetic factors in explaining variations in human height across ethnic frontiers. Data from the Habsburg monarchy provide a good example. In the mid-eighteenth century, average heights of various ethnic groups living in different provinces were within a five-centimeter range (see Table 1). Between 1740 and 1790, with the integration of the grain market in the monarchy and the subsequent convergence of food prices, the range of the height of adult males of various nationalities decreased. By the 1790s it was about two centimeters. This is an indication that environmental factors, rather than genetic ones, were the most important determinants of the final height of the population.

The average height of children during the growing years and the terminal height of a population is, at least in market economies, a rough proxy for family income in so far as food consumption depends on real income.[28] This proxy has the advantage of measuring net nutrition (net, that is, after debilitation from disease, physical exertion, and psychological stress has been calculated). Consequently, it is perhaps a better proxy of the standard of living during the early stages of industrialization than any other measure hitherto devised.[29]

---

Physical Stature," 425; James M. Tanner, *A History of the Study of Human Growth* (Cambridge, 1981); Robert Fogel, "Nutrition and the Decline of Mortality since 1700: Some Preliminary Findings" (typescript, Department of Economics, University of Chicago, 1984); Roderick Floud, "The Physical State of the British Working Class, 1860–1914: Evidence from Army Recruits" (typescript, Department of History, University of London, 1984); Robert W. Fogel *et al.*, "The Economics of Mortality in North America, 1650–1910: A Description of a Research Project," *Historical Methods*, 11 (1978): 83–89, and Fogel *et al.*, "Exploring the Uses of Data on Height," 416. Weight is another indicator of nourishment, and birth weight is a good indicator of the mother's nourishment during the last trimester of pregnancy. See W. Peter Ward and Patricia C. Ward, "Infant Birth Weight and Nutrition in Industrializing Montreal," *AHR*, 89 (1984): 326.

[27] Reynaldo Martorell, "Genetics, Environment, and Growth: Issues in the Assessment of Nutritional Status," in A. Velasquez and H. Bourges, eds., *Genetic Factors in Nutrition* (New York, forthcoming), as cited in John Mellor and Bruce Johnston, "The World Food Equation: Interrelations among Development, Employment, and Food Consumption," *Journal of Economic Literature*, 22 (1984): 546. Also see Reynaldo Martorell *et al.*, "Malnutrition, Body Size, and Skeletal Maturation: Interrelationships and Implications for Catch-Up Growth," *Human Biology*, 51 (1979): 371–89; and Lawrence Greene and Francis Johnston, eds., *Social and Biological Predictors of Nutritional Status, Physical Growth, and Neurological Development* (New York, 1980).

[28] Richard H. Steckel, "Height and Per Capita Income," *Historical Methods*, 16 (1983): 1–7.

[29] Floud, "Heights of Europeans," 5; and Fogel *et al.*, "Exploring the Uses of Data on Height," 406. Using height as an index of nourishment has hitherto produced some noteworthy studies in the field of American history. According to these analyses, slaves born in the United States were better nourished than African and West Indian blacks or urban British workers but not quite as well nourished as rural white Americans of the time. Historians have also demonstrated that there were cycles in the level of nourishment in the United States, with peaks in 1795 and 1825; after 1835 there was a decline in the

TABLE 1
Height (in Centimeters) of Soldiers
in the Habsburg Monarchy by Place of Birth[a]

| *Year of Birth* | *Hungary* | *Lower Austria* | *Galicia* | *Bohemia* | *Moravia* |
|---|---|---|---|---|---|
| 1740 | 170.1 | n.a. | n.a. | 165.3 | 166.3 |
| 1750 | 168.6 | 164.6 | 168.9 | 164.4 | 165.8 |
| 1760 | 167.0 | n.a. | 167.1 | 163.4 | n.a. |
| 1770 | 162.5 | 163.1 | 163.7 | 161.5 | n.a. |
| 1780 | 163.3 | 161.6 | 164.2 | 163.8 | n.a. |
| 1790 | 162.1 | 163.2 | n.a. | 160.8 | n.a. |
| 1800 | 163.0 | n.a. | n.a. | n.a. | n.a. |
| 1810 | 165.0 | n.a. | n.a. | n.a. | n.a. |
| 1820 | 163.7 | n.a. | 158.8 | n.a. | n.a. |
| 1830 | n.a. | n.a. | 160.1 | n.a. | n.a. |

[a] The table lists not the raw means but the estimated means after an adjustment for the minimum height requirement enforced by the army. The recruits were twenty-three years old or older, unskilled, and from rural households.

SOURCE: Vienna, Kriegsarchiv, Musterlisten and Standestabellen, Infantrieregimenter, Grundbücher.

The subsequent discussion is confined to the peasant population; data on artisans and urban dwellers await further archival research. The results reflect the standard of living of the lower classes from whose ranks army recruits were drawn. Because the army enforced a minimum height requirement for recruits, the data on the height of the adult male population must be corrected statistically. The estimation is complicated because this standard varied over time and was always flexible.[30]

ONE PATTERN THAT EMERGES from the seventy-five thousand observations extracted from the archives up to now is that stature of peasants was increasing for cohorts born in the 1740s. This feature was particularly noticeable for the sons of lower aristocrats who entered the military academy founded by Maria Theresa.[31]

general level of nutrition until about 1870. Yet soldiers during the American revolutionary war were as tall as those during the Civil War and the Second World War. Robert W. Fogel *et al.*, "Secular Changes in American and British Stature and Nutrition," *Journal of Interdisciplinary History* (hereafter, *JIH*), 14 (1983): 445–82; Gerald C. Friedman, "The Heights of Slaves in Trinidad," *SSH*, 6 (1982): 482–515; Robert Margo and Richard Steckel, "The Height of American Slaves: New Evidence on Slave Nutrition and Health," *SSH*, 6 (1982): 516–38, and "Heights of Native Born Whites during the Antebellum Period," *JEH*, 43 (1983): 167–74; Richard Steckel, "Slave Height Profiles from Coastwise Manifests," *EEH*, 16 (1979): 363–80; Fogel, "Nutrition and the Decline of Mortality"; and Floud, "Physical State of the British Working Class."

[30] I would like to thank Roderick Floud and Annabel Gregory of the University of London and Richard Steckel of Ohio State University for providing a computer program that adjusted the statistics. The method is described in detail by D. W. Wachter and James Trussell; see their "Estimating Historical Heights," *Journal of the American Statistical Association*, 77 (1982): 279–93.

[31] John Komlos, "Patterns of Children's Growth in East-Central Europe in the Eighteenth Century," *Annals of Human Biology* (forthcoming).

This finding is extremely important because the increase in stature coincided with the favorable agricultural conditions of the period and thereby affirms the relationship between height and the nutritional status of a wide segment of the population.[32] Stature and nutritional standards were increasing during the early eighteenth century in the British colonies of North America as well, where meat appears to have become a more important element in the diet.[33]

Cross-sectional analysis shows that before the middle of the eighteenth century the tallest men in the Habsburg monarchy were born in the economically least developed lands: Hungary and Galicia (see Table 1). These economies were overwhelmingly agricultural and, in the main, not yet market oriented. Although technologically backward, the peasants were self-sufficient and lived on productive land that was not densely populated. The nutritional gradient rose from west to east: while Hungarian peasants born in the 1740s reached an average height of 170 centimeters, men in the most advanced parts of the monarchy, Lower Austria and Bohemia, were about 165 centimeters tall, the same height as British soldiers at this time. The relationship changed once the grain market in the monarchy became integrated; agricultural exports to Austria had the effect of decreasing the nutritional status of the peasant populations at the lower end of the income distribution. A hundred years later, Polish and Hungarian army recruits were among the shortest in the monarchy, having declined by about eight centimeters in stature. In 1850 the economically isolated province of Dalmatia and in the early twentieth century the predominantly agricultural province of Bosnia-Hercegovina had the tallest men in the monarchy.[34]

Another noteworthy trend is the declining height of males born in the 1750s in Bohemia and Lower Austria, a trend that resulted from population growth. Diminishing returns to labor meant diminishing consumption per capita—a typically Malthusian pattern, which undermined the well-being of the population in these provinces, even though they were industrially the most advanced of the monarchy. The resource base was not expanding at the same rate as the population grew. The declining trend of heights did not stabilize until the generation of the 1790s, some of whom matured after the end of the Napoleonic Wars and benefited from the economic advances that followed the peace.

I would not feel confident in publishing even these preliminary results were it not that other data support the notion of a decline in nutrition. The estimates in Table 1 pertain only to men over the age of twenty-three, but the pattern is similar for younger recruits. In addition, Swedish recruits, poor London boys, and American slaves suffered a similar decline toward the end of the eighteenth

[32] E. L. Jones, *Agriculture and the Industrial Revolution* (New York, 1974); and Le Roy Ladurie, *Territory of the Historian*, 13, 89, 201, 232.

[33] Kenneth Sokoloff and Georgia C. Villaflor, "The Early Achievement of Modern Stature in America," *SSH*, 6 (1982): 453–81, 470; Robert Fogel, "Growth as a Measure of the Economic Well-Being of Populations: The Eighteenth and Nineteenth Centuries," in James M. Tanner, ed., *Human Growth* (forthcoming); and Sarah McMahon, "Provisions Laid Up for the Family: Toward a History of Diet in New England, 1650–1850," *Historical Methods*, 14 (1981): 22–30.

[34] *Militär-Statistisches Jahrbuch für das Jahr 1871* (Vienna, 1873), 48.

century.[35] The most significant corroborative evidence in support of my thesis comes from records of orphanages and army schools in the Habsburg monarchy (twenty-five thousand observations). These data have an advantage over muster-rolls. Orphanages did not enforce a minimum height requirement, and, consequently, no statistical adjustment is required to estimate mean population height from sample height. Confidence in the results is therefore increased considerably. These data show a pattern similar to that of the adult recruits: a decline in height of boys of all ages until nutrition improved in the post-Napoleonic era.[36]

Beginning in the 1750s, the Habsburg monarchs perceived this decline in the standard of living and adopted countermeasures. A subsistence crisis of immense proportions in 1770–71 and several jacqueries of an old-fashioned sort heightened their perception. Maria Theresa and her son, Joseph II, proceeded systematically to break the power of the guilds and to remove other institutional bottlenecks that stood in the way of economic development. In 1755 all persons, regardless of gender, living outside of towns were allowed to weave linen at home. In 1763 restrictions on nonguild production of cotton textiles were lifted; a similar measure for woolen textiles followed in 1764. As a result peasants could use industrial activity as a means of defending their standard of living. Then a number of agricultural reforms in the 1770s and 1780s guaranteed greater mobility to peasants and enabled them to enter the industrial labor force in greater number. The government also aided the peasantry at the expense of landlords by setting a ceiling on rents.

Several factors contributed to the success of the monarchs' economic program: the state had attained greater power, in particular, the means to intervene in the economy; it had sufficient wealth at its disposal to overcome some of the initial difficulties of industrialization; the social infrastructure as well as capital accumulation, both human and physical, had greatly expanded; and, finally, government advisers could adopt economic innovations from the West. By fostering industrialization, and thereby ameliorating the subsistence crisis, the state acted in self-defense; otherwise, its more efficient neighbors would have gained at its expense.[37]

In addition to concern for the well-being of the population, the Habsburg monarchy had another reason for taking action: the desire to prevent a crisis of ideology. A social system can function effectively only as long as its participants have so accepted established social norms that excessive monitoring of individual behavior is not required. Ideology is a substitute for monitoring. The ideology of

35 Richard Steckel, letter to the author. Steckel was kind enough to send me a revision of the data that he and Lars G. Sandberg published; see their "Soldier, Soldier, What Made You Grow So Tall?" *Economy and History*, 23 (1980): 91–105, 98, 103. Also see Floud and Wachter, "Poverty and Physical Stature"; and Steckel, "Slave Height from Coastwise Manifests." In Leipzig a nutritional decline seems to have taken place, perhaps because of the War of the Austrian Succession. The voices of teenage boys in Leipzig changed at a later age in the 1740s than in the 1730s, indicating that a decline in nutritional status that delayed the onset of puberty began earlier there than in the Habsburg monarchy. S. F. Daw, "Age of Boy's Puberty in Leipzig, 1727–49, as Indicated by Voice Breaking in J. S. Bach's Choir Members," *Human Biology*, 42 (1970): 87–89.

36 Komlos, "Patterns of Children's Growth."

37 Douglass North, *Structure and Change in Economic History* (New York, 1981), 44.

enlightened absolutism included the notion that the monarch is responsible for the well-being of his people. Therefore, a considerable decline in the standard of living of Habsburg peasants threatened the social contract. To prevent a breakdown of civil order, the monarchs opened the way for industrialization by their energetic intervention in the nation's economic life.

Questions of ideology aside, the monarchs did not welcome a decrease in their tax base. Maintenance of their standing army, and in turn their power, depended on the ability of the peasantry to pay taxes. A fiscal crisis could be expected if the standard of living of the peasantry continued to deteriorate. The institutional changes that took place in the decades after the onset of the nutritional decline can be understood only in light of the threat that this potential crisis posed to the power of absolute rulers. To maintain power, they had to impinge on the lopsided influence of the landed oligarchy, on the one hand, and the guilds, on the other, thus paving the way for economic development. That they succeeded in doing so was not only because of their perseverence and enlightened policies but also because of the external threat from Prussia that made the aristocracy more pliable than it otherwise might have been.

THE DATA FROM BOHEMIA AND LOWER AUSTRIA indicate that industrialization can be conceptualized as an antidote to the immiseration brought about by the beginning of the population explosion. That is not to say that industrialization led immediately to an improvement in living standards, but in East Central Europe, at least, it slowed the rate of deterioration until the trend could be reversed. In addition, industrialization allowed population growth to continue, even accelerate, by removing the Malthusian constraint of the declining marginal product of labor.[38] Had industrialization not proceeded as it did in these provinces, the Malthusian mechanism of diminishing returns to labor would have continued to impinge on the nutrition, and hence on the health, of the populace until population growth declined. In other words, if the industrial sector had not expanded sufficiently to provide an additional source of income, economic hardship would eventually have curtailed population growth. People could have adapted to the crisis by having fewer children, which would also have averted a demographic crisis. Without any of the above ameliorating factors, population growth would have led to the kind of disasters that occurred in earlier centuries, when population growth exceeded the growth of the resources necessary for its support. Population growth led to a decline in the standard of living, and this decline provoked institutional change, the growth of the industrial sector, and a redistribution of income that favored the peasantry over the landlords. These new economic and legal environments furthered population growth, because the industrial sector was not constrained by declining returns to labor to the same extent as the agricultural sector, but at the

[38] For a fuller account, see John Komlos and Marc Artzrouni, "A Simulation Model of Malthusian Population Dynamics in a Two-Sector Economy with Capital Accumulation between the First and Second Economic Revolutions" (typescript, Carolina Population Center, 1985).

same time capital accumulation and technical change proceeded fast enough to absorb the growing population, thereby breaking the Malthusian cycle.[39]

Another implication of this study is that economic development can lead to a decline in nutrition for part of a society undergoing rapid economic change. This happened to a substantial segment of the Galician and Hungarian peasantry because the more advanced parts of the monarchy achieved a better and more rapid integration of the market. This finding conforms to economic theories that stress the initial disadvantages of free trade to the less developed partner.[40] Even if the whole society gains from free trade, the rapid economic change that results from it imposes hardship as well. Although the gains might be greater than the burdens, especially in the long run, the costs and benefits are not distributed evenly among various segments of the population. The less advantageously placed members of the society bear the brunt of the burdens of economic development.

A recent study of birth weights in Montreal concludes that "malnutrition and [economic] development went hand in hand."[41] The same can be said of the Habsburg monarchy's less developed regions. In economically advanced provinces, such as Bohemia and Lower Austria, the spread of malnutrition preceded the rapid industrialization of the 1760s and 1770s. Maintenance of a minimum level of nutrition in these provinces was achieved with imports of agricultural products from Hungary and Galicia. The population in the western region was able to escape the Malthusian menace only by exporting its malnutrition to the peasantry of the two agricultural provinces.

The pattern in Bohemia and Lower Austria, where the diminution in real income preceded the beginning of industrial growth, is consistent with the thesis of those economic historians who begin their analysis of the "condition-of-England-question" with "some set of exogenous changes (mainly population growth) which would have had even more severe repercussions on the standard of living without the alternative of industrial development."[42] According to Arthur J. Taylor, "The most powerful short-term influence on economic prosperity in the eighteenth century was the pressure of a rapidly expanding population."[43] François Crouzet put it this way, "The decisive factor in the undoubtedly low standards of living and widespread poverty of the working classes in the eighteenth and early nineteenth centuries was the pressure of a rising population."[44] A recent demographic study confirms the Malthusian dynamic: "Before 1800, the situation

39 This is similar to the pattern found for eighteenth-century Flanders. "The expansion of the rural industry . . . occurred in response to the growth of population, which tended to outrun the means of subsistence. . . . The development of the cottage industry made possible the acceleration of the rate of population growth." Franklin F. Mendels, *Industrialization and Population Pressure in Eighteenth-Century Flanders* (New York, 1981), 3.

40 For a similar pattern in contemporary Central America, see Gretel Pel and Pertti Pelto, "Diet and Delocalization: Dietary Changes since 1750," *JIH*, 14 (1983): 2, 523.

41 Ward and Ward, "Infant Birth Weight," 326, 345.

42 R. M. Hartwell and Stanley Engerman, "Models of Immiseration: The Theoretical Basis of Pessimism," in Arthur J. Taylor, ed., *The Standard of Living in Britain in the Industrial Revolution* (London, 1975), 213.

43 Taylor, "Introduction," in Taylor, *Standard of Living*, lii.

44 F. Crouzet, *Capital Formation in the Industrial Revolution* (London, 1972), 63–64.

developed much as Malthus had insisted it must: the faster population grew, the faster the price of food rose, the lower the standard of living fell, and the grimmer the struggle to exist became. As Malthus had postulated, there were indeed long, slow oscillations in the rate of population growth and in the standard of living."[45]

In contrast, for the pessimists "the analysis generally begins with the proposition that the transition to rapid economic growth could have been achieved with less cost to the working classes."[46] That, too, is probably true. In hindsight one can argue that the government could have intervened even more energetically to better the lot of the underprivileged, but it was not prepared to respond quickly to their plight by redistributing income on a grand scale. Yet the most pressing problems of life and death were indeed solved, and in Austria they were solved without a revolution. The success of the government's economic program, given the entrenched power of the landed oligarchy, is in itself an admirable achievement.

Many contemporary observers pointed to the poverty of the hand-loom weavers as evidence of the suffering brought about by industrialization.[47] The pattern of the data on height suggests, however, that such weavers were not poor because they wove; rather, they wove because otherwise they would have been poorer. Gertrude Himmelfarb recently summed up this and related issues: "The 'standard-of-living' question . . . is generally discussed as if it were a question of industrialism alone, a measure of the social and economic costs and gains of mechanization and urbanization. In fact, . . . it was as much a product of such natural and political catastrophes as famine and war as of the industrial and agricultural revolution. It was also a product of the 'demographic revolution,' which itself was partly a consequence of industrialism and partly a cause or 'precondition' of it."[48] It was not industrialism that caused poverty; rather, the growth of the agricultural population caused poverty of sufficient magnitude to force many peasants to engage in industrial activity, even if, at first, on a part-time basis. Industrialization, in turn, enabled the labor force to emancipate itself from the Malthusian menace, which had threatened it since time immemorial.[49]

IN SUM, THE EFFECT OF INDUSTRIALIZATION on the standard of living was ameliorative in Bohemia and Lower Austria. The standard of living, as measured by height, had begun to deteriorate before the spread of industrialization, because of the growth of the agricultural labor force. The falling standard of living led the

45 Roger Schofield, "The Impact of Scarcity and Plenty on Population Change in England, 1541–1871," *JIH*, 14 (1983): 275.

46 Hartwell and Engerman, "Models of Immiseration," 213.

47 An observer of hand-loom weavers in England referred to them as a race "rigidly descending to the size of Liliputians"; Floud and Wachter, "Poverty and Physical Stature," 424.

48 Himmelfarb, *Idea of Poverty*, 135.

49 The Malthusian positive checks (population growth constrained by the availability of the food supply) no longer operated in sixteenth-century England and thereafter, but Malthusian negative checks—restraints voluntarily imposed in response to adverse economic developments—operated until the early nineteenth century; E. A. Wrigley and R. S. Schofield, *The Population History of England, 1541–1817* (Cambridge, Mass., 1981).

government to embark on a program of institutional change that enabled many peasants in the 1750s and thereafter to enter the industrial sector, at first as cottage workers in the employ of an outputter. Protoindustrialization in Bohemia helped prevent the utter collapse of the local economy. The standard of living fell until the end of the century, then stabilized, and after a few decades began to rise again.

The extent to which this pattern of nutrition will hold up for other societies must await further research. Those who plan to explore the relationship between the standard of living and industrialization in other regions should keep in mind that it is insufficient to show a correlation between industrialization and a decline in the standard of living in order to construct a case for causation from the former process to the latter. Rather, one must explore the contemporaneous changes in the well-being of the rural population as well and determine the beginning of the downturn in the standard of living. One should furthermore investigate the rate of change of the deteriorating standard of living and correlate that with the extent of industrialization. In other words, the standard of living might be declining on a grand scale even after the onset of industrialization, but it might be declining at a slower rate than just before industrialization.

The case of the Habsburg monarchy shows how widespread rural poverty was and that immiseration of the rural population preceded the Industrial Revolution. Without industrialization, the effect of immiseration on the health of the population could have been catastrophic. By the nineteenth century, or even earlier, the population of Lower Austria and Bohemia could have been weakened to such a degree that a demographic crisis similar to the one in mid-nineteenth-century Ireland might have occurred. Indeed, in the Habsburg lands, industrialization was a conscious response designed to ameliorate the pauperization that had begun to threaten this European society.

# XIII

# *The Height and Weight of West Point Cadets: Dietary Change in Antebellum America*

A decline in nutritional status is inferred from data on the height and weight of West Point cadets in the antebellum period. The decline was geographically widespread and affected farmers and blue-collar workers the most; middle-class cadets did not experience a decline in nutritional status until the Civil War. Nutritional status declined because meat output did not keep pace with population growth. Urbanization and the expansion of the industrial labor force increased the demand for food. However, the agricultural labor force grew at a slower pace, and productivity growth in food production was insufficient to redress the imbalance.

Economic historians have recently begun to analyze in detail the ways in which economic, demographic, and biological processes interacted in the eighteenth and nineteenth centuries. Anthropometric measures play an important role in this research because they enable one to quantify nutritional status, which had not previously been attempted. Birth weight, for instance, can be used to proxy maternal nutritional status, while stature measures cumulative net nutrition during the growing years of an individual.[1] Proxies for nutrition are important because nutrition affects fertility and mortality. Consequently anthropometric measures can have powerful explanatory value for demographic trends. In addition, because food intake in prior centuries was an important component of total expenditure, changes in stature can reflect changes in economic conditions and in the well-being of populations. Nutrition is an important element linking these processes.

*Journal of Economic History*, Vol. XLVII, No. 4 (Dec. 1987).  ISSN 0022-0507. By permission of Cambridge University Press.

The support of NICHD National Research Award No. 2732 HD07168 from the Center for Population Research and Grant No. 2900 from the National Bureau of Economic Research Program on the Development of the American Economy is gratefully acknowledged. The study was written while I was a postdoctoral fellow at the Carolina Population Center of the University of North Carolina at Chapel Hill. I am greatly indebted to Robert Gallman for helping me make sense of antebellum agricultural statistics, to Peter Coclanis, Robert Gallman, Claudia Goldin, Kenneth Sokoloff, Richard Steckel, Richard Oesterreicher, and James Tanner for commenting on the manuscript, and to Robert Fogel for interesting me in the history of human stature. However, I alone assume responsibility for the views expressed in this paper. I would like to thank Paul Mageli, Francesca Turchi, Amanda Lyerly, and Su Keun Kwak for their assistance in processing the data.

[1] John Komlos, "Patterns of Children's Growth in East-Central Europe in the Eighteenth Century," *Annals of Human Biology*, 13 (Jan./Feb. 1986), pp. 33–48; W. Peter Ward and Patricia C. Ward, "Infant Birth Weight and Nutrition in Industrializing Montreal," *American Historical Review*, 89 (Apr. 1984), pp. 324–45.

Its analysis, longitudinally and cross-sectionally, affords insights that have hitherto eluded the historian.[2]

Two major findings about the American population have emerged: that the colonial population achieved high nutritional standards by the eighteenth century and that even the stature of American slaves shows they were well nourished as adults, although not as children.[3] An anomaly did arise in the course of examining the extant data. Adult male stature in the United States began to decline among the birth cohorts of the 1830s, and the trend was not reversed until late in the century.[4] The pattern is puzzling because according to conventional indicators the American economy was expanding rapidly during the antebellum decades. For instance, between 1840 and 1870 per capita net national product increased by more than 40 percent and value added in the agricultural sector was growing between 2.3 and 4.2 percent per annum in the 1840s and 1850s.[5] The inference usually drawn from such rapid economic development is that the standard of living and hence food consumption, too, must have been increasing. Yet the diminution in the age-by-height profile of the male population implies that the average nutritional status was deteriorating amid growing prosperity. My aim is to resolve the conflict between these two images of the American economy at a crucial juncture in its development, by arguing that the two views are by no means inconsistent. The nutritional status of a population can decline even during prosperous times at the early stages of economic development, while the economy adjusts to sectoral shifts in production.

## HEIGHT AND WEIGHT OF WEST POINT CADETS

Data on the stature of cadets entering the West Point Military Academy from 1843 to 1894 were collected on the basis of annual

[2] John Komlos, "Stature and Nutrition in the Habsburg Monarchy: The Standard of Living and Economic Development in the Eighteenth Century," *American Historical Review*, 90 (Dec. 1985), pp. 1149–61.

[3] Kenneth L. Sokoloff and Georgia C. Villaflor, "Early Achievement of Modern Stature in America," *Social Science History*, 6 (Fall 1982), pp. 1–7; Richard H. Steckel, "A Peculiar Population: The Nutrition, Health, and Mortality of American Slaves from Childhood to Maturity," this JOURNAL, 46 (Sept. 1986), pp. 721–42.

[4] Robert W. Fogel, "Nutrition and the Decline in Mortality since 1700: Some Preliminary Findings," in Stanley L. Engerman and Robert E. Gallman eds., *Long Term Factors in American Economic Growth*, National Bureau of Economic Research, Studies in Income and Wealth, vol. 51 (Chicago, 1987). Robert A. Margo and Richard H. Steckel, "Heights of Native-Born Whites during the Antebellum Period," this JOURNAL, 43 (Mar. 1983), pp. 167–74. Birth weight of babies born in the Philadelphia alms house also declined between 1856 and 1866. I am indebted to Claudia Goldin and to Robert Margo for making these preliminary data available to me.

[5] Robert E. Gallman, "Changes in Total U.S. Agricultural Factor Productivity in the Nineteenth Century," *Agricultural History*, 46 (Jan. 1979), pp. 191–210. Robert E. Gallman, "The Pace and Pattern of American Economic Growth," in Lance E. Davis, Richard A. Easterlin, and William N. Parker eds., *American Economic Growth: An Economist's History of the United States*, (New York, 1972), pp. 15–60.

reports of medical examinations of candidates on admission.[6] Name, state of birth, age, and the state from which appointed were noted. During the latter decades of the period weight is also available.[7] The data collected number 4,180. Individuals younger than 16, older than 21, or born outside the United States were disregarded. The records were subsequently linked to some socioeconomic attributes of the parents.[8] Information is available on parental residence, income, and occupation. The compilation states residence as city, town, or rural, and divides parents into four income groups: independent, moderate, reduced, and indigent (Table 1). Criteria for inclusion into these categories were not specified.

The age-by-height observations are standardized. The average stature for a particular age is first calculated for the whole period (Table 2). Each observation is subsequently divided by the average for the relevant age category. The procedure enables the construction of a height index. An index number of 100.0 for a group of cadets means that their stature is equal to the average for their age.[9] A similar procedure is used to create the index of weight.[10] Unless otherwise specified, dates refer to the cadets' date of birth, not the date his height and weight were measured.

The frequency distributions of the height index for the 1820s to the 1870s show no evidence of the truncation introduced into some military samples by a minimum height requirement. At age 18, the West Point cadets were about 1 cm taller than comparable Austrian and French students, but were 3 cm shorter than Harvard students.[11] The 18-year-old cadets fall into the 25th percentile of today's height distribution.

Because diminution in stature in the antebellum period is evident for all age groups (Table 2), is pervasive across all regions (see Table 3), and in many of the socioeconomic categories (Tables 4 and 5), and corresponds to the pattern found among army recruits studied by other

[6] National Archives, Washington, D.C., Military Archives Division, Record Group 94, entries 103, 234. There are no blacks in the sample. In the antebellum period about 7.5 percent of the candidates were rejected. James Morrison, Jr., "*The Best School in the World*": *West Point, the Pre-Civil War Years, 1833–1866* (Kent, 1986), p. 65. Generally, applicants rejected on the basis of the physical examination do not appear on the lists.

[7] The physical description of the candidates was recorded upon entrance to the academy, consequently it reflects nutritional status prior to entering the academy.

[8] Archive of the West Point Military Academy, Record Group 404, no. 196. Lists relating to economic and social status of cadets' parents; Circumstances of Parents of Cadets, 1842–1910.

[9] Similarly, an index number of 99 means that the cadet was 1 percent below the average for his age. In this manner the index of height can be considered without regard to the age of the cadet.

[10] For this index average weight for age was used for those born in the 1860s.

[11] Jean Sutter, René E. Izac, and Tran Ngoc Toan, "L'Evolution de la Taille des Polytechniciens, 1801–1954," *Population*, 13 (July/Sept. 1958), pp. 373–406; Vienna, Austria, Kriegsarchiv, Kadettenschule; Gordon T. Bowles, *New Types of Old Americans at Harvard and at Eastern Women's Colleges* (Cambridge, Mass., 1932), p. 19.

TABLE 1
CHARACTERISTICS OF THE SAMPLE
(percentages)

| Cadets | | Parents | |
|---|---|---|---|
| Orphan | 3.2% | Income | |
| Mother head of household | 9.6 | Unknown | 31.9% |
| Age | | Independent | 3.0 |
| 16 | 8.4 | Moderate | 59.2 |
| 17 | 24.7 | Residence | |
| 18 | 23.3 | Missing | 31.9 |
| 19 | 20.2 | City | 22.1 |
| 20 | 14.9 | Town | 28.9 |
| 21 | 8.0 | Rural | 17.1 |
| Other | 0.5 | Occupation | |
| Birth decade | | Farmer | 15.7 |
| 1820 | 7.0 | Blue collar | 10.6 |
| 1830 | 17.8 | Merchant[a] | 15.7 |
| 1840 | 14.5 | Professional[b] | 16.9 |
| 1850 | 22.7 | Government[c] | 6.4 |
| 1860 | 31.9 | None[d] | 2.3 |
| 1870 | 5.8 | Elected official | 0.5 |
| Other | 0.2 | Unknown | 31.9 |
| Birthplace | | | |
| Mid-Atlantic | 26.3 | | |
| Midwest | 21.3 | | |
| Upper South | 17.4 | | |
| South | 14.7 | | |
| New England | 8.9 | | |
| West | 7.8 | | |
| Other or missing | 3.6 | | |
| State from which appointed | | | |
| Mid-Atlantic | 22.0 | | |
| Midwest | 21.0 | | |
| South | 14.2 | | |
| Upper South | 13.8 | | |
| West | 10.7 | | |
| New England | 7.7 | | |
| At large | 6.9 | | |
| Other or missing | 3.8 | | |

[a] Includes banker, cotton broker, insurance man.
[b] Includes clergyman, druggist, teacher.
[c] Includes military and elected officials.
[d] Includes out of work and gentleman.
*Note*: Number of observations = 4,180.
*Sources*: National Archives, Washington, D.C. Military Archives Division, Record Group 94, entries 103 and 234; Archive of the West Point Military Academy, Record Group 404, no. 196.

scholars, the diminution is not likely to be an artifact of the sample.[12] The number of observations in the 1820s is not large, but the decline in stature is not confined to the 1830s and continues into the 1860s. Hence the conclusion that stature declined does not depend on a difference in

[12] Fogel, "Nutrition and the Decline in Mortality."

TABLE 2

STATURE OF CADETS BY AGE AND DECADE OF BIRTH

(in centimeters)

| Birth | 16 | | | 17 | | | 18 | | | 19 | | | 20 | | | 21 | | | Index[a] | | Regular Recruits |
|---|---|---|---|---|---|---|---|---|---|---|---|---|---|---|---|---|---|---|---|---|---|
| Decade | $N$ | $\bar{X}$ | s.d. | $N$ | $\bar{X}$ | s.d. | $N$ | $\bar{X}$ | s.d. | $N$ | $\bar{X}$ | s.d. | $N$ | $\bar{X}$ | s.d. | $N$ | $\bar{X}$ | s.d. | $N$ | $\bar{X}$ | $\bar{X}$ |
| 1820s | 35 | 167.9 | 6.6 | 45 | 169.4 | 6.4 | 35 | 172.0 | 7.1 | 36 | 173.0 | 5.9 | 34 | 173.5 | 4.4 | | n.a. | | 185 | 100.4 | 173.0 |
| 1830s | 152 | 166.0 | 6.4 | 159 | 170.2 | 5.9 | 169 | 171.3 | 6.2 | 143 | 171.7 | 5.9 | 76 | 174.0 | 5.7 | | n.a. | | 699 | 100.1 | 173.3 |
| 1840s | 113 | 166.5 | 6.4 | 138 | 169.5 | 5.3 | 99 | 170.7 | 6.2 | 90 | 171.5 | 6.4 | 74 | 171.9 | 6.2 | 17 | 172.8 | 6.2 | 531 | 99.8 | 171.9 |
| 1850s | 28 | 168.1 | 5.0 | 228 | 169.2 | 6.1 | 185 | 170.5 | 6.1 | 160 | 170.4 | 5.7 | 146 | 170.8 | 6.3 | 98 | 171.3 | 5.9 | 845 | 99.5 | 170.9 |
| 1860s | | n.a. | | 324 | 169.5 | 5.8 | 377 | 170.4 | 6.1 | 336 | 171.1 | 6.1 | 231 | 171.7 | 6.0 | 198 | 172.1 | 6.2 | 1466 | 99.6 | 170.9 |
| 1870s | | n.a. | | 102 | 170.3 | 5.6 | 88 | 171.4 | 6.0 | 64 | 172.6 | 6.0 | 49 | 173.4 | 5.1 | 21 | 171.7 | 6.3 | 332 | 100.3 | 171.0 |
| Average | | 166.2 | | | 169.6 | | | 170.6 | | | 171.3 | | | 171.8 | | | 171.9 | | | 170.9 | |

[a] Index standardized for age.

*Sources*: See Table 1; for recruits see Robert W. Fogel, "Nutrition and the Decline in Mortality since 1700: Some Preliminary Findings," in Stanley L. Engerman and Robert E. Gallman, eds., *Long Term Factors in American Economic Growth*, National Bureau of Economic Research, Studies in Income and Wealth, vol. 51 (Chicao, 1987). In this source the number of observations is not given.

TABLE 3
INDEX OF STATURE BY REGION AND DECADE OF BIRTH OF CADET

| Birth Decade | New England N | New England $\bar{X}$ | Mid-Atlantic N | Mid-Atlantic $\bar{X}$ | South N | South $\bar{X}$ | Upper South N | Upper South $\bar{X}$ | Midwest N | Midwest $\bar{X}$ | West N | West $\bar{X}$ |
|---|---|---|---|---|---|---|---|---|---|---|---|---|
| 1820s | 17 | 99.9 | 48 | 100.2 | 35 | 101.2 | 52 | 100.4 | 23 | 100.1 | 4 | 101.5 |
| 1830s | 82 | 100.4 | 218 | 99.5 | 118 | 100.8 | 159 | 100.8 | 92 | 99.2 | 24 | 99.6 |
| 1840s | 53 | 99.7 | 163 | 99.3 | 67 | 100.2 | 89 | 100.6 | 109 | 99.9 | 32 | 99.3 |
| 1850s | 76 | 99.8 | 209 | 99.1 | 134 | 100.3 | 135 | 99.7 | 202 | 99.7 | 69 | 98.2 |
| 1860s | 98 | 99.3 | 347 | 99.5 | 249 | 100.1 | 224 | 100.1 | 372 | 99.2 | 143 | 99.9 |
| 1870s | 30 | 99.6 | 80 | 100.2 | 50 | 100.2 | 41 | 100.8 | 82 | 100.9 | 46 | 99.8 |
| Average | 356 | 99.7 | 1065 | 99.5 | 653 | 100.4 | 700 | 100.3 | 880 | 99.6 | 318 | 99.5 |

*Note*: Index is standardized for age.
*Sources*: See Table 1.

stature between 1820s and 1830s. The trend bottomed out in the 1860s and generally reversed in the 1870s. This was several decades before the stature of ordinary recruits recovered. Because cadets came from higher-than-average socioeconomic stratum, they experienced a smaller diminution in height than regular army recruits.[13] Between the birth cohorts of the 1820s and 1860s, the cadets deciined on average by 1.4 cm, whereas the regulars declined by 2.1 cm (Table 2).[14] However, class differences in America did not imply as large a difference in stature as in

TABLE 4
INDEX OF STATURE BY OCCUPATION OF FATHER AND DECADE OF BIRTH OF CADET

| Birth Decade | Farmer N | Farmer $\bar{X}$ | Blue Collar N | Blue Collar $\bar{X}$ | Middle Class: Merchant N | Merchant $\bar{X}$ | Middle Class: Professional N | Professional $\bar{X}$ | Middle Class: Government N | Government $\bar{X}$ | Middle Class: Total N | Total $\bar{X}$ | Harvard Students Index | Harvard Students cm | Harvard Students N |
|---|---|---|---|---|---|---|---|---|---|---|---|---|---|---|---|
| 1820s | 50 | 101.3 | 17 | 100.1 | 29 | 99.7 | 35 | 99.6 | 14 | 100.5 | 78 | 98.8 | | | |
| 1830s | 169 | 101.0 | 70 | 99.7 | 132 | 99.6 | 144 | 99.7 | 62 | 100.0 | 338 | 99.7 | 100.0 | 172.5 | 550 |
| 1840s | 88 | 100.0 | 57 | 98.9 | 93 | 100.4 | 106 | 100.1 | 50 | 100.4 | 249 | 100.3 | 100.5 | 173.3 | 1089 |
| 1850s | 107 | 99.5 | 98 | 98.7 | 114 | 99.2 | 143 | 99.8 | 48 | 99.9 | 305 | 99.6 | 100.2 | 172.9 | 335 |
| 1860s | 198 | 100.0 | 169 | 99.2 | 226 | 100.0 | 233 | 99.5 | 91 | 100.0 | 550 | 99.8 | 101.1 | 174.4 | 506 |
| 1870s | 35 | 100.0 | 26 | 99.9 | 60 | 100.7 | 41 | 100.2 | 20 | 101.7 | 121 | 100.7 | 101.8 | 175.6 | 307 |
| Average | 647 | 100.3 | 437 | 99.2 | 654 | 99.9 | 702 | 99.7 | 284 | 100.2 | 640 | 99.9 | | | |

*Note*: Index is standardized for age.
*Sources*: See Table 1. For Harvard students see Gordon T. Bowles, *New Types of Old Americans at Harvard and at Eastern Women's Colleges* (Cambridge, Mass., 1932), p. 25.

[13] After all, political connections were needed to gain admission to the school.

[14] The 1.4 cm decline refers to all age groups. This number is not reported in the tables presented in the essay, but the evidence is available from the author on request. The data on the recruits are from Fogel, "Nutrition and the Decline in Mortality," Table A.1.

TABLE 5
INDEX OF STATURE BY INCOME OF PARENT AND DECADE OF BIRTH OF CADET

| Birth Decade | Parental Income: Independent N | $\bar{X}$ | Moderate N | $\bar{X}$ | Reduced N | $\bar{X}$ | Indigent N | $\bar{X}$ |
|---|---|---|---|---|---|---|---|---|
| 1820s | 3 | 101.3 | 109 | 100.3 | 30 | 100.4 | 6 | 100.8 |
| 1830s | 43 | 100.2 | 539 | 100.1 | 26 | 99.0 | 1 | 98.0 |
| 1840s | 36 | 101.1 | 367 | 99.8 | 14 | 101.7 | 4 | 97.4 |
| 1850s | 25 | 99.5 | 456 | 99.5 | 39 | 99.1 | 11 | 98.6 |
| 1860s | 13 | 99.7 | 818 | 99.6 | 88 | 100.4 | 12 | 100.4 |
| 1870s | 2 | 100.2 | 165 | 100.5 | 15 | 99.3 | 0 | |
| Average | 122 | 100.3 | 2454 | 99.8 | 212 | 100.0 | 34 | 99.5 |

*Note*: Index is standardized for age.
*Sources*: See Table 1.

Europe. Middle-class cadets were only 1.2 cm taller than were those of the working class (Table 4), while aristocratic boys in Europe were 4 to 7 cm taller than middle-class boys. An extreme example is England, where the boys of gentry ancestry were 20 cm taller than London slum boys.[15]

Among the birth cohorts of the 1820s and 1830s, sons of farmers were the tallest (Table 4). Thereafter they remained taller than sons of blue-collar workers, but became shorter than cadets whose fathers were either merchants, professionals, or government officials, except for the decade of the 1860s. While the height of cadets born to farmers or blue-collar workers began to diminish in the 1830s and continued to do so until the 1860s, the height of middle-class cadets did not. In the 1840s their stature even increased, and only in the decade of the 1850s did their stature decline below the level of the 1830s. In the 1860s the process stabilized, and by the 1870s middle-class cadets show a substantial increase in stature. Compared with the 1820s their stature increased by 0.9 percent. In contrast, farmers and workers in the 1870s were still 1.3 percent and 0.2 percent below the stature attained by their counterparts in the 1820s. By the 1870s, the stature of cadets with parents in high-status occupations was greater than the stature of sons of farmers and blue-collar workers. All in all, middle-class cadets maintained their nutritional status throughout the period; a substantial decline was evident among them only in the 1850s. This pattern was consistent in all regions of the country.[16]

The stature of students enrolled in Harvard University correlates

[15] Komlos, "Patterns of Children's Growth"; Roderick Floud and Kenneth W. Wachter, "Poverty and Physical Stature, Evidence on the Standard of Living of London Boys, 1770–1870," *Social Science History*, 6 (Fall 1982), pp. 422–52.
[16] Data available from the author upon request.

TABLE 6
INDEX OF STATURE BY RESIDENCE OF PARENT AND DECADE OF BIRTH OF CADET

| Birth Decade | Parental Residence: City N | City $\bar{X}$ | Town N | Town $\bar{X}$ | Rural N | Rural $\bar{X}$ |
|---|---|---|---|---|---|---|
| 1820s | 19 | 99.9 | 77 | 99.8 | 51 | 101.4 |
| 1830s | 166 | 99.6 | 248 | 99.7 | 195 | 100.7 |
| 1840s | 124 | 99.8 | 183 | 100.1 | 114 | 99.9 |
| 1850s | 177 | 99.3 | 246 | 99.4 | 108 | 99.7 |
| 1860s | 342 | 99.5 | 384 | 99.6 | 206 | 100.2 |
| 1870s | 88 | 100.6 | 59 | 100.4 | 35 | 99.9 |
| Average | 916 | 99.7 | 1,197 | 99.7 | 709 | 100.3 |

*Note*: Index is standardized for age.
*Sources*: See Table 1.

positively with the stature of the middle-class cadets (Table 4). Harvard students also came from families with high-status occupations.[17] Like middle-class cadets, Harvard students declined in stature in the 1850s but increased continuously thereafter.

Most parents of cadets were classified in the upper-middle income group as "moderate."[18] The stature of cadets in this category began to diminish in the 1830s and reached a nadir in the 1860s. By the 1870s their stature reached the level of the 1820s. Because of the small number of observations in the other income groups, no significance can be attached to the decadal variations of stature in these categories (Table 5). Nonetheless, on average, stature of cadets in the highest income group was greatest and stature of cadets in the lowest income group was smallest. Moreover, stature of cadets whose parents were in the highest income category correlates positively with that of those in higher socioeconomic occupations. The category experienced a decline in stature only in the 1850s.

The height of cadets whose parents resided in rural areas declined consistently and similarly to that of cadets whose parents were farmers; by the 1870s neither group had recovered to the level of the 1820s (Table 6). The declining trend is more pronounced among rural than among urban residents. Town and city dwellers fared better than rural residents in the 1830s and 1840s; in the 1850s, however, their stature also declined, and did not begin to recover until the 1860s. By the 1870s urban cadets were actually taller than rural cadets. Therefore, the trend

[17] Bowles, *New Types of Old Americans*, p. 20. Only 10 percent of the Harvard fathers were manual workers.

[18] They were apparently asked to rank themselves with strong pressure to choose the moderate designation in order to forestall the criticism that the school was an elitist institution. I am indebted to Peter Karsten for this suggestion.

of the nutritional status of urban residents is quite similar to the pattern of middle-class cadets. The similarity is understandable insofar as the latter group must have had a higher propensity to reside in urban than rural areas. Rural residents were taller than urban residents in the 1820s, 1830s, 1850s, and 1860s.[19]

Regional trends adhere to the basic pattern (Table 3). The declining tendency was experienced generally and was reversed in the 1870s.[20] Cadets from the South were the tallest until the 1870s, when they were overtaken by cadets from the Midwest and the Upper South. Although the Civil War seems to have had a great impact on cadets of the Upper South, throughout the whole period cadets from the Upper South were perhaps least affected by the decline in nutritional status.[21] By the 1870s, their average stature was 0.4 percent greater than in the 1820s. Midwesterners also quickly recovered their nutritional status after the Civil War. In the 1870s they were 0.8 percent taller than in the 1820s.

The weight of the cadets indicates that by modern standards they were underweight for their height; the distribution of their weight, however, is not unusual. They weighed less than Harvard students.[22] Cadets born in the 1850s were about 2.5 pounds lighter than those born in the 1870s (Table 7). Income and occupation of the father did not have a significant effect on the weight of the cadets. Cadets from the New England states and from the Midwest and West were heavier than those from other regions. Rural residents were heavier than town or city dwellers.

## THE HYPOTHESIS OF ANTEBELLUM DIETARY CHANGE

There are a number of possible explanations for the decline in stature after 1830: 1) changes in the epidemiological environment, which could have affected stature to the extent that diseases lay claim to nutrients; 2) a decline in nutritional status because of immigration; 3) increased work intensity, which might have claimed sufficient nutrients to have caused the decline in stature; and 4) decline in the availability of nutrients because of rapid population growth coupled with urbanization and expansion of the non-nutrient producing sector.

The first possibility is not a very promising hypothesis partly because

[19] In the 1820s and 1830s Civil War army recruits born in rural areas were also taller (by 2.2 cm) than urban-born recruits. Margo and Steckel, "Height of Native-Born Whites," p. 169.

[20] The pattern of stature by residence, that is, the state from which the cadet was appointed, is very similar to the pattern of stature by birthplace. Averages for the whole period (1820–1870) by residence and birthplace did not diverge from one another by more than 0.2 percent in any region. Evidence available from the author on request.

[21] Tennessee was the only state and the South the only region whose cadets maintained an index of stature above 100.0 throughout the antebellum period. Cadets of all states showed some signs of deterioration in stature, but those born in Tennessee appear to have withstood the stresses of the period most successfully.

[22] Bowles, *New Types of Old Americans*, p. 25.

TABLE 7
DETERMINANTS OF THE WEIGHT OF CADETS

| | (1) | (2) | (3) | (4) | (5) | (6) | (7) |
|---|---|---|---|---|---|---|---|
| Constant | 124.8* | 126.2* | 127.9* | 126.1* | 124.9* | 126.6* | 129.4* |
| Age | | | | | | | |
| 18 | 0.3 | 0.4 | 0.3 | 0.3 | 0.3 | 0.3 | 0.4 |
| 19 | 4.5* | 4.6* | 4.4* | 4.5* | 4.5* | 4.3* | 4.5* |
| 20 | 4.3* | 4.6* | 4.2* | 4.4* | 4.4* | 4.0* | 4.4* |
| 21 | 7.0* | 7.4* | 6.8* | 7.2* | 7.2* | 6.4* | 6.9* |
| Birth decade | | | | | | | |
| 1850s | −2.3* | −2.6* | −2.7* | −2.6* | −2.3 | −2.5* | −2.7* |
| 1860s | 1.6 | 1.4 | 1.3 | 1.4 | 1.6 | 1.4 | 1.3 |
| Place of Birth | | | | | | | |
| Mid-Atlantic | 0.6 | | | | 0.5 | 0.9 | |
| Midwest | 2.5* | | | | 2.5* | 2.7* | |
| New England | 2.9* | | | | 2.9* | 3.3* | |
| South | −0.2 | | | | −0.2 | −0.3 | |
| West | 2.5* | | | | 2.6* | 2.7* | |
| Height[a] | 3.7* | 3.7* | 3.7* | 3.7* | 3.7* | 3.7* | 3.7* |
| Income | | | | | | | |
| Independent | | 1.1 | | | 1.2 | | |
| Reduced | | −1.9 | | | −1.9 | | |
| Indigent | | −1.9 | | | −2.1 | | |
| Residence | | | | | | | |
| City | | | −2.2* | | | −2.6* | −3.8* |
| Town | | | −1.8* | | | −1.8* | −3.1* |
| Occupation | | | | | | | |
| Farmer | | | | 0.4 | | | −2.0 |
| Blue collar | | | | −0.0 | | | 0.0 |
| $R^2$ | .4 | .4 | .4 | .4 | .4 | .4 | .4 |

* Significant at the 5 percent level.
[a] Weight increment in pounds for each inch in height above 67 inches and symmetrically for each inch below 67.
*Notes*: Dependent variable is weight in pounds. The intercept represents a 17-year-old cadet, born in the 1870s, 67 inches tall, and (1) born in the Upper South; (2) whose father was in the moderate income category; (3) whose parents were rural residents; (4) whose father had a Group II occupation; (5) born in the Upper South, whose father was in the moderate-income category; (6) born in the Upper South, whose parents were rural residents; (7) whose parents were rural residents and had a Group II occupation.

the decline in stature was geographically widespread. It is not likely that the disease environment would have changed so dramatically across such a large area. If the decline in stature had been caused by a change in the epidemiological environment, then one might expect that urban areas with higher population densities would have been affected first and most severely. Instead, cadets whose parents were living in rural areas experienced a greater decline in height than those in urban areas (Table 6). Moreover, in the 1840s the stature of both town and city dwellers increased. Should one conclude from this that the disease environment was actually improving in the cities while deteriorating in the countryside, although there is no evidence at this time for such a pattern?[23] In addition, middle-class cadets maintained or increased their stature throughout the 1830s and 1840s (Table 4), and there is no reason to think that they would have been immune to a deterioration in the disease environment.

Nor is it likely that the deterioration in nutritional status was caused by epidemics, because they were episodic: cholera struck, for instance, in 1832, 1849, and 1866.[24] Could such sporadic events give rise to a long-term trend? Besides, when it did strike, cholera affected relatively few individuals, and because of the high mortality rate, not many of those who contracted the disease would have enrolled in West Point.[25]

Another reason to discount this hypothesis is that during the first decades of the eighteenth century, when the disease environment was in many ways worse than a century later, the stature of American men rose, probably because meat consumption was increasing.[26] One could argue that in the antebellum era the disease environment was to some extent improving. Yellow fever and smallpox, for instance, were no longer as prevalent (in the North at least) as they had been earlier.[27] At the same time, malaria and hookworm persisted well into the twentieth century, long after the nutritional status of the population began to improve substantially.[28] Thus the incidence of these diseases did not prevent an upturn in nutritional status after the Civil War. In the face of these patterns, the first hypothesis can not have much appeal until changes in the disease environment are studied more systematically.

Although genetically immigrants were quite similar to the native white population, because both had originated in Europe, immigration could nonetheless cause a decline in stature because immigrants' nutritional status was below the American standard. The reason is that nutritional status is inheritable to some extent. Although cadets in the sample were all born in the United States, some could have been children of immigrants and shorter on that account.[29] However, circum-

[23] Richard A. Easterlin, "Population Issues in American Economic History: A Survey and Critique," in Robert E. Gallman ed., *Research in Economic History*, supplement 1, (1977), pp. 131–58.

[24] Charles Rosenberg, *The Cholera Years: The United States, 1832, 1849, 1866* (Chicago, 1962), p. 90.

[25] In New York City, for instance, only 4 percent of the population contracted the disease in 1849. Of those who did, 40 percent died.

[26] Fogel, "Nutrition and the Decline in Mortality"; Sarah McMahon, "Provisions Laid Up for the Family: Toward a History of Diet in New England, 1650–1850," *Historical Methods*, 14 (Winter 1981), pp. 22–30.

[27] Rosenberg, *The Cholera Years*, p. 1.

[28] Greer Williams, *The Plague Killers: Untold Stories of Three Great Campaigns Against Disease* (New York, 1969), pp. 94, 122.

[29] It is possible that not many first-generation Americans would have been nominated to West Point, given the political connections needed for appointment to the academy. Morrison, *The Best School*, p. 62.

stantial evidence tends not to support this hypothesis either. The influx of immigrants in the 1820s was only 1.3 percent of the white population and in the 1830s only 4.0 percent.[30] Thus, the decline in stature started at a time when the share of immigrants in the population was still very low. By 1850 the share of immigrants in the total population had risen to 12.9 percent. Yet the rise of this share could not have been the primary factor determining the decline in stature because the trend reversed in the 1870s, even though the share of immigrants in the population continued to grow. Hence, if the decline in stature were attributed to the increased share of the foreign born in the population, one could not explain the turning points in the 1830s and 1870s.

Another compelling reason for discarding the immigration hypothesis is that the decline in stature was pervasive across all regions of the United States but immigration was not. In 1850 only 4.5 percent of the white population was foreign born in the South in contrast to 18.5 percent in the Northeast. Yet the decline in stature between the 1830s and 1850s in the New England and Mid-Atlantic states (Table 3) was about the same as in the South. In fact, the decline in stature was the greatest in the Upper South, where the immigration rate was quite low. Hence, the immigration rate does not correlate well with the regional decline in stature either. Moreover, the rate of immigration must have been as high in the eighteenth century as it was in the antebellum period, yet the earlier wave of immigration did not affect the mean stature of the native population.

The third hypothesis, increased work effort, cannot be rejected, but because of the lack of systematic evidence it cannot be accepted either. The real question, in any case, is not if the work intensity of the labor force increased during the 1830s but if children and pregnant women began to work harder during this decade because only that change would have affected terminal height of this birth cohort.[31] Without further evidence, the most one might suggest is that changes in the work intensity could have contributed to the decline in nutritional status.

The fourth hypothesis remains. Indeed, an accounting of the average calorie and protein consumption per capita (Table 8) suggests that nutritional intake was declining in the late antebellum period.[32] The availability of nutrients declined because food output did not keep pace

[30] U.S. Bureau of the Census, *Historical Statistics of the United States: Colonial Times to 1970* (Washington, D.C., 1975), vol. I, pp. 12, 14, 15, 105.

[31] At least in Philadelphia the labor force participation of female heads of households declined to some extent during the antebellum period. Claudia Goldin, "The Economic Status of Women in the Early Republic: Quantitative Evidence," *Journal of Interdisciplinary History*, 16 (Winter, 1986), pp. 375–404.

[32] The decline was somewhat more pronounced if the caloric intake is converted to adult male equivalent. Evidence is available upon request.

TABLE 8
ESTIMATE OF DAILY PRODUCTION AND CONSUMPTION OF CALORIES AND PROTEIN PER CAPITA

| | Production | | | | | | Consumption[a] | | | | |
|---|---|---|---|---|---|---|---|---|---|---|---|
| | Lower Bound | | | | Upper Bound | | | | | | |
| | Meat | Vegetable[b] | Total | Extra[c] | Total | Net Exports[d] | Lower Bound | Index | Upper Bound | Index | Weight[e] (pounds) |
| Calories | | | | | | | | | | | |
| 1839 | 1,007 | 2,512 | 3,519 | 0 | 3,519 | 146 | 3,021 | 100 | 3,021 | 100 | 4.1 lbs. |
| 1849 | 869 | 2,217 | 3,086 | 64 | 3,150 | 81 | 2,696 | 89 | 2,754 | 91 | 3.5 |
| 1859 | 762 | 2,525 | 3,287 | 107 | 3,394 | 66 | 2,892 | 96 | 2,989 | 99 | 3.8 |
| 1869 | 525 | 2,842 | 3,367 | 77 | 3,444 | 138 | 2,892 | 96 | 2,962 | 98 | 3.6 |
| 1879 | 678 | 3,250 | 3,928 | 206 | 4,134 | 591 | 2,944 | 97 | 3,130 | 104 | 4.1 |
| 1880[f] | 750 | | | | | | 2,700 | | 3,100 | | |
| 1909 | 710[g] | | | | | | | | 3,132[h] | | 4.3 |
| 1910–1919 | 677[g] | | | | | | | | 3,064[h] | | |
| Protein (grams) | | | | | | | | | | | |
| 1839 | 36 | 84 | 120 | 0 | 120 | 5 | 103 | 100 | 103 | 100 | |
| 1849 | 31 | 74 | 105 | 1 | 106 | 3 | 92 | 89 | 92 | 89 | |
| 1859 | 28 | 86 | 114 | 2 | 116 | 5 | 98 | 95 | 99 | 96 | |
| 1869 | 20 | 95 | 115 | 2 | 117 | 10 | 93 | 90 | 95 | 92 | |
| 1879 | 25 | 118 | 143 | 5 | 148 | 29 | 100 | 97 | 104 | 101 | |

[a] 90 percent of production in order to allow for noningested items minus net exports.
[b] Includes dairy products.
[c] The meat equivalent of the unaccounted portion of corn output (see Appendix).
[d] Does not include alcohol, nuts, and cocoa imports.
[e] Weight of food consumed daily.
[f] Ingested foods only.
[g] Refers to consumption.
[h] 10 percent allowance made for noningested items.

*Notes*: Restricted to products enumerated in 1840 census except as specified. To convert into adult male equivalent, multiply by one of the following ratios: 1839—1.285, 1849—1.266, 1859—1.267, 1869—1.262, 1879—1.253.

*Sources*: See Appendix. Estimates for 1880 are from Merrill Bennett and Rosemond Peirce, "Change in the American National Diet, 1879–1959," *Food Research Institute Studies*, 2 (May 1961), pp. 95–119; 1909–1919 are from *Consumption of Food in the United States, 1909–48*, U.S. Department of Agriculture, Miscellaneous Publication No. 691 (Aug. 1949), p. 123.

with the demands placed upon it by the rapid growth of the urban-industrial sector.[33] Population adjusted for age structure grew by between 33 and 38 percent per decade, while the agricultural labor force expanded more slowly, in the 1840s, for instance, by 27 percent. In contrast, in the 1840s the labor force employed in the other sectors increased by 78 percent and the urban population almost doubled, only to expand by 75 percent in the 1850s.[34] There were some gains in labor productivity in grain production in the first half of the century, but these were very small, and in animal husbandry they were negligible.[35]

Hence, productivity advances would not have been able to offset the dramatic sectoral shifts of the labor force.[36] As a result, the number of persons without ties to the land and dependent on the farm population for sustenance was unprecedently large. Between 1800 and 1820 one agricultural worker produced enough food for about four individuals.[37] The ratio began to change in 1830, and by 1870 one agricultural laborer supported almost six people—an increase of 50 percent. It is not likely that labor productivity in the food-producing sector could have offset a shift of such magnitude.[38]

The finding that West Point cadets were underweight contributes to the argument that the decline in their nutritional status is related to

[33] The decline relative to 1840 was not caused by an unusually good harvest in 1840, since the 1840 Harvest appears to have been average for the period. Robert E. Gallman, "A Note on the Patent Office Crop Estimates, 1841–1848," this JOURNAL, 23 (June 1963), pp. 185–95.

[34] Stanley Lebergott, *The Americans: An Economic Record*, (New York, 1984), p. 66; *Historical Statistics of the United States*, vol. 1, pp. 12, 14, 15.

[35] Robert E. Gallman, "The Agricultural Sector and the Pace of Economic Growth: U.S. Experience in the Nineteenth Century," in David C. Klingaman and Richard K. Vedder, eds., *Essays in Nineteenth Century Economic History* (Athens, 1975), pp. 35–76, esp. p. 42. Gallman estimates that labor productivity in grain and cotton production rose at a rate of 0.3 percent per annum. This is far below the rate that would have been needed to maintain food intake at the level of the 1820s. In the 1840s value added per gainful worker actually declined by 5 percent, and the rate of growth between 1839 and 1869 was 0.6 percent per annum. Robert E. Gallman, "Commodity Output, 1839–1889," National Bureau of Economic Research, *Trends in the American Economy in the Nineteenth Century*, vol. 24, (Princeton, 1960), pp. 13–67.

[36] The impression is propagated in the literature that labor productivity increased considerably during the first half of the century. One finds in Stanley Lebergott's textbook, for instance, that the number of man-hours required to produce 100 bushels of wheat declined from 373 to 233. *The Americans*, p. 301. He quotes Robert McElroy, Reuben Hecht, and Earle Gavett, "Labor Used to Produce Field Crops, Estimates by States," U.S. Department of Agriculture, *Statistical Bulletin*, No. 346 (May 1964), p. 3. This publication, in turn, relies on Martin Cooper, Glenn Barton, and Albert Brodell, "Progress of Farm Mechanization," U.S. Department of Agriculture, *Miscellaneous Publications*, No. 630 (Oct. 1947) p. 3. The latter study, however, focuses mainly on the twentieth century, and its authors do not divulge their sources and methods for the nineteenth century, merely stating that "yields for 1800 and 1840 are estimates by the authors." This assertion might suffice if one's interest in the nineteenth century is peripheral, but the evidence is insufficient to understand the dynamics of antebellum agricultural development. Hence Gallman's productivity estimate appears to be the most plausible.

[37] Lebergott, *The Americans*, p. 66. Of course, this assertion is not strictly correct because some members of the agricultural labor force did not produce nutrients.

[38] Gallman, "The Agricultural Sector and the Pace of Economic Growth," p. 47.

dietary change.[39] Their weight for height was less than that of females today. According to the latest standard, 17-year-old boys are on average 176.3 cm tall and weigh 146 pounds. Even after accounting for differences in stature, by subtracting 5.5 pounds, one still finds that 17-year-old West Point cadets of the nineteenth century weighed 14 pounds less than the mean for today's youth.[40] That 17-year-old cadets were 6 cm shorter than today's 17-year-old boys also supports the notion that there was a binding caloric constraint on the biological growth process. Furthermore, one should also consider that human growth usually ceases today at age 18, but a century ago it did not. Compared with adult army recruits born in the 1830s, 20-year-old cadets were 0.9 cm shorter.[41] This is significant because the earlier growth ceases for a given height, the better the nutritional status of the individual. The anthropometric evidence given here contradicts the notion that an abundant food supply meant that there were no constraints on the food consumption of the population.

Estimates of calorie and protein intake are restricted to food products enumerated in the 1840 census with the addition of dairy products (see Appendix). Consequently they capture accurately only the trend in these variables, but may be imperfect estimates of the level of total protein or calorie intake of the population because fish, poultry, fruit and some vegetables are not included in the estimates and because the noningested portion of food production might be underestimated.[42] The estimated intake in 1849, of about 2,700 calories per capita or 3,400

[39] Frank Kidder, *The Architect's and Builder's Pocketbook* (New York, 1900), p. 721. White adults also weighed less than they do today. The average weight of men in Boston at the time of the Civil War was 141.5 pounds, 25 pounds below the standard of the 1950s. (In contrast, however, adult male slaves weighed about 157 pounds, much closer to modern standards.) Robert A. Margo and Richard H. Steckel, "The Height of American Slaves: New Evidence on Slave Nutrition and Health," *Social Science History*, 6 (Fall 1982), pp. 516–38.

[40] U.S. Department of Health, Education and Welfare, National Center for Health Statistics, "NCHS Growth Charts, 1976," *Monthly Vital Statistics Report*, 25, No. 3, supplement (June 22, 1976) (HRA 76–1120). The average weight at ages 18–24 of white women born in the 1950s was 138 pounds. "Weight and Height of Adults 18–74 Years of Age: United States, 1971–74," U.S. Department of Health Education, and Welfare Publication No. (PHS) 79–1659 (May 1979) p. 24; "Weight by Height and Age for Adults 18 to 74 Years: United States, 1971–74," U.S Department of Health, Education, and Welfare Publication No. (PHS) 79–1656 (Sept. 1979) p. 18. Sixteen-year-old black boys weighed about as much as 20-year-old cadets at West Point a hundred years earlier. Phyllis Eveleth, Evelyn Bowers, and Joan Schall, "Secular Change in Growth of Philadelphia Black Adolescents," *Human Biology*, 51 (May 1979), pp. 213–28. The Metropolitan Life Insurance Company's height and weight standards for men 172 cm (without clothing), medium frame, aged 25–59 in 1983 was 138–150 pounds. Taking the mean of this range would make our 20-year-old cadets about 14 pounds below today's norm. The British weight standard in 1966 was 140 pounds. This would be 9 pounds above the mean of the 19-year-old cadets and at the 26 centile of the British weight distribution.

[41] In the category of farmers 174.4 vs. 173.5 cm and among urban residents 172.1 vs. 171.2 cm. Margo and Steckel, "Height of Native-Born Whites," p. 169.

[42] However, the neglected items could not have been important because the estimates are reasonably close to the ones made for the 1880s and thereafter. The lower-bound estimates actually appear to be a couple of hundred calories too high.

calories per adult male, was quite adequate for ordinary work, but may have been insufficient to provide an excess supply of nutrients required for the physical activity of the time. The weight of the cadets indicates that the claims on their caloric intake must have been enormous.[43] The decline in caloric and protein intake was about 10 percent in the 1840s. (The decline per adult male equivalent was slightly greater.) Thereafter, consumption levels improved to some extent, but did not reach the level of 1840 until 1880. Whereas net exports of nutrients had not been important previously, in the late 1870s they increased manyfold. Consequently average nutrient intake in 1880 was still near the 1840 level.[44]

Indeed, after the 1840s the only food crop whose output kept pace with population growth was corn, which was used mostly as feed. Yet a rise in corn production did not necessarily lead to a large increase in meat production.[45] I calculate a lower-bound and an upper-bound estimate of caloric intake by varying the meat equivalent of a portion of the growth in corn output. (Details are in the Appendix.) Even the upper-bound estimate fails to redress the imbalance between population and meat availability after 1840. Of course, the destruction of livestock during the Civil War must have exacerbated the situation. To maintain per capita meat production at the 1840 level, either the dressed weight of cattle and swine would have had to increase by 41 and 134 percent respectively by 1870, or the slaughter ratio would have had to increase correspondingly (Table 9). Since such drastic changes in the productivity of the meat industry are implausible, there can be little doubt that per capita meat consumption declined. Per capita consumption of meat in the antebellum period must have been close to 200 pounds, but by the early twentieth century, when actual slaughtering statistics for the nation first appeared, meat consumption had fallen to 154 pounds per capita and continued to do so until the nadir of 131 pounds was reached

[43] I am indebted to James Tanner of the Institute of Child Health, University of London, for this suggestion.

[44] The divergence of consumption and production per adult male equivalent after 1870 provides a clue to why the stature of West Point cadets diverged from that of regular recruits after the Civil War. While the stature of cadets increased, that of regular soldiers did not. The export and import of nutrients may help to explain this pattern insofar as the origin and destination of the flow of nutrients through international trade affected the various elements of the population differently. While food exports were composed of such products as corn, ham, and bacon, eaten generally by the population, nutrient imports were made up of such products as sugar and exotic fruit, whose consumption in the population was not as widespread. It is understandable that the trend in the cadets' stature diverged from that of the regular soldiers because cadets were drawn from better situated families than those of the average recruits.

[45] For instance, in the 1870s and 1880s pork production stagnated even though corn output doubled. Although, beef production did double, the divergence between corn and pork production shows the dangers of drawing inference from one to the other. William N. Parker, "Agriculture," et al., eds., *American Economic Growth*, pp. 369–417.

TABLE 9
ESTIMATE OF PER CAPITA ANNUAL MEAT PRODUCTION AND CONSUMPTION IN THE UNITED STATES
(in pounds)

| | | Production | | | | Dressed Weight Needed to Maintain Per Capita Consumption Plus Exports at the 1839 Level | | | | | | | | |
|---|---|---|---|---|---|---|---|---|---|---|---|---|---|---|
| | | | | | | Beef | | | Pork | | | Mutton | | |
| | Consumption[a] | Beef | Pork | Mutton | Total | Weight | Increment | % | Weight | Increment | % | Weight | Increment | % |
| 1839 | 213 lbs. | 79 lbs. | 142 lbs. | 18 lbs. | 239 lbs. | 400 lbs. | | | 140 lbs. | | | 40 lbs. | | |
| 1849 | 194 | 72 | 121 | 14 | 207 | 441 | +41 | 10% | 166 | +26 | 19% | 49 | +9 | 23% |
| 1859 | 181 | 73 | 99 | 11 | 183 | 435 | +35 | 9 | 203 | +63 | 45 | 64 | +24 | 60 |
| 1869 | 130 | 56 | 61 | 12 | 129 | 565 | +165 | 41 | 327 | +187 | 134 | 61 | +21 | 53 |
| 1879 | 161 | 64 | 88 | 11 | 163 | 509 | +109 | 27 | 240 | +100 | 71 | 65 | +25 | 63 |
| | | | | | | Actual Values | | | | | | | | |
| 1909[a] | 154 | 70[b,c] | 71[c] | 13[c,d] | | 496 | | | 122 | | | | | |
| 1910–1919 | 142 | | | | | 472 | | | 123 | | | | | |
| 1920–1929 | 139 | | | | | 474 | | | 122 | | | | | |
| 1930–1939 | 131 | | | | | 460 | | | 129 | | | | | |

[a] Production minus 10 percent wastage, minus exports, plus the meat equivalent of the unaccounted portion of corn output.
[b] Includes veal.
[c] Consumption.
[d] Includes all other meat.
*Sources*: See Appendix. For 1899–1907 see: *Consumption of Food in the United States, 1909–48*, U.S. Department of Agriculture, Miscellaneous Publication No. 691 (Aug. 1949), p. 123; *Livestock and Meat Statistics, 1957*, U.S. Department of Agriculture, Statistical Bulletin No. 230 (July 1958); George K. Holmes, "Meat Situation in the United States," U.S. Department of Agriculture Report No. 109 (July 1916), p. 137.

TABLE 10
ANNUAL PRODUCTION AND CONSUMPTION OF FOOD PRODUCTS IN THE UNITED STATES IN 1879 PRICES

| | Production | | | Consumption |
|---|---|---|---|---|
| | Total (000,000) | Per Adult[a] | Change[b] (percent) | Per Adult (dollars) |
| 1839 | $582 | $43.8 | | $43.3 |
| 1849 | 733 | 40.0 | −8.7% | 39.6 |
| 1859 | 1,065 | 42.9 | −2.1 | 43.3 |
| 1869 | 1,318 | 43.6 | −0.5 | 43.7 |
| 1879 | 1,960 | 49.0 | +11.9 | 40.5 |

[a] Per adult male equivalent.
[b] Relative to the base year 1839.
*Sources*: Robert E. Gallman, "Commodity Output, 1839–1889," in National Bureau of Economic Research, *Trends in the American Economy in the Nineteenth Century* (Princeton, 1960), vol. 24, pp. 13–67; for exports see: *Historical Statistics of the United States from Colonial Times to 1970* (Washington, D.C., 1975) p. 890.

in the 1930s. The only real question is when the decline in meat consumption began.[46]

Food production measured in constant (1879) dollars conforms to the above pattern: output per adult male equivalent continued to decline through 1870 (Table 10). Imports to some degree compensated for this declining trend, not in 1850, but in 1860 and 1870, only to have a reverse effect in 1880. Because of exports, the value of food consumption per adult male equivalent in 1879 was still below that of 1839.

[46] That growth in the livestock population did not keep pace with expansion of the human population was noticed by George Holmes who pointed out that the number of cattle per capita of the population was 0.88 in 1840, but did not reach that level again until 1900. He did not think this to be a statistical artifact. He observed that 1840 "was a time when meat was a much larger fraction of national dietary than in 1900" and added that "a steady, persistent, and strong decline in per capita number of swine in the United States is observed from the first census year in which swine were enumerated to the latest one." He concluded that "the displacement of meat in the dietary by products of the vegetable kingdom advances slowly but surely." George K. Holmes, "Meat Situation in the United States," U.S. Department of Agriculture Report No. 109 (July 1916), part 1, pp. 22, 34; George Warren and Frank Pearson estimated food and feed crops production per capita stagnating between 1839 and 1849, rising to 1859, dipping somewhat during the Civil War, and rising thereafter; see *Prices* (New York, 1933), p. 44. However, their estimate is misleading because it does not contain meat production. See also Walter Baron von Richtofen, *Cattle Raising on the Plains of North America* (1st edn. 1885; 2nd edn., Norman, 1964), p. 10. Sam Hilliard, *Hogmeat and Hoecake: Food Supply in the Old South, 1840–1860* (Carbondale, 1972), p. 66; Hilliard thought that there was a decline in meat consumption accompanied by ". . . a marked improvement in the quality of the diet in the nation after about 1840 . . ." Another source suggested that the national average meat consumption between 1830 and 1880 was 175 pounds, of which 150 pounds was pork. Hilliard, *Hogmeat and Hoecake*, pp. 41, 130; and Richard Cummings, *The American and His Food: A History of Food Habits in the United States* (rev. edn.; Chicago, 1941), p. 258 as cited in James T. Lemon, "Household Consumption in Eighteenth Century America and Its Relationship to Production and Trade: The Situation among Farmers in Southeastern Pennsylvania," *Agricultural History*, 41 (Jan. 1967) pp. 59–70; *Live Stock and Meat Statistics* (New York, 1957) pp. 283, 284.

Increase in demand due to population growth, coupled with the slow growth of agricultural productivity, put upward pressure on food prices in the antebellum era. In nominal terms the price of grain products rose between 20 and 30 percent in the 1830s and 1850s (Table 11). According to one index, grain prices rose 15 percent faster than wages in the 1830s but then leveled off in the 1840s and 1850s.[47] According to another index, the wages of agricultural workers in Maryland relative to the price of foodstuffs peaked in the 1820s and fell thereafter.[48] In Vermont, the wages of farm laborers, deflated by the price of grain, fell by 13 percent in the 1830s relative to the 1820s, rose in the 1840s slightly above the 1820s level, and fell again in the 1850s.[49] In Massachusetts wages relative to farm prices fell by nearly 4 percent in the 1830s and rose by only 4.5 percent in the 1840s.[50] Wage indexes deflated by grain prices do corroborate, in the main, the notion that grain prices were rising faster than wages in most of the antebellum decades.

There were other reasons for consumers to decrease their food intake. Even an individual whose salary rose faster than food prices would have had an incentive to decrease food consumption, because the price of food rose relative to industrial prices. This would have been an inducement to substitute away from food and toward all other goods. For instance, grain prices relative to industrial prices (67 commodities) began to rise rapidly in 1830, and, with the exception of two years, remained high during the subsequent decades.[51] Between the 1820s and the 1850s, the price of grain rose 30 percent and the price of meat 22 percent relative to industrial products.[52] Also, in Vermont the rise in the relative price of grain was substantial. Relative to the price of clothing it more than doubled (Table 11).

[47] Paul A. David and Peter Solar, "A Bicentenary Contribution to the History of the Cost of Living in America," in Paul Uselding, ed., *Research in Economic History*, 2 (1977), p. 1–80.

[48] Donald Adams, Jr., "Prices and Wages in Maryland, 1750–1850," this JOURNAL, 46 (Sept. 1986), pp. 625–46.

[49] T. M. Adams, "Prices Paid by Farmers for Goods and Services and Received by Them for Farm Products, 1790–1871; Wages of Farm Labor, 1780–1937," University of Vermont and State Agricultural College, Vermont Agricultural Experiment Station (Feb. 1939). If the agricultural labor market had been in competitive equilibrium, then one would expect money wages in this sector to rise with the rise in the value of their marginal product. Thus wages should have risen with grain prices, unless they were offset by a decline in labor productivity. This is also an indication that labor productivity was not growing as quickly as is sometimes suggested because if it had agricultural wages probably would have risen faster.

[50] Winifred Rothenberg, "The Market and Massachusetts Farmers: A Paradigm of Economic Growth in Rural New England, 1750–1855" (Ph.D dissertation, Brandeis University, 1985), pp. 219, 224.

[51] Anne Bezanson, Robert Gray, and Miriam Hussey, *Wholesale Prices in Philadelphia 1784–1861* (Philadelphia, 1936), chap. 2.

[52] Rent and services are the only important components of the typical budget which are missing from this index. Regardless of how these missing prices behaved, my argument holds up as long as the cross-price elasticity between food and industrial products is stronger than between food and rent; 80 percent of a typical budget, is, however incorporated into these indexes. Jeffrey G. Williamson and Peter H. Lindert, *American Inequality: A Macro-economic History* (New York, 1980), p. 25.

TABLE 11
INDEX OF WHOLESALE PRICES IN THE UNITED STATES, 1820–1860

| | Philadelphia | | | | | | | | Cincinnati | Vermont | |
|---|---|---|---|---|---|---|---|---|---|---|---|
| | | | | | | | | | Corn/Pork Ratio | Grain | |
| | Grains | Grain Products | | Meat Products | | Grain and Meat Products | | | | | |
| 1820s | 100.0[a] | 100.0[a] | 100.0[b] | 100.0[a] | 100.0[b] | 100.0[a] | 100.0[b] | 100.0[c] | 100.0[d] | 100.0[a] | 100.0[e] |
| 1830s | 130.3[a] | 127.8[a] | 137.7[b] | 111.0[a] | 123.5[b] | 119.4[a] | 130.5[b] | 115.2[c] | 114.4[d] | 132.0[a] | 173.0[e] |
| 1840s | 109.6[a] | 100.1[a] | 119.6[b] | 82.8[a] | 102.1[b] | 91.5[a] | 110.8[b] | 121.0[c] | 132.5[d] | 113.0[a] | 189.0[e] |
| 1850s | 137.2[a] | 116.7[a] | 133.0[b] | 119.5[a] | 140.6[b] | 118.1[a] | 136.8[b] | 97.8[c] | 126.0[d] | 141.0[a] | 273.0[e] |

[a] Base year 1820–29.
[b] Relative to industrial products; base year 1820–29.
[c] Price of grain relative to meat; base year 1820–29.
[d] Price of corn relative to pork; base year 1820–29.
[e] Price of grain relative to cloth; base year 1820–29.

*Sources*: Anne Bezanson, Robert D. Gray, and Miriam Hussey, *Wholesale Prices in Philadelphia 1784–1861* (Philadelphia, 1936), pp. 392, 394; Thomas Berry, *Western Prices Before 1861* (Cambridge, Mass., 1943); T. M. Adams, "Prices Paid by Farmers for Goods and Services and Received by Them for Farm Products, 1790–1871," University of Vermont and State Agricultural College, Vermont Agricultural Experiment Station (Feb. 1939).

To the extent that the relative price of food rose, one might argue that at least the income of farm operators must have increased. An increase in their income would increase their demand for food, conceivably offsetting their inclination to substitute away from food and toward industrial products. However, the price elasticity of demand for meat was probably greater than its income elasticity, given their twentieth-century values of 0.92 and 0.35, respectively.[53] Thus, if income and relative prices rose by about the same percentage, the substitution away from meat consumption would have been considerably greater than the inclination to increase consumption on account of the rise in income. Of course, farmers' income may not have risen at all. It is possible that middlemen received much of the gains from the increased prices.

Because of the vagaries of constructing an accurate index of real wages, the trend of this variable in the antebellum period is by no means unambiguous. Paul David and Peter Solar have it rising continuously through the 1850s and falling only in the 1860s.[54] According to Jeffrey Williamson and Peter Lindert, however, the real wages of unskilled urban workers declined between 1844 and 1856 as well as between 1844 and 1880 at a rate of 1.65 and 0.13 percent per annum.[55] George Warren and Frank Pearson, too, have constructed a real wage series that falls somewhat between the 1840s and 1860s.[56] In any event, falling per capita food consumption could have occurred even had real wages not been falling, because increases in the relative price of food could have sufficed to induce people to rearrange their diet. Of course, they were probably unaware that by doing so they were impinging on their nutritional status and health, because the importance of a balanced diet was not generally known by the populace.

The rise in the price of grain relative to meat in the 1830s and 1840s was probably caused by the fact that the demand for meat was more elastic than that of grain.[57] Hence, a smaller increase in the price of meat sufficed to discourage consumption sufficiently to equilibrate the market.[58] A desire to obtain cheaper calories led consumers to substitute grain for meat and contributed to the rise in the relative price of grain.[59]

[53] Henry Schultz, *Theory and Measurement of Demand* (Chicago, 1983); and M. Wold, *Demand Analysis* (New York, 1953), p. 265 as cited in Edwin Mansfield, *Microeconomics, Theory and Applications* (2nd edn. New York, 1975), p. 109.

[54] David and Solar, "A Bicentenary Contribution," pp. 1–80.

[55] Williamson and Lindert, *American Inequality*, p. 131.

[56] Warren and Pearson, *Prices*, p. 197.

[57] In the twentieth century, the price elasticity of beef is said to be about 0.92, while the price elasticity of wheat is 0.08. Mansfield, *Microeconomics*, p. 109.

[58] Another reason was that immigrants were primarily grain consumers; hence they were more likely to increase the demand for grain than for meat products. Bezanson, et al., *Wholesale Prices*, p. 24.

[59] In the 1880s, for instance, 25 cents could obtain 870 calories if used to purchase beef (sirloin), but 4,000 calories if used to purchase wheat bread. Massachusetts, Bureau of Statistics and Labor, *Seventh Annual Report*, Public Document No. 15 (Mar. 1886) p. 254.

In Europe, pressure on the food supply repeatedly led to a fall in the price of meat relative to grain.[60]

The degree to which cadets, with parents in higher status occupations, escaped relatively unscathed supports the argument that the decline in nutritional status was related to changes in economic conditions. While the sons of blue-collar workers experienced the greatest cumulative decline in stature, the sons of professionals maintained their stature with only slight fluctuations. This result is in keeping with the notion that the income distribution was becoming more skewed in antebellum America beginning in the 1810s.[61] Skewing of income distribution has been linked to declining nutritional status.[62] Consequently, changes in the average caloric intake do not fully reflect the incidence of dietary change among all social groups. With the skewing of the income distribution, the range and variance of caloric and protein intake increased as entitlements to food supplies were redistributed.

Urbanization, commercialization, as well as intraregional specialization, separated individuals from sources of food, which increased the number of people who had to pay not only for the production of food but also for transportation costs and the efforts of middlemen in bringing the food within reach. After 1840 ". . . the number of swine on Massachusetts farms fell off so sharply that by the 1855 census, swine are not even counted in 15 of the 31 towns."[63] The retail price of food was at least twice as much as the farm-gate price but was often even greater.[64] In addition, losses of animals in transit were significant on cattle trails as well as on railroads; therefore as interregional trade accelerated so did the waste of nutrients.[65] Moreover, animals consumed feed while transported, and when slaughtered provided extra-edibles, or offal, not consumed to the same degree in an urban environment as in a rural one.[66] Thus, as consumption shifted from rural to urban areas, nutrient

[60] Wilhelm Abel, *Agrarkrisen und Agrarkonjunktur* (1935; 2d edn., Hamburg, 1966).

[61] This pattern is supported by data which show that the cost of living of the poor relative to the cost of living of the rich increased somewhat in the 1840s and 1850s. There was considerable annual variation in this index, but between 1844 and 1855 the ratio of the two indexes rose by 13 percent.

[62] Steckel, "Height and Per Capita Income"; Williamson and Lindert, *American Inequality*, pp. 68–70.

[63] Winifred Rothenberg, "A Price Index for Rural Massachusetts, 1750–1855," this JOURNAL 39 (Dec. 1979), pp. 975–1001. By 1830 farmers in Massachusetts were buying flour rather than growing grain. Winifred Rothenberg, "The Market and Massachusetts Farmers, 1750–1855," this JOURNAL, 41 (June 1981), pp. 283–304.

[64] *Consumption of Food in the United States, 1909–1948*, U.S. Department of Agriculture Miscellaneous Publication No. 691 (Aug. 1949), p. 99.

[65] According to one source, losses in shipment might have been about 1 percent. Robert James McFall, *The World's Meat* (New York, 1927), p. 116; *Consumption of Food in the United States, 1909–1948*, p. 129.

[66] Holmes, *Meat Situation*, p. 118. Blood and offal, almost certainly consumed on farms, were turned into fertilizer by some meat packers. The fat of meat often went to the soapmaker. The loss in nutrients for human consumption must have been enormous because blood and offal made up 18 percent of the weight of a swine. Margaret Walsh, *The Rise of the Midwestern Meat Packing*

costs increased considerably. Moreover, wild game was an important source of nutrition for pioneers.[67] With urbanization and rising population densities, this source of nutrients diminished. Even farmers began to specialize and to buy certain processed foods whose nutritional value, unbeknown to them, could have been inferior.

It is true that the index of grain and meat prices relative to industrial prices,which rose 30 percent in the 1830s, fell in the 1840s, only to rise once again in the 1850s. This is not precisely the mirror image of the trend in stature because, at least in some groups of the society, stature was monotonically decreasing. There are two reasons why one would not expect the two variables to correlate perfectly. One is that stature is a cumulative measure of nutritional status. Thus the birth cohorts of the 1840s might have initially benefited from the temporary fall in food prices in the 1840s. But in the 1850s and 1860s, when they were teenagers and at a time when growth is as sensitive to protein intake as it is during early childhood, they would have suffered nutritional assaults and therefore would not have realized their growth potential. The other reason why the food-price index may not correlate perfectly with the stature index is that nutritional status can be transmitted to some extent from one generation to another. This means that although the birth cohorts of the 1840s did benefit from the falling food prices, they were born to parents who, as teenagers, had experienced the more adverse times of the 1830s. In sum, more than a decade of respite is needed for the stature of a population to recover.

## CONCLUSION

The diminution of stature of West Point Cadets during the antebellum period indicates a decline in their nutritional status. I have presented evidence that the decline is related to dietary changes brought about by the processes of rapid economic development. An examination of the diet of the population indicates that, at least after 1839, average calorie and protein intake declined and did not reach its earlier level again until the 1870s.

The average calorie and protein intake of the American population does appear to indicate an adequate diet. One might be misled into thinking that the population was consuming above the point at which a decline of a few hundred calories would make a substantial impact on its nutritional status. Yet evidence on the weight of the cadets contradicts the notion that they had unlimited access to food supplies. In fact, they were quite underweight: many weighed between 100 and 120 pounds. This finding is important because the weight of the cadets is the only

*Industry* (Lexington, 1982), p. 83. Massachusetts. Bureau of Statistics and Labor, *Seventh Annual Report*, Public Document, No. 15 (Boston, 1886), p. 260.

[67] Hilliard, *Hogmeat and Hoecake*, pp. 13, 71, 78.

incontrovertible evidence on the contemporaneous nutritional intake of the cadets relative to the claims on those nutrients, and for the period no estimate of caloric consumption can be raised to the same level of accuracy.

The daily caloric intake of adult males, implied by the estimates in Table 8, was almost 3,900 in 1839. Although such a diet was adequate for moderate work, it might not have been adequate for heavy work.[68] Thus even a loss of 400 calories during the next decade could easily have impinged on the number of calories available to satisfy the body's requirements for growth. Admittedly, the body's ability to absorb nutrients was influenced by the epidemiological environment. Still, changes in disease environment are not likely to have played a major role because there is no evidence that epidemiological changes would have been more potent among certain socioeconomic groups than in others.

Very rapid population growth coupled with urbanization and the expansion of the industrial sector increased the demand for food considerably. Because the agricultural labor force grew more slowly and because advances in labor productivity in the agricultural sector were still minuscule, the supply response, although admirable, did not suffice to maintain calorie and protein intake at the level of the 1820s. Particularly noticeable was the decline in meat consumption. Insofar as immigration was partly the cause of rapid population growth, it was indirectly responsible for the decline in nutritional status through the pressure it exerted on food supplies. The problem was exacerbated by the disruptions in the flow of nutrients during the Civil War and by the destructions it caused.

The quick pace of urbanization during the antebellum period and the ongoing process of specialization even within the agricultural sector meant that an ever increasing share of the population was separated from the source of food supply, in whole or in part. Transportation to urban markets might have doubled food prices and, in addition, meant an enormous loss of nutrients. Transportation prices fell during the course of these decades, but as the price index shows, urban food prices did not. Food prices did not rise as much as they would have in the absence of transportation improvements. The fall in transportation costs and the increase in urban demand also meant that food was brought to urban areas from greater distances than they were earlier. This, of course, also fostered the integration of markets.[69]

[68] The reader should be cautioned that no estimate exists for the calorie requirements of the American population in the nineteenth century. In order to arrive at such an estimate one would have to first calculate the work intensity of the various groups in the society.

[69] Diane Lindstrom, "American Economic Growth before 1840: New Evidence and New Directions," this JOURNAL, 39 (Mar. 1979), 289–302; Douglass North, *The Economic Growth of the United States, 1790 to 1860* (Englewood Cliffs, 1961), pp. 106, 251; Rothenberg, "The Market and

For farm operators new market opportunities meant an increase in income, and, at the same time, a fall in the relative price of industrial products which induced them to substitute away from the consumption of nutrients, particularly meat. Because the price elasticity of demand for meat was greater than its income elasticity, they reduced their meat intake voluntarily in exchange for other goods whose relative price had declined. The upshot was a decline in nutritional status even for this socioeconomic group.

For urban workers, as well as for farmhands, whose wages did not keep up with food prices, a decrease in food consumption was a natural reflex. Only the middle class appears to have overcome the problem of adjusting to the new economic conditions. For this group the Civil War, alone, interrupted its customary consumption of food. Perhaps their income rose fast enough to counteract the substitution effect; being better educated than other groups, perhaps they were more aware of the body's nutritional needs and were consequently more reluctant to substitute away from food consumption.

Dietary changes may very well have contributed to the decline in life expectancy during the antebellum period.[70] Of course, even if this turns out to be the case, it should not be interpreted to mean that the population consciously chose to become less healthy, because the nature and importance of a balanced diet was not yet widely known.

The above argument suggests that the human biological system can experience periods of stress even as aggregate output per capita grows significantly. Thus anthropometric measurements do not appear to be perfect proxies for the material well being of a population. Rather, they might be considered a component of the biological standard of living, which under certain conditions might diverge in significant ways from such conventionally defined standards as output per capita. During early stages of rapid industrialization there are reasons to believe that the diet of populations undergoes changes as the consumption of meat declines. There is nothing counterintuitive about the supposition that even in otherwise prosperous times the per capita consumption of meat decreases. Of course, to the extent that the dietary change had unforeseen health effects, one must consider this a hidden cost of economic development.

If the above argument holds up to further scrutiny, one will also have to revise estimates of growth of national product for the first half of the nineteenth century because these have been predicated on the assump-

---

Massachusetts Farmers, 1750–1855," pp. 283–304; Christopher Clark, "The Household Economy, Market Exchange and the Rise of Capitalism in the Connecticut Valley, 1800–1860," *Journal of Social History*, 13 (Winter 1979), pp. 169–90.

[70] Clayne L. Pope, "Native Adult Mortality in the U.S.: 1770–1870," in R. W. Fogel, ed., *Long-Term Changes in Nutrition and the Standard of Living* (Berne, 1986), pp. 76–85.

tion that food output per capita remained constant between 1800 and 1840.

These results for antebellum America, coupled with my findings concerning the Austro-Hungarian experience of a hundred years earlier, begin to form a pattern showing how rapid population growth, market integration, and urbanization in the early stages of industrialization could initially exert downward pressure on the nutritional status of a large segment of the population.[71] If this was the case in the United States, with its abundant fertile land and absence of customs and institutions that might have been inimical to economic expansion, one can easily imagine the strain experienced by European societies in prior centuries during episodes of rapid population growth. Although the caloric and protein intake of the American population did decline, it does not reflect badly on the responsiveness of the American agricultural sector. On the contrary, that the decline was relatively small in face of such rapid population growth and sectoral shifts and that the subsequent recovery turned out to be as quick as it was, should be considered a reflection of the successful adaptation to new circumstances on the part of the American economy.

# *Appendix*

NOTES TO TABLE 8

Calculation of calorie and protein production is restricted to products enumerated in the 1840 census, with allowance for dairy products. Seed to yield ratios are those of Strauss and Beane, most of which have become standard in the literature on value added in American agricultural production.[72] The output of cheese and butter in 1840 is estimated by multiplying the number of milk cows by 49 to obtain the quantity of butter produced per annum and by 16.67 to obtain the amount of cheese produced per annum (in pounds). Milk cows in 1840 were assumed to be 35 percent of the total cattle (without calves).[73] Milk production was estimated by multiplying the number of milk cows by the probable output per cow.[74] Milk consumption was estimated by subtracting the milk-equivalent of cheese and butter output from milk production.

*Animal Stock*

According to the 1860 census, approximately 10 percent of hogs, 7 percent of sheep, and 13 percent of cattle were kept in urban areas. Since the census regularly reported only the number of animals kept on farms, the census figures were increased by these

[71] A similar pattern is evident in industrializing Montreal. Ward and Ward, "Infant Birth Weight."

[72] Frederick Strauss and Louis Beane, "Gross Farm Income and Indices of Farm Production and Prices in the United States 1869–1937," U.S. Department of Agriculture, Technical Bulletin No. 703 (Dec. 1940); Gallman, "Commodity Output," p. 52.

[73] *Seventh Census*, "Report of the Superintendent of the Census" (Washington, D.C., 1853).

[74] Fred Batemen, "Improvement in American Dairy Farming, 1850–1910: A Quantitative Analysis," this JOURNAL, 28 (June 1968), pp. 255–73.

percentages in other census years. Since the livestock kept in urban areas was probably declining during the period, this assumption introduces a slight upward bias in meat consumption estimates toward the end of the period.

### *Beef*

Beef production was estimated by assuming a 20 percent slaughter ratio (the share of stock slaughtered in any one year) and a dressed weight of 400 pounds per head of cattle.[75] Northern cattle were thought to average 450 pounds dressed weight, but southern cattle were likely to have been closer to 300 pounds.[76] Changes in the assumed ratios are difficult to ascertain for the period covered by this study. By the early twentieth century, the dressed weight of cattle was about 500 pounds, and the trend was declining because cattle were being brought to market younger. Within a decade, the average decline in the dressed weight of cattle was about 25 pounds.[77] Yet, the slaughter ratio was increasing sufficiently so that the yield of meat per animal in stock rose from 100 pounds in 1899 to 1908, to 110 pounds in 1905 to 1909, to 121 pounds in 1910 to 1914.[78]

### *Mutton and Pork*

For mutton a 36 percent slaughter ratio and a 40-pound dressed weight was assumed.

For pork there is considerable variance in estimates of dressed weight and slaughter ratios. Estimates of dressed weight are in the range of 120 to 144 pounds. Estimates of the slaughter ratio range from 50 to 83 percent. Robert Gallman assumed a slaughter ratio of 0.5 and a net weight of 120 pounds, while Richard Sutch assumed a net weight of 144 pounds.[79] Average dressed weight of hogs on one southern plantation in the 1850s was 133 pounds.[80]

The average dressed weight of 3,494 hogs slaughtered in the early 1840s was 133 pounds, but in the 1850s it rose to 153 pounds for 7,718 animals. The overall average was 146 pounds. Commercial hogs in Cincinnati were said to weigh about 200 pounds live. Animals slaughtered on the farm weighed less than hogs slaughtered commercially.[81] I assume a slaughter ratio of 0.6 and a dressed weight of 140 pounds. This is greater than the 120-pound average dressed weight obtained in the early twentieth century.[82]

[75] Town and Rasmussen, "Farm Gross Product," p. 283. Calves are not included in the statistics.

[76] Dale Evans Swan, "The Structure and Profitability of the Antebellum Rice Industry: 1859," (Ph.D. dissertation, University of North Carolina, Chapel Hill, 1972), p. 230; Hilliard, *Hog Meat and Hoecake*, p. 129; U.S. Department of Agriculture, "Conversion Factors and Weights and Measures for Agricultural Commodities and Their Products" (Washington, D.C., May 1952).

[77] *Historical Statistics of the United States, Colonial Times to 1957*, p. 137; *Livestock and Meat Statistics, 1957*, pp. 148, 196.

[78] *Livestock and Meat Statistics*, 1957, p. 2.

[79] Richard Sutch, "The Care and Feeding of Slaves," in Paul A. David et al., eds., *Reckoning with Slavery* (New York, 1976), pp. 246–47; Robert W. Fogel and Stanley L. Engerman, *Time on the Cross: Evidence and Methods—A Supplement* (Boston, 1974), p. 95; Gallman, "Self Sufficiency."

[80] Swan, "The Structure and Profitability," pp. 212, 215. Slaughter ratio in the South was assumed to be 0.5.

[81] Hilliard, *Hog Meat and Hoecake*," pp. 101, 102, 261; Berry, *Western Prices*, p. 231.

[82] *Historical Statistics of the United States, Colonial Times to 1957*, p. 291.

*Corn Production*

There is considerable disagreement on the allocation of corn. Direct evidence is nonexistent. Corn output grew greatly after the 1840s, but is is not clear how much of this increase was used for human consumption. Town and Rasmussen assumed humans consumed 4.4 bushels per capita—about 17.5 percent of total product. According to the first actual tabulation undertaken in 1900, only 15.5 percent of output entered directly into value added.[83] According to Strauss and Beane, in the early twentieth century 18.5 percent of corn output entered directly into value added.[84] Yet not all of this was used as food, because in 1910 average human consumption of corn was between 1.0 and 1.5 bushels per capita.[85] To keep near these bounds, I assumed that per capita corn consumption was 4 bushels. This amounted to 18 percent of production in 1840, 16 percent in 1850, 15 percent in 1860, 20 percent in 1870, and 12 percent in 1880. These are relatively small shares of total product, but they must still be considered as upper bounds, because it would have been physically impossible for individuals to eat more. With this estimated diet, daily per capita food consumption amounted to more than 3 pounds of food (Table 8), more than 4 pounds per adult male, and this does not even include all vegetables or game. More food could not possibly have been ingested. In 1910 per capita food consumption was 4.4 pounds daily, but this included items not ingested because of waste or spoilage.[86]

The primary use of corn was for animal feed. Estimates of the quantities consumed per farm animal have a wide range. Hogs in Ohio were said to consume between 10 and 15 bushels, and in Illinois about 15 to 20 bushels of corn; in the South the average was nearer to 4 bushels.[87] The average was perhaps 8.25 bushels or 460 pounds of corn per hog, but in the absence of systematic evidence it is futile to attempt even educated guesses about the true corn consumption of hogs in the 1840s or during the subsequent decades.[88]

The great increment in corn production took place in the Midwest. Corn production more than doubled between 1840 and 1850 in Iowa, Missouri, and Illinois, almost doubled in Indiana, and increased by 75 percent in Ohio. In 1850 these states accounted for 215 million bushels or 36 percent of total corn production. The output was probably not shipped out of the state, or even out of the county where it was produced, unless the farm was near water transportation or a railroad.[89] Midwestern corn output must have been fed to local livestock, but this may not have resulted in much weight gain, because in the 1830s and 1840s cattlemen depended on the range for grazing. "Before the settlers encroached on the prairies, they were the domain of the herds. . . ." Thereafter,

83 Town and Rasmussen, "Farm Gross Product," p. 297.

84 Strauss and Beane, "Gross Farm Income," p. 37; according to "Consumption of Food, 1909–1949," p. 177, humans consumed only 8 percent of the corn crop.

85 Merrill Bennett and Rosemond Peirce, "Change in the American National Diet, 1879–1959," *Food Research Institute Studies*, 2 (May 1961), pp. 95–119; *Consumption of Food, 1909–1949*, p. 85.

86 Bennett and Peirce, "Change in the American National Diet," p. 115.

87 Hilliard, *Hog Meat and Hoecake*, p. 101; Allan Bogue, *From Prairie to Corn Belt* (Chicago, 1963), pp. 105, 112. Hogs were often fattened on acorns in the woods until five or six weeks before slaughter and were then turned into the cornfields; Thomas Berry, *Western Prices Before 1861* (Cambridge, Mass., 1943), p. 230.

88 McFall, *The World's Meat*, p. 141. According to one twentieth-century experiment it took 18 bushels of corn to produce a 225-pound hog; William Smith, *Pork Production* (New York, 1937), p. 435.

89 Bogue, *From Prairie to Corn Belt*, pp. 130, 234; Hilliard, p. 15; ". . . corn could stand the transportation costs for a moderate distance by canal, river or railway, but could not yet be economically brought from many farms to the local depots." The perishability of corn also predicated against long distance trade in it. Berry, *Western Prices*, p. 203.

stockmen concentrated increasingly on corn feeding.[90] Corn could have been a substitute feed for the prairie grasses and forage crops grown extensively in the Midwest without making a contribution to net meat production.

During the course of the century the taste for a leaner pork developed, which had the effect of lowering slaughter weights.[91] In the 1840s swine in Illinois were fed corn only for a few weeks prior to slaughter. With the disappearance of open grazing, corn consumption of pigs might have increased without causing their weight to increase. This, in fact, is plausible, because evidence from Cincinnati does not indicate a substantial increase in the slaughter weight of hogs.[92] It is difficult to confirm how quick the transformation from a low to a high corn feeding regime was. Nonetheless, in 1850 the 215 million bushels of corn produced in the Midwest could have been consumed by hogs. After a 10 percent deduction for seed and waste and further deductions for human consumption and commercial use, 21 bushels remain per swine. This is not outside the feasible corn consumption allowance, particularly because in the early twentieth century the ratio of the price of corn to the price of hog was such that 11 bushels of corn could be used economically to produce 100 pounds of pork.[93]

Even less is known about corn consumption by other animals. According to anecdotal accounts, a steer could eat as much as 75 bushels of corn.[94] In Ohio between 1840 and 1850 fat three-year-old cattle were fed 10 to 20 bushels of corn in March and April and allowed to graze during the summer and fall. Then for four or five months they were fed all the corn they could eat, about a half-bushel per day. This amounted to 80 bushels of corn per head slaughtered.[95] By 1860 consumption per cow was supposed to have declined to 25 to 40 bushels.

The problem of allocating corn to its various uses is complicated because, according to some accounts, there could have been such overproduction of corn that much of the crop was "left to rot," or used as fuel.[96] Corn was often not harvested, but hogs were allowed into the cornfield to feed on it, which led to much waste. More research is needed to understand corn consumption in the antebellum period.

The lower-bound estimates of calorie and protein consumption assume that none of the increment to corn production after 1840 was translated into meat production (Table 8). This supposes that the increased use of corn as feed was a substitute for declining pasture and forest lands used to feed livestock. The upper-bound estimates are predicated on the assumption that the increases in corn output after 1840 which cannot be accounted for by increases in the human or animal population was translated into meat production by increasing the yield of meat per animal (Table 12).

### *Weight of Livestock*

Circumstantial evidence does not support the notion that hogs were brought to market fatter in the late antebellum period than earlier. According to one source, the live weight of hogs in Cincinnati fluctuated around 200 pounds throughout the antebellum era.[97] On Massachusetts farms hogs had the following average live weight in the antebellum

[90] Bogue, *From Prairie to Corn Belt*, pp. 94, 101.
[91] Ibid., p. 109.
[92] The 1862 figure is most likely an outlier.
[93] McFall, *The World's Meat*, p. 143.
[94] Bogue, *From Prairie to Corn Belt*, p. 94.
[95] U.S. Census Office, *Agriculture of the United States in 1860: Eighth Census* (Washington, D.C., 1864), p. cxxxi.
[96] Diane Lindstrom, "Southern Dependence upon Interregional Grain Supplies: A Review of the Trade Flows, 1840–1860," *Agricultural History*, 44 (Jan. 1970), pp. 101–13; Bogue, *From Prairie to Corn Belt*, p. 131.
[97] Berry, *Western Prices*, p. 231.

TABLE 12
CORN PRODUCTION AND ITS PROBABLE USES
(millions of bushels)

| | 1840 | 1850 | 1860 | 1870 | 1880 |
|---|---|---|---|---|---|
| Production | 378 | 592 | 839 | 761 | 1755 |
| Seed | 19 | 30 | 42 | 38 | 88 |
| Human foods[a] | 68 | 93 | 126 | 152 | 225 |
| Hog feed[b] | 237 | 273 | 302 | 226 | 429 |
| Cattle feed[c] | 40 | 48 | 67 | 62 | 93 |
| Whiskey[d] | 13 | 35 | 55 | 70 | 75 |
| Exports | 1 | 7 | 4 | 2 | 98 |
| Unaccounted | 0 | 106 | 243 | 211 | 747 |
| Waste, spoilage | 0 | 10 | 24 | 21 | 75 |
| Substitute for other feed | 0 | 10 | 24 | 21 | 75 |
| Extra feed for hogs and cattle | 0 | 86 | 195 | 169 | 597 |

[a] Assuming 4 bushels consumed per capita, 4.5 in 1880.
[b] Assuming 9 bushels consumed per swine.
[c] Assuming 2.6 bushels consumed per head of cattle.
[d] Three bushels of corn yield one gallon whiskey, see Thomas Berry, *Western Prices Before 1861* (Cambridge Mass., 1943).
*Sources*: See text.

decades: 329.5 pounds in 1820s, 370.5 pounds in 1830s, 322.6 pounds in 1840s, and 384.9 pounds in 1850s.[98] A crude estimate of the slaughter weight of hogs in Cincinnati can be made by dividing the shipment of pork and bacon out of Cincinnati by the number of hogs brought to the city. According to such a calculation the slaughter weight increased by 12 percent in the 1860s and by another 8 percent in the 1870s.[99] This is considerably below the 16 percent extra weight allowed in the upper-bound estimates in these two decades (Table 8). To determine upper-bound estimates, I allocated all of the corn crop of 1839 among its various uses (Table 12). I then applied these ratios to the livestock and population totals subsequent to 1840. The quantities of corn production which remain unaccounted for ranged from 86 to 597 million bushels per annum after a 10 percent reduction was made for waste and another 10 percent was assumed to have served as substitutes for other feed.

To calculate the hypothetical increment to meat production brought about by this unaccounted portion of corn output, one can assume that it was used equally to produce pork and beef. I assume a marginal cost of 5 pounds of corn to produce 1 pound live weight of hog and 11 pounds of corn to produce 1 pound live weight of cattle. A dressing ratio of 70 and 50 percent was assumed.[100] In this respect, too, the evidence is not without contradictions. While the above-quoted marginal product of corn was broadly valid in a number of experiments, other experiments showed that corn without proper nutritional supplements was actually not a very good feed for swine.[101] At times swine suffered protein deficiency and their growth was stunted on an exclusive diet of corn.

[98] Personal correspondence with Winifred Rothenberg whose research of Massachusetts farm account books has yielded livestock weights. In addition to the data cited she found some eighteenth-century hog weights which indicate a large increase in weight between the eighteenth and nineteenth centuries.

[99] Walsh, *The Rise of the Midwestern Meat Packing Industry*, pp. 20, 34.

[100] Smith, *Pork Production*, p. 143. It took about a pound of corn per 100 live weight to keep a hog at constant weight, pp. 148, 173.

[101] Ibid., pp. 150, 172, 435.

Moreover, a corn diet increased fat more than the meaty parts, and the meat had an abnormally high component of water.[102] This might explain why the protein and calorie content of pork and beef was said to be less in the nineteenth century than that established by modern research.[103]

### *Nutritional Content of Food*

Modern sources were used to establish the nutritional content for all food products except meat.[104] The calorie content of meat has increased since the nineteenth century because of quality changes. Therefore nineteenth century conversion factors were used for beef, pork, and mutton.[105] A 10 percent allowance was made for noningested items.

[102] Ibid., pp. 171, 175.

[103] Holmes, *Meat Situation*, p. 16. The average caloric content of a pound of meat was said to be 1,040.

[104] Bernice Watt and Annabel Merrill, *Composition of Foods, Raw, Processed, Prepared*, U.S. Department of Agriculture, Agriculture Handbook No. 8 (Dec. 1983).

[105] Massachusetts, Bureau of Statistics and Labor, *Seventh Annual Report*.

XIV

Peter A. Coclanis
John Komlos

# Nutrition and Economic Development in Post-Reconstruction South Carolina

## An Anthropometric Approach

The determination of anthropometric historians to unearth the broad patterns of human biological well-being is now too well known to need reiteration. From Richard Steckel's exploratory essays, which could be taken as the launching manifesto of the discipline, to the most recent publications, many hundreds of thousands of records from nearly all continents of the globe have been examined (Mascie-Taylor 1991; Steckel 1979). All of this effort notwithstanding, we are still in the initial phases of this research

*Social Science History* 19:1 (spring 1995). 

program, with many important features of the agenda remaining unsettled (Floud, Wachter, and Gregory 1990; Fogel 1993; Komlos 1989; Mokyr and O'Grada 1988). One such question is related to the most startling discovery to date, namely, that as part of its adaptability to its socioeconomic environment, human physical stature has undergone cyclical fluctuations during the course of the last quarter of a millennium (Fogel 1986; Komlos 1987, 1994a, 1994b; Nicholas and Oxley 1993).

On American soil a major downturn in human stature has been identified beginning among the birth cohorts of the 1830s, though its geographic and social correlates have not yet been investigated in detail. It is possible, for instance, that African Americans led the cycle (Komlos 1992). As with business cycles, moreover, the nature and causal connections of height cycles remain controversial, even if their existence is no longer questioned (Komlos 1993a, 1993b). The downswing in height of the 1830s, for instance, is considered an anomaly inasmuch as the American economy—and the West in general—was at the time undergoing one of the most powerful and sustained economic expansions in history. Nonetheless, it now seems clear that the biological standard of living failed to keep pace with the overall growth in per capita income. If scholars cannot yet fully explain this anomaly, research is continuing to broaden our knowledge about such phenomena. The present study is an effort in this direction: it explores the upturn of the American nutritional cycle at the end of the nineteenth and the beginning of the twentieth centuries. So far only two studies have been devoted to the analysis of the secular trend in human stature in the late nineteenth and early twentieth centuries (Steckel and Haurin 1994; Wu 1994). Both of these studies are geographically limited (to Ohio and to Pittsburgh), therefore we need to expand our knowledge in this area to other regions of the country. The goal here is to identify as precisely as possible when physical stature began to increase, and to explore the socio-economic correlates of the upturn.

## Data

The records of the Alumni Association of The Citadel, The Military College of South Carolina, serve as the source of the data included in this study.[1] The Citadel, located in Charleston, was established by an act of the General Assembly of South Carolina in December 1842 and opened its doors in March of the following year (Act to Convert, 1842, 244–45). The school provided military and academic training to small numbers of student cadets during the remainder of the antebellum period and throughout most of the Civil War, before suspending operations in February 1865 after Federal troops occupied Charleston and seized the institution's buildings and grounds.

These troops remained stationed at The Citadel until April 1879; the school did not reopen until the fall of 1882. The records analyzed here begin with the school's reopening after this seventeen-year hiatus (Bond 1936: 17–104; Thomas 1879: 2–73; Act to Authorize, 1879–82, 727–29).

Before moving on to the data and results, it is important to say a few words about the social composition of the corps cadets at The Citadel during its early years. While higher education in the nineteenth-century South, generally speaking, was solely an upper- and middle-class prerogative, The Citadel drew its student cadets from a broader social spectrum[2] (Coulter 1928: vii–viii; Hollis 1951–56, 1: 255–70). This was so for several reasons. Because student cadets in the early years were as much militia auxiliaries and guardians of the state's arms as they were scholars, and because of the heavy curricular emphasis on engineering and military arts, The Citadel, it seems, appealed to students from a more diverse pool than did nineteenth-century institutions with more "elevated" belletristic concerns. This point is underscored by the fact that this public military school allowed numerous young men to enroll *gratis* as "beneficiaries" or scholarship students "selected from those not able to bear their own expenses" (Bond 1936: 24; statutes at Large, 1873: 244–45).[3] So-called beneficiaries were not necessarily destitute; most, in fact, came from "respectable" yeoman families and from *petit bourgeois* groups in the cities and small towns of the Palmetto State. To be sure, the relative proportion of beneficiaries—less wealthy, and possibly less physically robust, cadets—declined as the school's enrollment increased in the first three decades of this century (Bond 1936: 156, 161, 162, 166, 167, 185, 189, 208, 209, 214). We are unable to isolate the group of beneficiaries from our sample as a whole and, hence, cannot estimate precisely their impact on observed trends. It is unlikely, however, that beneficiaries constituted a socioeconomic subgrouping sufficiently distinct from other cadets as to make a meaningful difference in our results. Clearly, though, The Citadel was one institution of higher learning in the South that catered to more than a narrow elite.

Table 1 Characteristics of the sample

| | % | N | % | N |
|---|---|---|---|---|
| **Birth decade** | | | | |
| 1860s | 3.3 | 218 | | |
| 1870s | 4.6 | 299 | | |
| 1880s | 5.1 | 336 | | |
| 1890s | 9.7 | 638 | | |
| 1900s | 6.8 | 449 | | |
| 1910s | 1.3 | 87 | | |
| 1920s | 49.4 | 3234 | | |
| 1930s | 19.7 | 1289 | | |
| Total | 100.0 | 6550 | | |
| **Age** | | | | |
| 15 | 3.5 | 227 | | |
| 16 | 13.9 | 913 | | |
| 17 | 31.4 | 2054 | | |
| 18 | 30.9 | 2023 | | |
| 19 | 11.7 | 769 | | |
| 20 | 4.6 | 300 | | |
| 21 | 1.6 | 106 | | |
| 22–25 | 2.4 | 158 | | |
| **Geographic distribution: Regional breakdown** | | | | |
| | Place of birth | | Place of residence | |
| North | 17.8 | 1164 | 15.2 | 994 |
| New England | 2.3 | 149 | 1.8 | 119 |
| Mid Atlantic | 10.8 | 705 | 9.7 | 634 |
| Midwest | 4.7 | 310 | 3.7 | 241 |
| South | 54.6 | 3578 | 77.0 | 5042 |
| South Carolina | 28.5 | 1867 | 50.6 | 3314 |
| Upper South | 11.7 | 765 | 11.3 | 740 |
| Lower South | 14.4 | 946 | 15.1 | 988 |
| West | 1.9 | 125 | 1.1 | 72 |
| Puerto Rico | 1.6 | 103 | 1.4 | 94 |
| Unknown | 24.1 | 1580 | 1.3 | 84 |
| **South Carolina breakdown** | | | | |
| Charleston | 19.9 | 364 | 18.7 | 599 |
| Upper Piedmont | 17.7 | 324 | 17.3 | 554 |
| Lower Piedmont | 11.7 | 215 | 12.6 | 405 |

**Table 1** Continued

| | % | *N* | % | *N* |
|---|---|---|---|---|
| Sand Hills | 13.1 | 239 | 13.0 | 417 |
| Upper Pine Belt | 23.0 | 421 | 23.6 | 757 |
| Lower Pine Belt | 11.4 | 209 | 10.6 | 338 |
| Coast | 3.2 | 58 | 4.1 | 131 |
| Total | 100.0 | 1830 | 99.9 | 3201 |

The data, compiled from Citadel Alumni Association records, consist of observations on 6,550 cadets for the period from the early 1880s through the 1950s[4] (Table 1). Throughout this period, the minimum-height requirement for admission into The Citadel was five feet; there were no rules pertaining to maximum height.[5] Because the information on each cadet was taken upon his initial enrollment at the school, our data pertain to birth cohorts stretching from the 1860s to the 1930s. Two other attributes of the data set should be noted at this time: 55 percent of the students were born in the South, and almost 80% resided there at the time of application to the school, and for unknown reasons the school failed to maintain these records during the late 1930s and early 1940s. As a consequence the files we obtained included only 87 observations for those born in the 1910s, which limits somewhat the precision with which we can date the upturn in physical stature among the cadets.[6] However, the broad trends are unaffected by this anomaly of the sample.

## Results

We constructed an index by setting the average height in each age category equal to 100 for the whole sample, and then averaging these age-specific indices in order to obtain the overall average (Table 2). The trend in physical stature indicates little or no improvement among students born in the late nineteenth century. This generalization is true not only for the age-standardized composite index, but also in the main, for the separate age groups.

To be sure, a slight rebound is apparent among those born during the early phase of Reconstruction, but this was reversed, in the main, by the turn of the century (Figure 1). A local height nadir was experienced by

**Table 2** Height by age of students, 1860–1939

| | Age | | | | | | Index | |
|---|---|---|---|---|---|---|---|---|
| Birth decade | 15 | 16 | 17 | 18 | 19 | 20 | All | SC[a] |
| 1860s | 64.8 | 66.7 | 67.5 | 68.5 | 68.7 | — | 98.0 | 97.8 |
| 1870s | 65.3 | 66.5 | 68.0 | 68.4 | 69.5 | — | 98.4 | 99.0 |
| 1880s | 65.4 | 66.9 | 67.9 | 68.2 | 69.1 | — | 98.3 | 97.8 |
| 1890s | 66.6 | 66.4 | 67.4 | 67.8 | 68.2 | 68.2 | 97.8 | 97.6 |
| 1900s | 66.1 | 66.9 | 67.3 | 68.1 | 69.2 | — | 98.0 | 97.6 |
| 1910s | — | 68.4 | 69.2 | 70.1 | 70.3 | 69.3 | 99.8 | 97.6 |
| 1920s | 69.6 | 69.6 | 69.9 | 70.1 | 70.2 | 70.4 | 100.8 | 100.9 |
| 1930s | — | 69.3 | 70.2 | 70.5 | 70.3 | 70.5 | 101.0 | 101.2 |
| Before 1910 | 65.6 | 66.6 | 67.5 | 68.1 | 68.9 | 68.8 | 98.1 | 97.9 |
| After 1910 | — | 69.5 | 69.9 | 70.3 | 70.3 | 70.3 | 100.9 | 101.0 |
| All | 65.9 | 67.8 | 69.3 | 69.8 | 70.0 | 70.1 | 100.0 | 100.3 |

[a]Born in South Carolina.

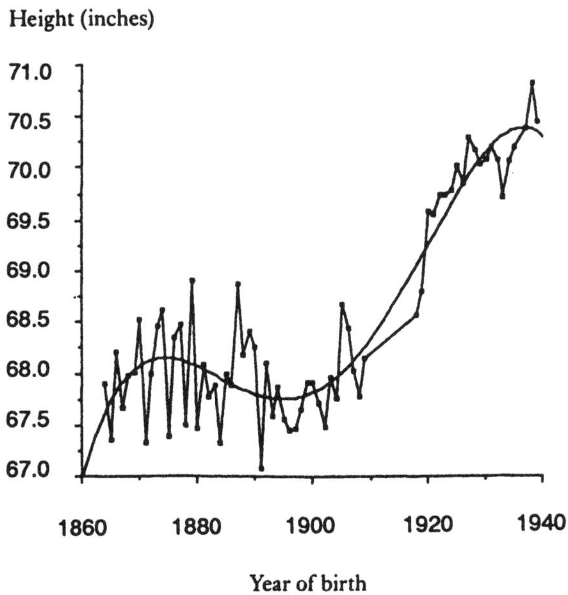


**Figure 1** Annual index of physical stature, and a fourth-degree polynomial estimate

Table 3 Growth increments of students

| Decade | 16 | 17 | 18 | 19 | 20 | Cumulative |
|---|---|---|---|---|---|---|
| 1860s | 1.9 | 0.8 | 1.0 | 0.2 | | 3.9 |
| 1870s | 1.2 | 1.5 | 0.4 | 1.1 | | 4.2 |
| 1880s | 1.5 | 1.0 | 0.3 | 0.9 | | 3.4 |
| 1890s | −0.2 | 1.0 | 0.4 | 0.4 | | 1.6 |
| 1900s | 0.8 | 0.4 | 0.8 | 1.1 | | 3.1 |
| 1920s | 0.0 | 0.6 | 0.2 | 0.1 | 0.2 | 1.1 |
| 1930s | n.a. | 0.9 | 0.3 | −0.2 | 0.2 | 0.8 |

the birth cohorts of the 1890s.[7] In contrast, immense improvements were registered by the third decade of the new century. Between the 1900s and 1920s the average physical stature increased by some 2.8%—nearly two inches. Precisely when this improvement occurred is not ascertainable on the basis of this evidence because of the aforementioned dearth of observations pertaining to the 1910s, but it is clearly evident by 1920.

During the late-nineteenth century, physical growth continued well into the late teenage years, with about four inches of growth taking place even after age 16 (Table 3). This prolongation of growth is in itself an indication of less than optimal nutritional status. By the 1920s, however, terminal height was reached by age 18 or 19, the pattern obtained in today's population. The fact that cumulative growth after age 16 diminished over time is indicative of an improvement in nutritional status.

The increase in nutritional status among the post-1910 birth cohort was most noticeable among the fifteen- and sixteen-year-olds, whose stature increased by 3.5 and 2.7 inches respectively, an obvious indication of an earlier completion of the adolescent growth spurt (Table 3). In the earlier epoch, the growth profile is quite steep in the late teenage years, a sign of late maturation[8] (Figure 2). At age 19 the difference in favor of the 1920s cohort was still 1 inch, even though this cohort lived through the hard times of the Great Depression.

Evidently, there was not a relapse to nineteenth-century conditions among the 1930s birth cohort. Instead, nutritional status appears to have leveled off among southerners during this decade. Although the decadal index shows minor improvement (Table 2), a more detailed examination re-

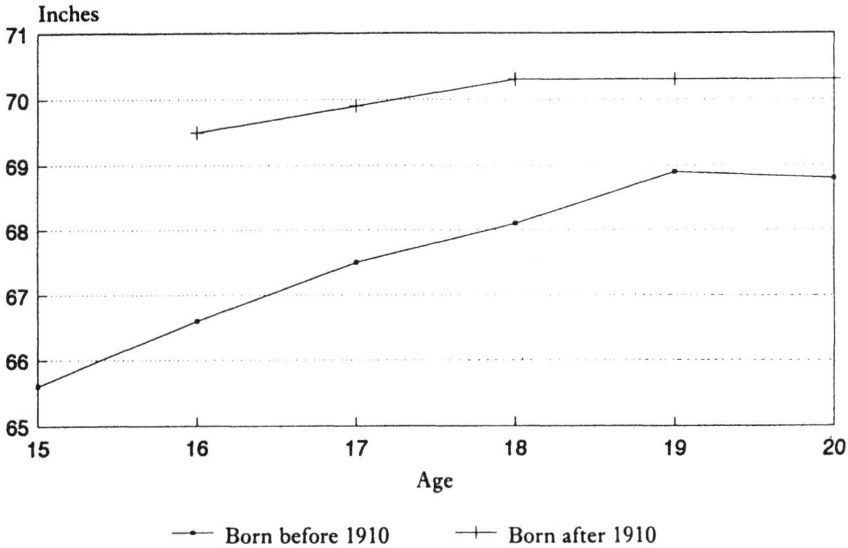


**Figure 2** Age by height profile of students

veals that the steep increase of the early 1920s had tapered off by the middle of the 1920s (Figure 1). At this point it seems useful to reiterate that the increase in height among the 1920s cohort was not caused by changes in the minimum height requirement for entrance into the school. Moreover, it should be noted that the height distributions of this decade remain approximately normal without any signs of truncation.[9]

The trend among those born in South Carolina was not appreciably different from the rest of the sample (Table 2). However, until 1920 the nutritional status among those born in that state was, in the main, below that of the rest of the country. By the 1920s this nutritional disadvantage had disappeared and South Carolinians were about as tall as the other students at The Citadel.

The data set does not contain sufficient information to permit us to explore regional variation within the United States prior to 1920. In this sample there does not appear to have been much variation in height by the 1920s (Table 4). This is in part due to the fact that many students born elsewhere were residents of South Carolina or other parts of the South for varying lengths of time prior to enrollment in the school (Table 1). Thus, they had

**Table 4** Index of height by region within the United States

| | 1920s | 1930s |
|---|---|---|
| North | 100.8 | 101.1 |
| Mid Atlantic | 100.8 | 100.9 |
| New England | 100.5 | 101.5 |
| Midwest | 101.0 | 101.4 |
| South | 100.9 | 101.2 |
| South Carolina | 100.9 | 101.2 |
| Deep South | 101.1 | 101.0 |
| Upper South | 100.9 | 101.4 |
| West | 100.4 | 101.0 |
| Puerto Rico | 98.9 | 98.8 |

a common experience both in terms of family earning opportunities and in terms of disease exposure experience. Only persons born in Puerto Rico stand out as having been much shorter than average; they were 1.4 inches shorter than South Carolinians (Table 5).

Comparison with the heights of northern white men found in other data sets indicates that the southern students at The Citadel were only slightly taller than cadets at the West Point Military Academy (Table 6 and Figure 3). In other words, by the late nineteenth century, southerners had lost the substantial height advantage they had enjoyed earlier in the century.

Within South Carolina the regional distribution of heights among the post-1910 cohorts is the same whether the data are examined by place of residence or by place of birth, with the single exception of the coastal region (Figure 4). Those who were born in this group of counties were substantially taller than those born outside of the coastal region, but who resided there. Students from the Upper Piedmont were the tallest and those from the Upper Pine Belt—an extremely poor section of South Carolina—were the shortest, a result borne out by regression analysis[10] (Table 7).

While the ranking is not identical among those born prior to 1910, it is similar to the one obtained for the post-1910 birth cohort. Those born or residing in the Piedmont or Sand Hills tended to be taller than average, but in the late nineteenth century students from the Upper Pine Belt were not among the shortest students within South Carolina. Among the pre-1910 birth cohort, too, the pattern is essentially the same whether the geographic

**Table 5** Regression estimates of students' height and BMI[a]

| | By birthplace | | By residence | |
|---|---|---|---|---|
| | Height | BMI | Height | BMI |
| Constant | 68.0* | 20.2* | 68.0* | 20.2* |
| Region | | | | |
| Mid Atlantic | 0.1 | 0.5* | 0.1 | 0.6* |
| Midwest | 0.3* | 0.6* | 0.1 | 0.7* |
| New England | 0.1 | 0.7* | −0.1 | 1.0* |
| Puerto Rico | −1.4* | −0.2 | −1.5* | −0.2 |
| Lower South[b] | 0.2 | −0.1 | 0.1 | 0.1 |
| Upper South | 0.2* | 0.1 | 0.2* | 0.1 |
| West | −0.1 | −0.2 | −0.2 | 0.2 |
| Age | | | | |
| 15 | −2.2* | −0.4* | −2.2* | −0.4* |
| 16 | −1.0* | −0.5* | −1.0* | −0.5* |
| 17 | −0.3* | −0.4* | −0.3* | −0.4* |
| 19 | 0.1 | 0.4* | 0.2 | 0.4* |
| 20 | 0.2 | 1.0* | 0.2 | 1.0* |
| Birth decade | | | | |
| 1860s | −0.2 | —[c] | −0.2 | —[c] |
| 1870s | −0.1 | 0.2 | −0.1 | 0.2 |
| 1880s | 0.0 | −0.0 | 0.0 | −0.0 |
| 1890s | −0.3** | −0.4* | −0.3** | −0.4* |
| 1910s | 1.4* | 1.3* | 1.4* | 1.2* |
| 1920s | 2.1* | 1.5* | 2.1* | 1.4* |
| 1930s | 2.4* | 2.0* | 2.4* | 2.0* |
| $R^2$ | 0.2 | 0.2 | 0.2 | 0.2 |
| F | 104.5* | 63.0 | 104.6 | 63.7 |
| *N* | 6285 | 5770 | 6285 | 5770 |

Note: Students older than twenty were not included in the regressions. The constant refers to an eighteen-year-old student born in the decade of 1900 and born in or resident of South Carolina.

*Significant at the 5 percent level.

**Significant at the 10 percent level.

[a]Weight divided by height squared in metric units.

[b]Without South Carolina.

[c]Not included in the regressions.

**Table 6** Comparison of physical stature in the U.S.

| Birth decade | Citadel[a] | West Point[a] | Pittsburgh[b] | Ohio Guard |
|---|---|---|---|---|
| 1860s | 68.7 | 67.4 | | 68.5 |
| 1870s | 69.5 | 68.0 | | |
| 1880s | 69.1 | | 68.0 | |
| 1890s | 68.2 | | 68.4 | |
| 1900s | 69.2 | | 68.7 | |
| 1910s | 69.3 | | 69.2 | |
| 1920s | 70.4 | | 69.6 | |
| 1930s | 70.5 | | 70.2 | |

Sources: See Table 2, and Komlos 1987; Wu 1994; Steckel and Haurin 1994.
[a]At age 19 or 20.
[b]Adult heights of white men.

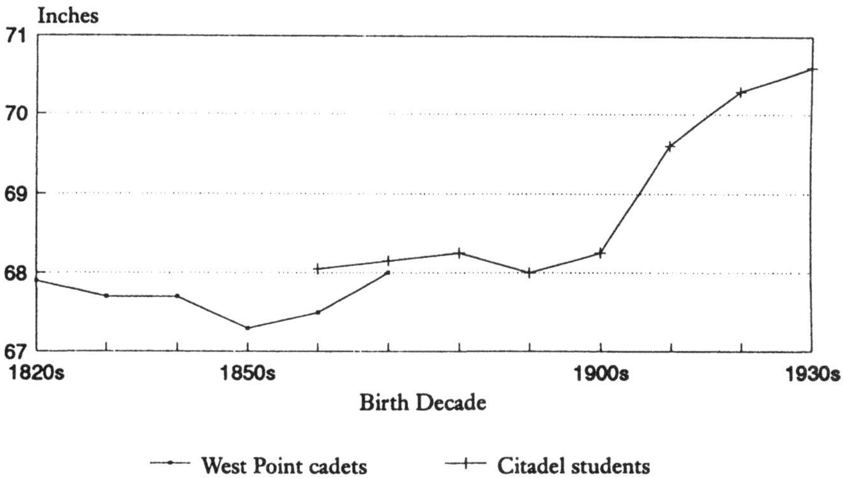


**Figure 3** Heights of American students at age 19, 1820–1940

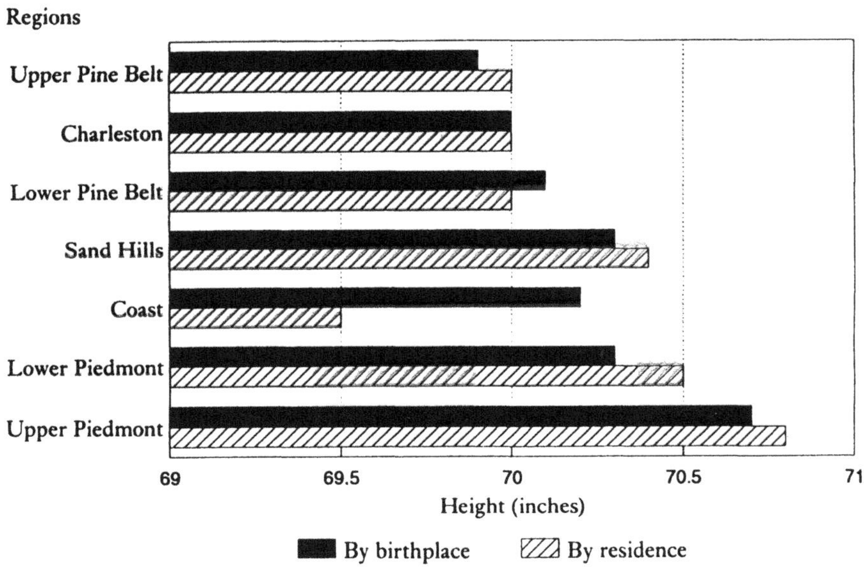


**Figure 4** Height by region, 1920s birth cohort, 18-year-olds

distribution of physical stature is examined by place of birth or by place of residence.

The body-mass index (BMI) is another measure used by auxologists to assess nutritional status. In contrast to physical stature, which is an historical record of nutritional status, BMI is an indicator of contemporaneous nutrient intake. (It is calculated by dividing the weight of an individual in kilograms by his height squared in meters.) The BMI is related to the well-being of an individual insofar as it has been found to be a good predictor of mortality (Fogel 1993; Costa 1993). More specifically, the optimal value of the BMI index lies between 20 and 26. Mortality risk rises with both people who are too slight as well as those who are obese as indicated by their BMI values. The BMI values of Citadel students were about the same as those found among West Point cadets, but considerably below modern standards (Cuff 1994). In fact, until the turn of the century, the mean of the BMI index was below the lower bound of the above-mentioned optimal range. In other words, more than half of The Citadel students were so underweight that their mortality risk was increased as a consequence. The BMI of students at The

Table 7 Variation of height and BMI[a] within South Carolina

| | By birthplace | | By residence | |
|---|---|---|---|---|
| | Height | BMI | Height | BMI |
| Constant | 67.6* | 20.5 | 67.9* | 20.2* |
| Region | | | | |
| Upper Piedmont | 0.7* | −0.2 | 0.8* | −0.1 |
| Lower Piedmont | 0.3** | −0.1 | 0.5* | −0.1 |
| Sand Hills | 0.3 | −0.2 | 0.4* | 0.1 |
| Upper Pine Belt | −0.1 | −0.2 | −0.0 | 0.0 |
| Lower Pine Belt | 0.1 | −0.1 | 0.0 | 0.1 |
| Coast | 0.2 | 0.5 | −0.5* | 0.5* |
| Age | | | | |
| 15 | −1.6* | −0.8** | −2.2* | −0.6* |
| 16 | −1.1* | −0.6* | −1.2* | −0.6* |
| 17 | −0.4* | −0.5* | −0.6* | −0.4* |
| 19 | 0.2 | 0.3 | 0.3* | 0.3** |
| 20 | 0.5 | 0.9* | 0.2 | 0.6* |
| Birth decade | | | | |
| 1860s | 0.1 | —[b] | −0.1 | —[b] |
| 1870s | 0.5 | 0.1 | −0.1 | 0.3 |
| 1880s | 0.1 | −0.4 | −0.1 | 0.0 |
| 1890s | −0.2 | −0.5 | −0.3** | −0.3** |
| 1910s | 1.9* | 1.1** | 1.6* | 1.9* |
| 1920s | 2.4* | 1.4* | 2.1* | 1.5* |
| 1930s | 2.8* | 1.7* | 2.4* | 1.9* |
| $R^2$ | 0.2 | 0.1 | 0.3 | 0.1 |
| F | 27.5* | 12.2 | 68.9 | 28.5 |
| *N* | 1780 | 1779 | 3315 | 2903 |

Note: The constant refers to an eighteen-year-old student born in the decade of 1900 and born or resident in Charleston.
*Significant at the 5 percent level.
**Significant at the 10 percent level.
[a]Weight divided by height squared in metric units.
[b]Not included in the regressions.

Citadel exhibits the same secular trend as height (Tables 5, 7–9). The setback of the 1890s is evident, as well as the absence of an overall improvement in the late nineteenth century. Like stature, the BMI increased markedly in the 1920s. The fact that the increases in both BMI and physical stature took place at the same time supports the notion that these increases mirror nu-

**Table 8** Weight (pounds) by age of students, 1865–1939

| | Age | | | | | |
|---|---|---|---|---|---|---|
| Birth decade | 15 | 16 | 17 | 18 | 19 | 20 |
| 1870s | 118.5 | 125.0 | 133.9 | 134.2 | 144.4 | — |
| 1880s | 115.6 | 125.2 | 134.0 | 133.5 | 138.8 | — |
| 1890s | 127.3 | 121.4 | 125.2 | 129.8 | 133.0 | 132.1 |
| 1900s | 118.1 | 125.7 | 127.6 | 134.9 | 139.3 | — |
| 1920s | — | 146.0 | 148.3 | 152.5 | 156.1 | 162.4 |
| 1930s | — | 146.9 | 153.1 | 157.9 | 160.0 | 167.5 |
| All | 124.9 | 133.8 | 143.9 | 151.0 | 154.1 | 158.8 |

**Table 9** Body mass index[a] of students, 1860–1939

| | Age | | | | | | Index | |
|---|---|---|---|---|---|---|---|---|
| Birth decade | 15 | 16 | 17 | 18 | 19 | 20 | All | SC[a] |
| 1870s | 19.5 | 20.0 | 20.2 | 20.2 | | — | 98.4 | 99.0 |
| 1880s | 19.0 | 19.6 | 20.4 | 20.2 | 20.4 | — | 98.3 | 97.8 |
| 1890s | 20.2 | 19.4 | 19.4 | 19.9 | 20.1 | 20.0 | 97.8 | 97.6 |
| 1900s | 19.1 | 19.8 | 19.8 | 20.5 | 20.5 | — | 98.0 | 97.6 |
| 1920s | 22.1 | 21.2 | 21.4 | 21.8 | 22.3 | 23.0 | 100.8 | 100.9 |
| 1930s | — | 21.6 | 21.9 | 22.3 | 22.8 | 23.6 | 101.0 | 101.2 |
| Before 1910 | 19.6 | 19.6 | 19.8 | 20.2 | 20.4 | 20.0 | | |
| After 1910 | 22.3 | 21.2 | 21.5 | 22.1 | 22.5 | 23.1 | | |
| All | 19.9 | 20.3 | 21.0 | 21.7 | 22.2 | 22.7 | 100.0 | 100.3 |

[a]Weight divided by height squared in metric units.

tritional reality and were not caused by some statistical artifact. Northerners were somewhat heavier for their height than southerners (Table 5). Within South Carolina there does not appear to have been much regional variation in the BMI, except that those individuals born or residing on the coast were somewhat heavier for their height than average (Table 7).

## Discussion

The short-term cycles in nutritional status apparent among the students of The Citadel are practically identical to the pattern found among soldiers of the Ohio National Guard (Steckel and Haurin 1994).[11] An increase in physical stature in the 1870s was followed by a decline in the 1880s and 1890s and a stabilization by the turn of the century. Thus, there is no evidence among either Ohio soldiers or mostly southern students of any permanent improvement in nutritional status during the second half of the nineteenth century. On the contrary, it appears that moderate advances were followed by setbacks in nutritional status during the 1890s. Physical stature did not begin to improve appreciably until the second or third decade of the century.

In contrast to the stagnation in physical stature found among students at The Citadel, the nutritional status of men born in Pittsburgh did increase continuously between about 1890 and 1920. Yet, there are similarities as well between the two samples, insofar as the nutritional experience of the 1920s and 1930s are concerned. In addition, the effect of the Great Depression and of World War II on the nutritional status of both groups of men was relatively minor. It is now clear that the American population did not suffer a reversal in its nutritional status even during the most severe downswings in economic activity of the twentieth century (Wu 1994). This finding is consistent with what we know of the improvements in public health during the period, as well as with the fact that food consumption was a relatively minor share of the consumer's budget by the 1930s. While disposable income declined by some 30 percent between 1929 and 1933, expenditure on food decreased by half as much. That meat consumption did not change at all, and that the consumption of other food items was only marginally affected, is perhaps even more telling (U.S. Department of Agriculture 1949: 136–40). In other words, with the income levels reached in the twentieth century, temporary economic downturns did not influence consumption patterns sufficiently to affect permanently the nutritional status of the U.S. population.

To what can one attribute the broad changes in nutritional status revealed in the Citadel data? More specifically, how can one explain the declining physical stature of cadets born in the 1880s and 1890s and the increase in stature among those born in the second and third decades of the twentieth century? Given the data limitations discussed earlier, it is impossible to issue categorical answers to such questions at this time. Enough is known, however, about both the South in general and South Carolina in particular during this period to suggest some of the variables that were likely involved.

We have long known, for example, that the late nineteenth century was a time of extreme economic duress in the South. Richard Easterlin's classic studies on interregional income have shown that the South Atlantic census region was the poorest in the U.S. in both 1880 and 1900 (though it is revealing that this was no longer true by 1920)[12] (Easterlin 1960: 73–140; 1961: 525–47). Clearly, the collapse of short-staple cotton prices, the concomitant rise in cotton-corn ratios, the decline in southern livestock herds, and the chronic shortage of capital in the region together could have created conditions sufficient for nutritional decline not just in the South Atlantic states but throughout the South (James 1981: 443–54; Ransom and Sutch 1977; Wright 1978: 158–84, 1986: 3–197; United States Department of Agriculture 1916 209–10). It should be noted, moreover, that South Carolina, home of The Citadel, faced additional economic hardship because of the demise of two local mainstays: the rice and long-staple cotton industries. Indeed, a number of scholars have argued that economic difficulties were so severe in the Palmetto State in the postbellum period that real per capita income and wealth were lower in 1880 than in 1860 and reached the 1840 per capita average of $56 only in 1900.[13]

Still other factors seem to have been involved in bringing about the decline in stature among Citadel cadets born in the late nineteenth century. According to most historians of medicine and public health, the "disease environment" of the South, which long was characterized by high levels of morbidity and mortality, became even more lethal in the decades after the Civil War[14] (Breeden 1988). It is not altogether surprising, of course, that a region devastated by four years of combat, traumatized by emancipation and the ensuing transformation of social relations, pinched by decades of economic hardship, and enervated by a severe lack of capital, would find itself in difficult epidemiological straits. With poverty widespread, urban population and pathologies increasing, and a minimal regional commitment to matters of public health, most of the South's "traditional" diseases—malaria, typhus, diphtheria, smallpox, etc.—were surging in the late nineteenth century, and "newer" diseases, most notably tuberculosis, were becoming deadly as well. Indeed, of the South's great killers, only yellow fever seems to have abated in the late nineteenth century, and, even then, only after the great epidemic of 1878[15] (Breeden 1988, Duffy 1988, Carrigan 1988, Waring 1964–71).

The diseases mentioned above led not only to higher levels of mortality in the region, but, in many cases, to chronic health problems among those fortunate enough to survive them. Such problems, in turn, were often compounded by the dietary regime characteristic of most parts of the South in the late nineteenth century. This regime, of course, was dominated by the so-called "three M's"—meat, meal, and molasses—which meant a diet very high in fats and carbohydrates, but low in minerals, vitamins, and (because of the poor quality of pork, the South's principal meat) proteins[16] (Fite 1984: 30–47; Hilliard 1972: 62–69). For our purposes here it is sufficient to note that the epidemiological environment, characteristic dietary practices, and economic and social conditions prevailing in the late-nineteenth-century South resulted in a variety of deficiencies and wasting diseases—chronic malaria, hookworm, pellagra, and rickets in particular—among a sizable proportion of the southern population (Beardsley 1987: 11–61, Duffy 1981, Etheridge 1972, Ettling 1988, Fite 1984: 30–47, Marcus 1988).

The small and slight stature of Citadel cadets born in the 1880s and 1890s was likely related to the factors mentioned above. This hypothesis seems particularly plausible in light of the fact that a sizable proportion of cadets at the time, as we have seen, were so-called beneficiaries, drawn from socioeconomic groups at greater risk to suffer from the debilitating problems present in the late nineteenth-century South. One last point is also instructive in this regard: there was an extremely high rate of medical rejections among both white and black South Carolinians mobilized for the Spanish-American War. The most common reason for rejection among men examined in Columbia, the state capital, was failure to meet the minimum weight requirement of 125 pounds[17] (Bailey, Jr. 1991; Beardsley 1987: 11–61; Cosmas 1971: 127–28; Edgar 1992: 19).

This dreary picture at long last began to brighten in the first quarter of the twentieth century. During this period, solid economic gains and significant strides in public health resulted in progress against, if not victory over, the harsh epidemiological and nutritional regime prevalent in the late nineteenth-century South. Cotton prices in the South, for example, were generally much higher in the first 25 years of the twentieth century than they had been earlier, and the rapid pace of industrial expansion in the region left many southerners—whites in the piedmont of the Carolinas, Georgia, and Alabama especially—much better off in material terms[18] (Easterlin 1960; Lee 1957–60: 685–97, 729–33; Wright 1986: 60–64, 117–23, 124–97).

If the general pace of southern economic life quickened in the first quarter of this century, the region remained well below national economic norms even in 1925. Far more important in explaining the changes in southern stature during this period were the widespread advances made in public health. As a result of a host of public and private initiatives—efforts by the U.S. Public Health Service and the Rockefeller Sanitary Commission were particularly important in this regard—considerable progress was made against several of the South's leading killers and debilitators[19] (Cowdrey 1983: 127–46; Edgar 1992: 35–37; Grantham 1983: 310–18).

The greatest successes were achieved against malaria and hookworm, the vitality-sapping diseases whose emaciated and sallow-skinned victims long were associated with, indeed seemed to embody, the notion of the "Lazy South." To be sure, progress proved slower against pellagra, the other great "wasting" disease in the region; not until the late 1920s and 1930s with the widespread employment of brewer's yeast against the disease would the "butterfly caste"—victims of pellagra—begin to dwindle in membership. But "two out of three ain't bad," and progress against malaria and hookworm (and against certain other enteric diseases as well) in the first quarter of the twentieth century certainly can help to explain the improvement in stature and nutritional well-being among many southerners (Duffy 1982; Etheridge 1972; Ettling 1981).

It is possible, of course, that still other variables were involved. The setback experienced by those cadets born in the 1890s and 1900s could have resulted from the rise in food prices during World War I.[20] If the nutritional status of those born in the 1910s rose substantially, this hypothesis would be weakened, but the present sample contains few observations from this decade. More importantly, all of those from this decade who are included were actually born after the war ended in 1918, so they would not have been directly affected by wartime conditions. Thus, something of a puzzle remains.

We should note in closing that the height profile of Citadel cadets born in the late nineteenth century was quite similar to that found by Steckel (1987) among slaves born during an earlier time period. This "pre-modern" growth pattern could indicate that nutritional deprivation was experienced by both white and black children in the nineteenth-century South. The peculiar (cross-sectional) growth profile was possibly the result of a delay in the adolescent growth spurt among a large segment of the population caused by a synergistic relationship between nutritional well-being and the epidemiological environment.

In sum, this study contributes to our understanding of the nature of the American anthropometric cycles in the late nineteenth and early twentieth centuries. However, it remains for future research to ascertain (a) the extent to which the above trends are representative of the national average; (b) whether the decline in physical stature of the 1890s was related to the recession of 1893; (c) if improvements in physical stature had already begun in the 1910s; and finally, (d) just when white southerners lost the height advantage they had enjoyed earlier in the nineteenth century. At this point it is already evident that the difficult economic restructuring experienced by the South after the Civil War meant a stagnation not only in per capita income, but also in the biological standard of living until the turn of the century. Not until the 1920s, almost a full century after the beginning of the decline in nutritional status, did the population of the South appear to enjoy a marked and permanent improvement in its nutritional status in the wake of public health improvements, and in conjunction with a rise in per capita income. Yet more research must be conducted on the South—the region David M. Potter called a "sphinx on the American land"—before the intricacies and causal interconnections of its anthropometric history can be fully mapped out (Potter 1968: 4).

## Notes

Peter Coclanis is associate professor of history at the University of North Carolina at Chapel Hill. His book *The Shadow of a Dream: Economic Life and Death in the South Carolina Low Country, 1670–1920* (Oxford University Press, 1989) won the Allan Nevins Prize of the Society of American Historians.

John Komlos is professor and chair of economic history at the University of Munich, Germany. His book *Nutrition and Economic Development in the Eighteenth-Century Habsburg Monarchy: An Anthropometric History* (Princeton University Press, 1989) won the book prize of the Center for Austrian Studies at the University of Minnesota, as well as the National Book Award from the Phi Alpha Theta International Honor Society in history.

We would like to take this opportunity to thank Paul Mageli for collecting the data set, Deborah Coclanis and Allen Van Wyk for helping with the laborious task of coding it, and Richard Steckel and Timothy Cuff for their helpful comments on an earlier version of the paper.

1 The Citadel has gone through several name changes over the years. Between 1861 and 1910, for example, the official name of the institution was the South Carolina

Military Academy. In 1910 its name was officially changed to The Citadel, The Military College of South Carolina. Through most of its history, however, the institution has generally been referred to simply as The Citadel (Bond 1936: 167–70; *Bulletin of The Citadel* 1989–1990: 9–10).

2 The total yearly cost for attending South Carolina College (University of South Carolina) in 1835 was $350 (Godbold 1944: 78–83, 185–201).

3 During the academic years between 1892 and 1902, for example, "beneficiaries" comprised on average about 53% of opening-day enrollments and about 61% of last-day enrollments (Bond 1936: 128). On the curricula at institutions of higher education in the nineteenth-century South, see Coulter 1928; Hollis 1951–56; Thomas 1879.

4 We were left with 6,550 cases after the elimination of 255 unusual observations. Observations on cadets younger than 15 years of age or older than 25 were not considered. Observations on those born prior to 1864 or after 1940 were also disregarded, as were a small number of cases containing obvious typographical errors.

Note that the original records of the Alumni Association, which we used in 1985 and 1986 for this study, have regrettably been discarded since that time. We possess copies of these records, however, and have deposited them in the Southern Historical Collection, University of North Carolina, Chapel Hill, NC, as well as with the Inter-university Consortium for Political and Social Research (ICPSR), Ann Arbor, Michigan.

5 On The Citadel's admissions regulations regarding height, see materials located at the Citadel Military Academy Archives, Charleston, SC (Circular of Information, 1900: 17; Bulletin of The Citadel 1945–46, 11).

6 Our sample contains virtually no records for cadets born between 1911 and 1917. Since the Alumni Association records were kept in alphabetical rather than chronological order, we would not have missed a birth cohort during sampling. It is perhaps more likely that the Alumni Association stopped collecting systematic information on students in the late 1930s and early 1940s.

7 The graph depicts the terminal height of students residing in the South at the time of enrollment. A fourth degree polynomial is fitted on the annual data.

8 These growth velocity profiles are crossectional ones; i.e., they are based on measurements of different individuals at various ages. In comparison to longitudinal growth velocity profiles, which refer to the same individual throughout the growth experience, cross-sectional velocity estimates have a higher variance for a given sample size. In addition, most of the observations are for the late teenage years, so that the adolescent growth spurt is not fully identified. At the beginning of the sample period we can, nonetheless, observe the tail end of the spurt (Tanner 1978: 11–12).

9 These figures are not reported here.

10 For data on personal income per capita in the Upper Pine Belt counties of South Carolina in the 1920s, see Robert E. Graham 1971: 37–38).

11 Our sample does not allow us to determine the turning point accurately, and the

Ohio records are not helpful in this regard either, inasmuch as they do not extend beyond those born in 1910 (Steckel and Haurin 1994).

12 Easterlin's (1961: 528) figures on personal income per capita in the South (and the subregions thereof) as a percentage of the U.S. average in 1860, 1880, 1900, 1920, and 1940 are reproduced below.

| | Year | | | | |
|---|---|---|---|---|---|
| Regions | 1860 | 1880 | 1900 | 1920 | 1940 |
| United States | 100 | 100 | 100 | 100 | 100 |
| South | 72 | 51 | 51 | 62 | 65 |
| South Atlantic | 65 | 45 | 45 | 59 | 69 |
| East South-Central | 68 | 51 | 49 | 52 | 55 |
| West South-Central | 115 | 60 | 61 | 72 | 70 |

Source: Easterlin, *American Economic History* 528.

13 On the demise of the rice and long-staple cotton industries in the South Carolina low country, see Coclanis, 1989: 111–54, 1993; Kovacik and Mason, 1985: 77–104.

14 This claim is made, it should be noted, by scholars fully cognizant of the fact that disease reporting and public health improved over time. For a good overview of the epidemiological history of the South see Breeden (1988: 1–28).

15 Note that the South's last yellow fever epidemic occurred in 1905. On the medical and epidemiological history of South Carolina specifically in the late nineteenth century, see Waring (1964–71: 2, 161–79).

16 Note that Hilliard argues that the meat consumed in the South between 1890 and 1930 was even poorer in quality than the poor-quality meat consumed in the region during the antebellum era.

In part because of the South Atlantic region's poverty and its diminished livestock herds, per capita meat consumption in the area was roughly 6% lower than the U.S. average in 1921 (Holmes 1916: 209–10; McFall, 1927: 118).

17 For example, according to Bailey, only 64 of the 171 men examined in Columbia in May 1898 were judged fit for military service. 67% of those rejected failed to meet the minimum weight requirement of 125 pounds (Bailey 1991: 203). According to Trask (1981: 127) the rejection rate nationwide was about 50%. The current archivist at The Citadel, Jane Yates, recently recounted to one of the authors of this essay that when putting together an exhibit on uniforms at The Citadel, the institution's curatorial staff had trouble finding present-day forms (mannequins) sufficiently small to display uniforms preserved from the late nineteenth century; even children's forms were too large for many of the uniforms.

18 Cotton production and both nominal and real cotton prices rose markedly in the first three decades of the twentieth century (U.S. Department of Commerce 1975, 1:

200–201, 517–18). On the early industrialization of South Carolina specifically, see Carlton (1982). For data demonstrating the increase in the average annual earnings of South Carolina textile workers in the first three decades of the twentieth century, see Olin D. Johnston Papers, Departmental Commissions, South Carolina Department of Archives and History, Columbia, SC. The authors are indebted to Bryant Simon for bringing these earnings data to our attention.

19 On the relationship between the growth in per capita income and human physical stature, see Steckel (1983).

20 On the magnitude of the rise in food prices during the war, see U.S. Department of Commerce (1975: 1 200).

## References

Act to Convert the Arsenal at Columbia, and the Citadel and Magazine in and near Charleston, into Military Schools, December 20, 1842, South Carolina Statutes at Large 11 (1873): 1838–49.

Act to Authorize the Reopening of the South Carolina Military Academy, January 30, 1882, South Carolina Statutes at Large 27.

Bailey, Jr., Harris Moore (1991) "The splendid little forgotten war: The mobilization of South Carolina for the war with Spain." South Carolina Historical Magazine 92: 189–214.

Beardsley, Edward S. (1987) A History of Neglect: Health Care for Blacks and Mill Workers in the Twentieth-Century South. Knoxville, TN: University of Tennessee Press.

Bond, Oliver J. (1936) The Story of The Citadel. Richmond, VA: Garrett and Massie.

Breeden, James O. (1988) "Disease as a factor in southern distinctiveness," in Todd L. Savitt and James Harvey Young, eds., Disease and Distinctiveness in the American South. Knoxville, TN: University of Tennessee Press: 1–28.

Bulletin of The Citadel: The Military College of South Carolina, Catalogue Issue 1989–1990. Charleston, SC: The Citadel.

Carlton, David L. (1982) Mill and Town in South Carolina, 1880–1920. Baton Rouge, LA: Louisiana State University Press.

Carrigan, Jo Ann (1988) "Yellow fever: Scourge of the South," in Todd L. Savitt and James Harvey Young, eds., Disease and Distinctiveness in the American South. Knoxville, TN: University of Tennessee Press: 55–78.

Coclanis, Peter A. (1989) The Shadow of a Dream: Economic Life and Death in the South Carolina Low Country, 1670–1920. New York: Oxford University Press.

Cosmas, Graham A. (1971) An Army for Empire: The United States Army in the Spanish American War. Columbia, MO: University of Missouri Press.

Costa, Dora L. (1993) "Height, wealth, and disease among the nativeborn in the rural, antebellum North." Social Science History 17: 355–84.

Coulter, E. Merton (1928) College Life in the Old South. New York: Macmillan.

Cowdrey, Albert E. (1983) This Land, This South: An Environmental History. Lexington, KY: University of Kentucky Press.

Cuff, Timothy (1994) "The body mass index of West Point cadets in the nineteenth century." Historical Methods 27.

Duffy, John (1988) "The impact of malaria on the South," in Todd L. Savitt and James Harvey Young, eds., Disease and Distinctiveness in the American South. Knoxville, TN: University of Tennessee Press: 29–54.

Easterlin, Richard A. (1960) "Interregional differences in per capita income, population, and total income, 1840–1950," National Bureau of Economic Research, Conference on Research in Income and Wealth, Studies in Income and Wealth 24, Trends in the American Economy in the Nineteenth Century. Princeton, NJ: Princeton University Press.

——— (1961) "Regional income trends, 1840–1950," in Seymour E. Harris, American Economic History. New York: McGraw-Hill: 525–47.

Edgar, Walter B. (1992) South Carolina in the Modern Age. Columbia, SC: University of South Carolina Press.

Etheridge, Elizabeth W. (1972) The Butterfly Caste: A Social History of Pellagra in the South. Westport, CT: Greenwood.

——— (1988) "Pellagra: An unappreciated reminder of southern distinctiveness," in Todd L. Savitt and James Harvey Young, eds., Disease and Distinctiveness in the American South. Knoxville, TN: University of Tennessee Press: 100–19.

Ettling, John (1981) The Germ of Laziness: Rockefeller Philanthropy and Public Health in the New South. Cambridge, MA: Harvard University Press.

Fite, Gilbert C. (1984) Cotton Fields No More: Southern Agriculture 1865–1980. Lexington, KY: University of Kentucky Press.

Floud, Roderick, Kenneth Wachter and Annabel Gregory (1990) Height, Health and History: Nutritional Status in the United Kingdom, 1750–1980. Cambridge: Cambridge University Press.

Fogel, Robert W. (1986) "Nutrition and the decline in mortality since 1700: some preliminary findings," in Stanley L. Engerman and Robert E. Gallman, eds., Long-Term Factors in American Economic Growth, National Bureau of Economic Research, Conference on Research in Income and Wealth, Studies in Income and Wealth 51: Chicago: University of Chicago Press: 439–555.

——— (1993) "New sources and new techniques for the study of secular trends in nutritional status, health, mortality, and the process of aging." Historical Methods 26: 5–43.

Godbold, Albea (1944) The Church College of the Old South. Durham, NC: Duke University Press.

Graham, Jr., Robert E. (1971) "Personal income in South Carolina by type, source, and geographical areas, 1929–1969." Essays in Economics 24: Columbia, SC: Bureau of Business and Economic Research, College of Business Administration, University of South Carolina.

Grantham, Dewey W. (1983) Southern Progressivism: The Reconciliation of Progress and Tradition. Knoxville, TN: University of Tennessee Press.
Hilliard, Sam B. (1972) Hog Meat and Hoecake: Food Supply in the Old South, 1840–1860. Carbondale, IL: Southern Illinois University Press.
Hollis, Daniel Walker (1951–56) Sesquicentennial History of the University of South Carolina. Columbia, SC: University of South Carolina Press.
Holmes, George K. (1916) U.S. Department of Agriculture, Bureau of Crop Estimates, Meat Situation in the United States, Part I, Report No. 109. Washington, DC: U.S. Government Printing Office.
James, John A. (1981) "Financial underdevelopment in the postbellum South." Journal of Interdisciplinary History 11: 443–54.
Johnston, Olin D., Papers, Box 39, Departmental Commissions, South Carolina Department of Archives and History, Columbia, SC.
Komlos, John (1987) "The height and weight of West Point cadets: Dietary change in antebellum America." Journal of Economic History 47: 897–927.
——— (1989) Nutrition and Economic Development in the Eighteenth-Century Habsburg Monarchy: An Anthropometric History. Princeton, NJ: Princeton University Press.
——— (1992) "Toward an anthropometric history of African-Americans: The case of the free blacks of antebellum Maryland." Claudia Goldin and Hugh Rockoff, eds., Strategic Factors in Nineteenth Century American Economic History: A Volume to Honor Robert W. Fogel: National Bureau of Economic Research Conference Report. Chicago: University of Chicago Press: 297–329.
——— (1993) "The secular trend in the nutritional status of the population of the United Kingdom, 1730–1860." Economic History Review 46: 115–44.
——— (1993b) "A Malthusian crisis revisited: The stature of runaway indentured servants in Colonial America." Economic History Review 46: 768–82.
——— (1994a) "The stature of runaway slaves in colonial America," in John Komlos, ed., The Standard of Living and Economic Development: Essays in Anthropometric History. Chicago: University of Chicago Press: Chapter 6.
——— (1994b) "The nutritional status of the students of the École Polytechnique, 1770–1869." Journal of Interdisciplinary History 24 (Winter) 3: 493–509.
Kovacik, Charles F. and Robert E. Mason (1985) "Changes in the South Carolina sea island cotton industry." Southeastern Geographer 25: 77–104.
Lee, Everett S. et al. (1957–60) "Population redistribution and economic growth, 1870–1950." Memoirs of the American Philosophical Society 45 and 51: Philadelphia: The American Philosophical Society.
Marcus, Alan I. (1988) "The South's native foreigners: Hookworm as a factor in southern distinctiveness," in Todd L. Savitt and James Harvey Young, eds., Disease and Distinctiveness in the American South. Knoxville, TN: University of Tennessee Press: 79–99.
Mascie-Taylor, C. G. Nicholas (1991) "Biosocial influences on stature: A review." Journal of Biosocial Science 23: 113–28.

McFall, Robert James (1927) The World's Meat. New York and London: D. Appleton and Company.

Mokyr, Joel and Cormac O'Grada (1988) "Poor and getting poorer? Living standards in Ireland before the famine." Economic History Review 41: 209–35.

Nicholas, Stephen and Deborah Oxley (1993) "The living standards of women during the Industrial Revolution, 1795–1820." Economic History Review 46: 723–49.

Potter, David M. (1968) The South and the Sectional Conflict. Baton Rouge, LA: Louisiana State University Press.

Ransom, Roger L. and Richard Sutch (1977) One Kind of Freedom: The Economic Consequences of Emancipation. New York: Cambridge University Press.

South Carolina Military Academy, Charleston, SC (1990) Circular of Information. Charleston, SC; The Citadel, The Military College of South Carolina, Bulletin of The Citadel, The Military College of South Carolina: General Information Announcements for 1945–6. Charleston, SC.

South Carolina Statutes at Large, 1879–82, 1838–49.

Steckel, Richard H. (1987) "Growth depression and recovery: The remarkable case of American slaves." Annals of Human Biology 14: 111–32.

——— (1979) "Slave height profiles from coastwise manifests." Explorations in Economic History 16: 363–80.

Steckel, Richard H., and Donald R. Haurin (1994) "Health and nutrition in the American midwest: Evidence from the height of Ohio National Guardsmen, 1850–1910," in John Komlos (ed.), The Standard of Living and Economic Development: Essays in Anthropometric History. Chicago: University of Chicago Press: Chapter 7.

Tanner, James (1978) Foetus into Men: Physical Growth from Conception to Maturity. Cambridge, MA: Harvard University Press.

Thomas, John P. (1879) Historical Sketch of the South Carolina Military Academy. Charleston, SC: Walker, Evans & Cogswell.

Trask, David F. (1981) The War with Spain in 1898. New York: Macmillan.

U.S. Department of Agriculture (1949) Consumption of Food in the United States 1909–1948. Washington, DC: Bureau of Agricultural Economics.

U.S. Bureau of the Census (1975) Historical Statistics of the United States: Colonial Times to 1970. Washington, DC: U.S. Government Printing Office.

Waring, Joseph I. (1964–1971) A History of Medicine in South Carolina. Columbia, SC: South Carolina Medical Association 2.

Wright, Gavin (1978) The Political Economy of the Cotton South: Households, Markets, and Wealth in the Nineteenth Century. New York: W. W. Norton.

——— (1986) Old South, New South: Revolutions in the Southern Economy Since the Civil War. New York: Basic Books.

Wu, Jialu (1994) "The effects of the Great Depression on the material standard of living: Anthropometric evidence from Pittsburgh and Allegheny County, Pa., 1890–1950," in John Komlos (ed.), The Standard of Living and Economic Development: Essays in Anthropometric History: Chicago: University of Chicago Press: Chapter 8.

# XV

# Toward an Anthropometric History of African-Americans

## The Case of the Free Blacks in Antebellum Maryland

Economic historians have been increasingly interested in the interaction between economic and biological processes, particularly as it relates to economic development.[1] Indicators such as height, weight, body-mass, age at menarche, and morbidity are all related to nutritional status and consequently to demographic variables such as life expectancy, with a feedback effect on the economy through their impact on labor productivity.[2] Within this context, the importance of the anthropometric history of African-Americans is accentuated by the debate over their material standard of living, especially their food consumption while in bondage. Calculations based on agricultural censuses indicate that the calorie and protein content of the slave diet was adequate on average. For example, slaves are said to have consumed circa 1,000

The author is greatly indebted to Peter Coclanis, Timothy Cuff, Seymour Drescher, Stanley Engerman, Claudia Goldin, Richard Steckel, and James Tanner for commenting on the manuscript. Data collection was supported in part by a grant from the University of Pittsburgh's Office of Research and Development and was processed by Philip Sidel of the university's Social Science Computer Research Institute.

1. The notion of using heights to gain insights into the biological maturation of human beings and its socio-economic implications was stimulated by the debate over the diet, health, and age at menarche of American slaves. For early papers see James Trussell and Richard H. Steckel, "The Age of Slaves at Menarche and Their First Birth," *Journal of Interdisciplinary History*, 8 (Winter 1978), pp. 477–505; Richard H. Steckel, "Slave Height Profiles from Coastwise Manifests," *Explorations in Economic History*, 16 (Oct. 1979), pp. 363–80. For a recollection of the inception of the nutrition project see "An interview with Robert W. Fogel," *The Newsletter of The Cliometric Society*, 5 (July 1990), p. 3ff.

2. Robert W. Fogel, "Nutrition and the Decline in Mortality since 1700: Some Preliminary Findings," in Stanley L. Engerman and Robert E. Gallman, eds., *Long-Term Factors in American Economic Growth* (Chicago, 1986), pp. 439–555; Roderick Floud, Kenneth Wachter, and Annabel Gregory, *Height, Health and History: Nutritional Status in the United Kingdom, 1750–1980* (Cambridge, 1990). For a theory of anthropometric history see John Komlos, *Nutrition and Economic Development in the Eighteenth-Century Habsburg Monarchy: An Anthropometric History* (Princeton, 1989), chap. 1.

The University of Chicago Press, Chicago 60637
The University of Chicago Press, Ltd., London

calories more daily than European peasants.[3] The disadvantage of these estimates is that they cannot be decomposed by sex and by age, and, moreover, do not reveal temporal changes.

Anthropometric indexes were developed to overcome the limitations of these approaches. The method has become standard in assessing nutritional adequacy in third-world countries.[4] Several studies have explored the age-by-height profile of slaves to quantify their nutritional status and material well-being using two major sources of data, slave manifests and Civil War muster rolls.[5] The system of manifests was designed to discourage smuggling after the prohibition of slave imports in 1807. These shipping documents included physical descriptions used to identify slaves transported along the coast. The muster rolls pertain to black soldiers in the Union Army in the early 1860s. Both sources provide valuable information on the physical characteristics of African-Americans, but both have limitations.

Slaves, whose heights appear on the manifests, were transported in interregional trade, and thus they may not be representative of all slaves. Similarly, the military source contains information only on soldiers. Because it excludes females and youths, it cannot corroborate some of the results obtained from the manifest sample. Even with regard to males, the Civil War sample has no information on birth cohorts of the early nineteenth century and, consequently, provides only a limited trend in slave stature.[6] Moreover, the trend

3. On slave consumption see Robert W. Fogel and Stanley L. Engerman, *Time on the Cross: The Economics of American Negro Slavery* (Boston, 1974), p. 109; and for corroboration see Roger Ransom and Richard Sutch, *One Kind of Freedom: The Economic Consequences of Emancipation* (Cambridge, 1977), p. 244; for consumption of European peasants see Komlos, *Nutrition and Economic Development*, p. 101. The extent to which the diet was adequate in terms of vitamins and minerals, given the genetic make-up of slaves, is more controversial. See, for instance, Kenneth F. Kiple, "A Survey of Recent Literature on the Biological Past of the Black," *Social Science History*, 10 (Winter 1986), pp. 343–68.

4. Phyllis B. Eveleth and James M. Tanner, *Worldwide Variation in Human Growth* (2d edn., Cambridge, 1990).

5. Stanley Engerman, "The Height of U.S. Slaves," *Local Population Studies*, 16 (Spring 1976), pp. 45–50; Robert W. Fogel, Stanley L. Engerman, James Trussell, Roderick Floud, Clayne L. Pope, and Larry T. Wimmer, "The Economics of Mortality in North America, 1650–1910: A Description of a Research Project," *Historical Methods*, 11 (Spring 1978), pp. 75–108; Robert Margo and Richard H. Steckel, "The Heights of American Slaves: New Evidence on Slave Nutrition and Health," *Social Science History*, 6 (Fall 192), pp. 516–38; Richard H. Steckel, "A Peculiar Population: The Nutrition, Health, and Mortality of American Slaves from Childhood to Maturity," *Journal of Economic History*, 46 (Sept. 1986), pp. 721–41; Richard H. Steckel, "Birth Weights and Infant Mortality among American Slaves," *Explorations in Economic History*, 23 (Apr. 1986), pp. 173–98; Richard H. Steckel, "Growth Depression and Recovery: The Remarkable Case of American Slaves," *Annals of Human Biology*, 14 (Mar.–Apr. 1987), pp. 111–32; Robert W. Fogel, *Without Consent or Contract: The Rise and Fall of American Slavery* (New York, 1989), pp. 138–47.

6. In addition, the Civil War sample is not representative of the whole South because 43 percent of the black recruits were born in Tennessee and Kentucky; see Margo and Steckel, "The Heights of American Slaves." These two states had the tallest white men in the United States at that time. Therefore, the nutritional status of the black population born in the Upper South could have been above average.

obtained from the military sample differs from that of slaves shipped in interregional trade.

The trend for African-Americans obtained from Civil War military records essentially follows that of the whites. The results of the manifest sample, however, are more difficult to interpret. An early result actually resembled the one obtained from white soldiers.[7] Subsequent estimates, however, do not show a downward tendency prior to the Civil War.[8] If anything, they show the opposite.[9] The result is suspect, however, because African-born slaves, who were shorter than American-born slaves, might have been included in the early part of the sample, and the positive trend in height could be indicative of their changing share in the sample, instead of an improvement in nutritional status.

These limitations notwithstanding, major findings have emerged from the anthropometric evidence. The stature of slaves in nineteenth-century America indicates that they were well nourished as young adults, although not as children. In spite of their early nutritional deprivation, male slaves reached a terminal height exceeding 67 inches—within an inch of northern-born whites and well above contemporary African and European norms. In fact, their physical stature was closer to that of European aristocrats than to that of peasants.[10]

The evidence on slave children's nutritional status is more controversial than that on adults. Data in the manifests indicate that slave children were even shorter than children in the poorest third-world countries today.[11] They were below the first centile of modern height standards for industrialized countries. By comparison, the average slum child of Lagos, Nigeria, attains the twelfth centile. Although short children in developing countries grow into short adults, small slave children in America, according to the existing evidence, apparently grew up to be relatively tall adults, reaching the 25th centile of modern standards. The slaves' remarkable growth pattern has been attributed to a rise in food intake after a period of severe deprivation.[12] Previously, this pattern of "catch-up" growth has been observed only for shorter periods of deprivation than alleged for slaves.[13]

7. Steckel, "Slave Height Profiles," p. 377.

8. Margo and Steckel, "The Heights of American Slaves," p. 523.

9. This later result is the trend being accepted by the profession. See, for example, David Eltis, "Welfare Trends among the Yoruba in the Early Nineteenth Century: The Anthropometric Evidence," *Journal of Economic History*, 50 (Sept. 1990), pp. 521–40.

10. John Komlos, "Height and Social Status in Eighteenth-Century Germany," *Journal of Interdisciplinary History*, 20 (Spring 1990), pp. 607–21.

11. Eveleth and Tanner, *Worldwide Variation in Human Growth*.

12. Steckel refers to the degree of catch-up growth as "remarkable" in Steckel, "Growth Depression and Recovery," pp. 115, 129.

13. Technically it is incorrect to speak of catch-up growth here because the term usually is reserved for cases in which the nutritional deprivation is temporary. Slaves, in contrast, suffered prolonged deprivation. It is perhaps better to describe the pattern as moving through the centiles

Richard Steckel, using the manifest data, observed that slave children showed the first signs of improvement in height prior to adolescence, when they began to enter the labor force. As workers, Steckel reasons, they must have received greater allotments of food.[14] A weakness in this interpretation is that the initial signs of recovery among the children were fairly weak. Not much catch-up growth took place at the ages of ten and eleven, when many slaves entered the labor force and before the modern reference population experiences the adolescent growth spurt. At age eleven, slave children were still in the second or third centile of the modern height standards (Table 10.2).[15] During the early teenage years, they fall back through the centiles because the modern reference population reaches the adolescent growth spurt earlier than did slaves. But even after the adolescent growth spurt, at age seventeen the slave boys were still below the fifth centile (Figures 10.1 and 10.2). Hence the acceleration in the rate of growth appears to have been confined to a few years in the late teens.[16] Consequently, the juncture between the timing of the children's entrance into the labor force and their becoming taller relative to the modern reference population seems tentative. Why significant catch-up growth occurred earlier among females than among males is another unaddressed issue.[17] Because the degree of catch-up growth appears extraordinarily large, one might ask whether the results are a consequence of some peculiarity of the manifest sample as much as of the children's labor force participation rates. The military sample cannot illuminate the issue because it contains no evidence on slave children. Hence, the nature of the available data restricts our knowledge of slave nutritional experience in two respects: the nutritional experience of slave children and the secular trend in slave stature. It is fortunate that new data, such as the Certificates of Good Character of Louisiana and the Maryland manumission records, have been found.[18]

---

of the modern standard. See James Tanner, *Foetus into Man: Physical Growth from Conception to Maturity* (2d. edn., Cambridge, Mass., 1990), p. 161.

14. Steckel, "Growth Depression and Recovery," p. 130. Short stature is also indicative of deprivation in utero, brought about by women working in the cotton fields until close to delivery. Labor force participation of pregnant females could have varied by crop specialization, and thus could have been different in the tobacco fields of Maryland. Slave children older than twelve were more easily separated from their parents. Some states forbade the importation of children under ten years of age without their mothers unless they were orphans. Perhaps these factors had an impact on which slave children ended up in the manifest sample. See Frederic Bancroft, *Slave Trading in the Old South* ([1931]; New York, 1959), pp. 197, 202, 212.

15. This means that 97 or 98 percent of today's youth are taller than the average slave child.

16. Slaves in the tobacco fields of Maryland, at least, were working full time by the age of ten (Allan Kulikoff, *Tobacco and Slaves: The Development of Southern Cultures in the Chesapeake, 1680–1800* [Chapel Hill, 1986], pp. 373, 377).

17. Children were reared by their families. It is possible that parents (and kin) supplemented children's rations from their own, particularly since slaves did produce some of their own food, cooked it mostly themselves, and often consumed it together (Kenneth M. Stampp, *The Peculiar Institution: Slavery in the Ante-Bellum South* [New York, 1967], p. 287). Some slave owners discouraged maternal neglect; see Bancroft, *Slave Trading in the Old South*, p. 86.

18. The system of certifying the good character of imported slaves was in effect between 1829 and 1831 in an attempt to keep troublemakers out of Louisiana (Herman Freudenberger and Jona-

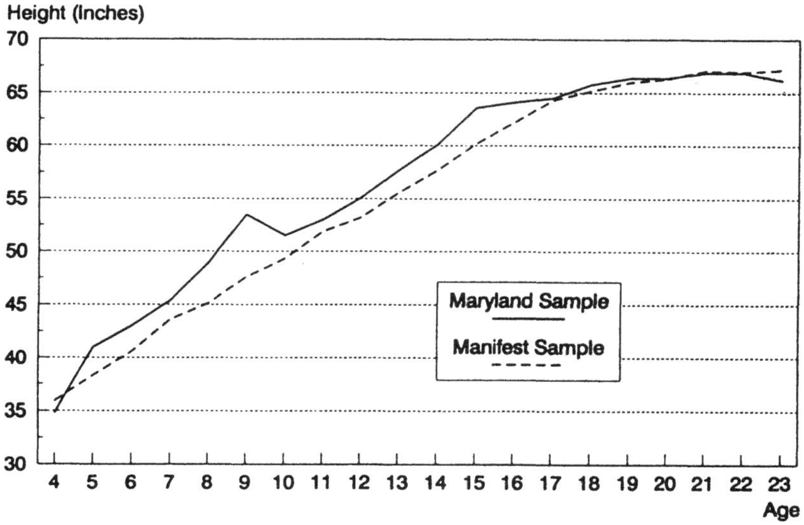


**Fig. 10.1 Heights of Male African-American Youth**
*Source:* Table 10.2.

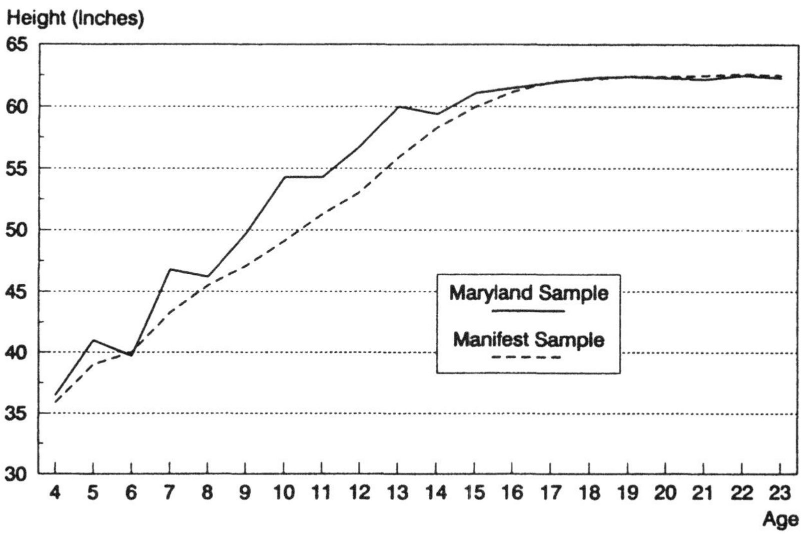


**Fig. 10.2 Height of Female African-American Youth**
*Source:* Table 10.2.

## 10.1 The Maryland Certificates of Freedom

The Maryland Certificates of Freedom were compiled between 1806 and 1864 to provide identification papers for blacks no longer bound to a master. The records contain information on the individual's height, age, sex, date of birth, color, county of residence and of birth, and whether the person was born free.[19] The number of records collected is 14,665; 35 percent of the females and 24 percent of the males were born free (Table 10.1).[20]

Current age, recorded to the nearest year, is not indicative of age at manumission because certificates were usually obtained when the former slave wanted to travel outside the immediate vicinity to an area where his or her status would not be common knowledge.[21] Although all ages are represented, few children are in the sample, probably because they were less mobile than adults (Table 10.1). The destination of the migrants was predominantly urban, evidenced by the fact that 39 percent of Maryland's free blacks lived in Baltimore in 1850. Like other migrant populations, most of the black migrants would have been young and probably unmarried when they applied for the certification. This explains why people in their twenties constitute nearly half of the sample and why there are so few children in the sample. There are more observations for teenage girls than boys. This could indicate that girls left home to marry or to work as domestics at a younger age than boys. Because adult migrants are frequently the more enterprising and possibly healthier, the sample may not be representative of all free blacks in Maryland.

Because physical descriptions were crucial for identification purposes, the height records are expected to be reliable. That heights were generally given to the nearest quarter of an inch supports their reliability. There was, however, some rounding to the nearest inch or half inch as indicated by the distribution of the height observations (Table 10.1). Unless otherwise stated, references to time indicate date of birth, not date of measurement, and references to Baltimore mean both the city and county.

---

than B. Pritchett, "The Domestic United States Slave Trade: New Evidence," *Journal of Interdisciplinary History*, 21 [Winter 1991], pp. 447–78).

19. Certificates of Freedom, Maryland State Archives, Annapolis, Maryland. The records from Allegany, Calvert, Cecil, Charles, Howard, and Montgomery counties were not available. Among these, only the Howard County records are said to be extant but have not yet been microfilmed. All other records are included in the sample. A few of the documents contain information on where the person grew up. Information on the age of manumission, also available in some instances, was not collected. Similar records exist in Virginia.

20. In 1850 there were about 75,000 free blacks in the state (*Abstract of the Seventh Census of the United States: 1850* [Washington, D.C., 1853], p. 150). Females may outnumber males in this sample because females may have been more likely to migrate either to work or to marry.

21. Jeffrey R. Brackett, *The Negro in Maryland: A Study of the Institution of Slavery* (Baltimore, 1889), pp. 89, 163. Only 1.7 percent of the sample pertains to those under the age of fifteen. Authorities feared that the manumission of children would burden public relief, and in 1858 it was forbidden in Maryland. See James M. Wright, *The Free Negro in Maryland, 1634–1860* (New York, 1921), p. 65.

**Table 10.1 Characteristics of the Sample of Maryland Certificates of Freedom**

Part A: By County

| County | Certificate[a] | Birthplace | Grew Up[b] |
|---|---|---|---|
| Anne Arundel | 13.7% | 13.4% | |
| Baltimore City | 4.2 | 7.7 | |
| Baltimore County | 16.4 | 7.1 | |
| Caroline | 7.6 | 6.7 | 5.5% |
| Cecil | 0.2 | 0.2 | |
| Charles | 0.3 | 0.0 | |
| Dorchester | 11.6 | 11.0 | |
| Frederick | 7.9 | 2.9 | |
| Harford | 2.5 | 2.0 | |
| Kent | 3.1 | 3.0 | |
| Prince George | 5.1 | 1.8 | 1.2 |
| Queen Ann | 3.2 | 0.4 | 0.3 |
| St. Mary's | 6.5 | 6.5 | |
| Somerset | 7.3 | 7.5 | |
| Talbot | 8.7 | 7.1 | 2.7 |
| Washington | 1.4 | 0.8 | |
| Other | | 1.0 | 0.5 |
| Unknown | | 20.2 | 89.7 |

Part B: By Legal Status at Birth

| | Male | Female | Total |
|---|---|---|---|
| Born free | 1,591 (24%) | 2,802 (35%) | 4,393 (30%) |
| Not born free | 6,107 (76%) | 5,165 (65%) | 10,272 (70%) |
| Total | 6,698 | 7,667 | 14,665 |

Part C: By Decade of Birth, Age, and Color

| Decade of Birth | | Age | | Color | |
|---|---|---|---|---|---|
| 1750 | 0.5% | 1–9 years | 0.6% | Black | 14.8% |
| 1760 | 1.7 | 10–14 | 1.1 | Brown | 5.6 |
| 1770 | 5.0 | 15–19 | 11.8 | Dark | 23.4 |
| 1780 | 10.4 | 20–24 | 28.9 | Light | 12.7 |
| 1790 | 15.0 | 25–29 | 19.4 | Yellow | 7.1 |
| 1800 | 18.0 | 30–34 | 14.2 | Copper | 1.3 |
| 1810 | 18.7 | 35–39 | 10.0 | Chestnut | 11.4 |
| 1820 | 15.1 | 40–49 | 10.6 | Mulatto | 10.0 |
| 1830 | 13.0 | 50–59 | 2.6 | Bright | 1.6 |
| 1840 | 2.3 | ≥60 | 0.6 | Unknown | 12.0 |
| 1850 | 0.0 | | | | |
| Unknown | 0.1 | | | | |

*(continued)*

**Table 10.1** (continued)

| Part D: By Precision of Height Measurement in Fractions of an Inch | |
|---|---|
| 0.0 inch | 46.6 |
| 0.25 | 11.6 |
| 0.50 | 30.0 |
| 0.75 | 11.0 |
| Other | 0.7 |

*Source:* Certificates of Freedom, Maryland State Archives, Annapolis, Maryland.
[a]County in which certificate was obtained.
[b]County in which person grew up.

## 10.2 Anthropometric Evidence from the Maryland Sample

### 10.2.1 Children

Although the Maryland sample corroborates the notion that slave children of both sexes were undernourished, they were, however, less malnourished than the manifest sample indicates (Table 10.2).[22] Free black children were several inches (about 3.5 percent) taller than the slave children transported on water to the Lower South (Table 10.3, rows 5 and 13; Figures 10.1 and 10.2 above). After age seventeen the distinction between the two results diminishes or even vanishes, even when the number of observations remains small, as in the case of seventeen-year-old boys. There is considerable variation in the centile of modern standards reached until the late teens because the number of observations in the free black sample is small. On average, children reached about the tenth centile of modern heights. Catch-up growth is evident and begins at about the same age as in the manifest sample, but it is less abrupt, and thus resembles more closely conventional growth profiles. Relative to their modern counterparts, free black girls were somewhat taller than boys (Table 10.2). Children born enslaved tended to be taller than those who were born free (Tables 10.3 and 10.4).

### 10.2.2 Time Trend

The time trend in stature is analyzed in three age categories for both sexes. Turning first to males, sample sizes are insufficient to explore the trend for those under age sixteen. Regression analysis on youth (ages 16 to 20) indicates that their height increased until approximately the birth cohorts of the 1820s and subsequently declined (Tables 10.5 and 10.6; Figure 10.3). Among

22. Here and elsewhere comparison is made with the hitherto published results of the manifest sample. Richard Steckel has now enlarged the sample to about 28,000 manifests, including all those housed in the National Archives. His preliminary investigation seems to indicate that slave children in Maryland and Virginia reached about the fifth centile of modern standards and were taller than those born in the Lower South. Hence, the yet unpublished results seem to be closer to the height of the free black children of Maryland.

**Table 10.2 Height by Age Profile of African-American Youth**

Part A: Height (in inches)

| | Male | | | | Female | | | |
|---|---|---|---|---|---|---|---|---|
| | Free | | Slave Manifests | | Free | | Slave Manifests | |
| Age | N | Height | N | Height | N | Height | N | Height |
| 1 | | | | | 1 | 27.0 | | |
| 2 | 1 | 33.5 | | | 7 | 30.1 | | |
| 3 | | | | | 7 | 34.7 | | |
| 4 | 2 | 34.8 | 195 | 35.9 | 6 | 36.5 | 206 | 35.9 |
| 5 | 1 | 41.0 | 169 | 38.3 | 8 | 41.0 | 200 | 39.0 |
| 6 | 2 | 43.0 | 218 | 40.6 | 5 | 39.7 | 262 | 40.0 |
| 7 | 4 | 45.4 | 200 | 43.6 | 14 | 46.8 | 241 | 43.3 |
| 8 | 4 | 48.9 | 281 | 45.1 | 8 | 46.2 | 337 | 45.5 |
| 9 | 3 | 53.5 | 266 | 47.6 | 11 | 49.7 | 306 | 47.1 |
| 10 | 1 | 51.5 | 557 | 49.3 | 11 | 54.3 | 528 | 49.1 |
| 11 | 5 | 53.0 | 347 | 51.9 | 10 | 54.3 | 443 | 51.3 |
| 12 | 6 | 55.1 | 751 | 53.2 | 23 | 56.8 | 736 | 53.1 |
| 13 | 6 | 57.7 | 470 | 55.6 | 30 | 60.0 | 556 | 55.9 |
| 14 | 18 | 60.1 | 732 | 57.7 | 54 | 59.4 | 765 | 58.3 |
| 15 | 12 | 63.6 | 571 | 60.2 | 89 | 61.1 | 812 | 60.0 |
| 16 | 30 | 64.1 | 709 | 62.2 | 245 | 61.5 | 113 | 61.2 |
| 17 | 42 | 64.5 | 655 | 64.3 | 352 | 61.9 | 871 | 62.0 |
| 18 | 61 | 65.8 | 1,142 | 65.2 | 439 | 62.3 | 1,268 | 62.2 |
| 19 | 96 | 66.4 | 900 | 66.0 | 375 | 62.4 | 594 | 62.4 |
| 20 | 168 | 66.4 | 1,527 | 66.3 | 412 | 62.3 | 1,764 | 62.4 |
| 21 | 746 | 66.9 | 944 | 67.1 | 367 | 62.2 | 337 | 62.5 |
| 22 | 685 | 66.9 | 1,374 | 67.0 | 439 | 62.5 | 664 | 62.6 |
| 23 | 466 | 66.2 | 795 | 67.2 | 339 | 62.3 | 404 | 62.5 |
| Total | 2,359 | | 12,806 | | 3,253 | | 11,807 | |

Part B: Standard Deviations and Centiles of Modern Height

| | Standard Deviations of Heights | | | | Centiles of Modern Height | | | |
|---|---|---|---|---|---|---|---|---|
| | Male | | Female | | Male | | Female | |
| Age | Free | Slave Manifests | Free | Slave Manifests | Free | Slave Manifests | Free | Slave Manifests |
| 2 | | | 2.2 | | 7.6 | | 0.02 | |
| 3 | | | 3.5 | | | | 1.7 | |
| 4 | 6.7 | 5.5 | 2.5 | 5.5 | 0.01 | 0.3 | 0.7 | 0.5 |
| 5 | 0.0 | 5.3 | 1.8 | 5.2 | 6.8 | 0.3 | 10.6 | 1.6 |
| 6 | 2.1 | 5.3 | 1.9 | 5.7 | 5.7 | 0.5 | 0.2 | 0.4 |
| 7 | 1.1 | 5.0 | 3.6 | 5.5 | 7.5 | 1.5 | 27.8 | 1.8 |
| 8 | 2.8 | 5.0 | 1.7 | 5.4 | 21.2 | 0.9 | 3.8 | 2.2 |
| 9 | 1.6 | 5.1 | 2.7 | 5.0 | 60.3 | 1.7 | 11.9 | 1.4 |
| 10 | 0.0 | 5.8 | 3.9 | 5.7 | 9.3 | 1.6 | 40.5 | 1.4 |

*(continued)*

**Table 10.2** (continued)

Part B: Standard Deviations and Centiles of Modern Height

| | Standard Deviations of Heights | | | | Centiles of Modern Height | | | |
|---|---|---|---|---|---|---|---|---|
| | Male | | Female | | Male | | Female | |
| Age | Free | Slave Manifests | Free | Slave Manifests | Free | Slave Manifests | Free | Slave Manifests |
| 11 | 3.4 | 4.7 | 4.3 | 4.9 | 7.4 | 3.6 | 13.8 | 2.1 |
| 12 | 3.3 | 5.0 | 3.1 | 4.7 | 8.4 | 2.4 | 12.7 | 0.9 |
| 13 | 3.3 | 4.8 | 2.5 | 4.7 | 10.8 | 3.0 | 20.3 | 0.9 |
| 14 | 3.4 | 4.5 | 2.6 | 4.1 | 8.5 | 2.1 | 5.0 | 1.7 |
| 15 | 3.7 | 3.9 | 2.3 | 3.4 | 12.9 | 1.3 | 12.9 | 5.6 |
| 16 | 2.6 | 3.7 | 2.6 | 3.1 | 5.7 | 1.2 | 16.6 | 13.3 |
| 17 | 3.0 | 3.1 | 2.3 | 2.8 | 5.5 | 4.6 | 21.2 | 21.5 |
| 18 | 3.0 | 3.0 | 2.4 | 3.4 | 12.7 | 8.9 | 26.4 | 24.5 |
| 19 | 2.7 | 3.2 | 2.5 | 3.1 | 18.1 | 14.5 | 28.1 | 27.4 |
| 20 | 2.8 | 3.0 | 2.3 | 3.0 | 18.1 | 17.6 | 26.4 | 26.8 |
| 21 | 2.6 | 2.9 | 2.7 | 2.8 | 23.6 | 26.1 | 25.1 | 28.4 |
| 22 | 2.6 | 3.0 | 2.7 | 3.0 | 23.6 | 24.8 | 29.5 | 29.5 |
| 23 | 2.6 | 3.0 | 2.5 | 3.4 | 23.6 | 26.8 | 26.4 | 28.4 |

*Note:* N = number of observations.

*Sources:* See Table 10.1. Slave heights are derived from the manifest sample, see Richard H. Steckel, "Growth Depression and Recovery: The Remarkable Case of American Slaves," *Annals of Human Biology*, 14 (Mar.–Apr. 1987), pp. 111–32. Modern height standards are from J. M. Tanner, R. H. Whitehouse, and M. Takaishi, "Standards from Birth to Maturity for Height, Weight, Height Velocity, and Weight Velocity: British Children, 1965, Part II," *Archives of Disease in Childhood*, 41 (Dec. 1966), pp. 613–35.

adults the trend is flatter prior to the 1820s, the coefficients being either small or insignificant. The increase of the 1820s and the decline in height among the birth cohorts of the 1830s is as evident as among the youth (Tables 10.6 to 10.8). The decline of the 1830s is more pronounced among the urban than the rural population. In spite of the decline in physical stature, adults of the 1830s were only marginally shorter than those of the late eighteenth and early nineteenth centuries. The few observations available for the 1840s suggest that the downward trend continued.

Regional differences in the extent of the decline are noteworthy as well. In Baltimore the birth cohorts of the 1830s were 0.87 inches shorter than those of the 1820s, while those for the rest of Maryland were only 0.22 inches shorter (Table 10.8). Only among those born in Baltimore was mean height in the 1830s markedly below the level that prevailed at the turn of the century. This might indicate that the decline in nutritional status among the males was initially, at least, an urban phenomenon.

If the notion of "urban" is expanded to include Anne Arundel County, be-

**Table 10.3** **Height Indexes of Black Children in Antebellum America**[a]

| | | Male | | Female | | Total | |
|---|---|---|---|---|---|---|---|
| Sample | Characteristics | N | Height | N | Height | N | Height |
| 1) Maryland | Rural, born slave | 81 | 101.0 | 336 | 100.5 | 417 | 100.6 |
| 2) Maryland | All, born slave | 91 | 100.7 | 373 | 100.4 | 464 | 100.5 |
| 3) Maryland | Rural, all | 120 | 100.7 | 592 | 100.3 | 712 | 100.4 |
| 4) Maryland | Rural, free | 39 | 100.0 | 256 | 100.0 | 295 | 100.0 |
| 5) Maryland | All, free | 163 | 100.0 | 860 | 100.0 | 1,025 | 100.0 |
| 6) Maryland | All, free | 72 | 99.3 | 487 | 99.7 | 559 | 99.6 |
| 7) Louisiana | Imported | 365 | 98.8 | 350 | 100.3 | 715 | 99.5 |
| 8) Maryland | Urban, born slave | 10 | 98.5 | 37 | 99.6 | 47 | 99.4 |
| 9) Maryland | Urban, free | 33 | 98.4 | 231 | 99.4 | 264 | 99.3 |
| 10) Maryland | Urban, all | 43 | 98.4 | 268 | 99.4 | 311 | 99.3 |
| 11) Manifests | New Orleans | 866 | 98.9 | 999 | 98.9 | 1,865 | 98.9 |
| 12) Manifests | New Orleans | 1,603 | 97.9 | 1,558 | 98.7 | 3,161 | 98.3 |
| 13) Manifests | Shipped | 7,263 | 96.6 | 7,366 | 96.3 | 14,639 | 96.5 |
| 14) Manifests | Shipped | 383 | 96.1 | 313 | 94.2 | 696 | 95.2 |

*Notes:* 1) manumitted, not born free, rural; 2) manumitted, not born free, rural and urban; 3) manumitted, free born, rural and urban; 4) free born, rural; 5) manumitted, free born, rural and urban; 6) free born, rural and urban; 7) transported to Louisiana overland and by ship; 8) manumitted, urban; 9) free born, urban; 10) manumitted, free born, urban; 11) shipped to New Orleans and sold there; 12) all slaves disembarked at New Orleans; 13) all slaves shipped in interregional trade; 14) slaves disembarked at New Orleans but not sold there.

*Sources:* See Tables 10.2 and 10.4; Jonathan Pritchett and Herman Freudenberger, "A Peculiar Sample: The Selection of Slaves for the New Orleans Market" (manuscript, Tulane University, 1990).

[a]Males include children between the ages of 4 and 18; females, between the ages of 4 and 17. The index is standardized for age. The calculation is made by setting the heights in the Maryland sample (row 5) equal to 100.

cause of its proximity to Baltimore, the case is strengthened. If the state is divided in this manner, one finds that urban heights began to decline in the 1820s, that is, a decade before rural areas (Table 10.8). In sum, the small number of observations, as well as the considerable regional variation in height, make it difficult to determine the turning point in the trend of the men's physical stature. It is probable that the urban population suffered a decline in height, and thus in nutritional status, earlier than the rural population (Figure 10.3). Moreover, regressions (3) and (4) in Table 10.7 indicate that the antebellum decline in nutritional status could have been greater among those who were not born free than among those who were, even though those born free tended to be shorter prior to the 1840s.

Among females the number of observations is sufficient to ascertain the trend for youth, adults, and children. The trends for all three strongly support the inference of a decline in nutritional status in the antebellum period (Tables 10.6 and 10.9; Figure 10.3). The change in average height among females after the 1820s is quite similar to that experienced by men. The decrease for females born in bondage was not greater than that of the free born. The female

**Table 10.4** **Height of African-American Children by Status and Place of Birth, Antebellum Maryland**

| | Males | | | | | | | | | | |
|---|---|---|---|---|---|---|---|---|---|---|---|
| | Born in Baltimore | | | | | Born Outside Baltimore | | | | | |
| | Born Free | | Born Slave | | | Born Free | | Born Slave | | | |
| | (1) | (2) | (3) | (4) | | (5) | (6) | (7) | (8) | | |
| Age | N | Height | N | Height | (2) − (4) | N | Height | N | Height | (6) − (8) | ΔH |
| 5 | | | | | | | | | | | |
| 6 | | | | | | | | | | | |
| 7 | | | | | | | | | | | |
| 8 | | | | | | | | | | | |
| 9 | | | | | | | | | | | |
| 10 | | | | | | | | | | | |
| 11 | | | | | | | | | | | |
| 12 | | | | | | | | | | | |
| 13 | | | | | | | | | | | |
| 14 | 5 | 60.5 | | | | 2 | 58.4 | 11 | 60.2 | −1.8 | −0.3 |
| 15 | 1 | 60.5 | 2 | 61.1 | −0.6 | 4 | 64.4 | 5 | 64.6 | −0.2 | 0.0 |
| 16 | 3 | 64.9 | 1 | 63.0 | +1.9 | 8 | 63.6 | 18 | 64.2 | −0.6 | −0.2 |
| 17 | 10 | 63.2 | 2 | 63.1 | +0.1 | 8 | 63.9 | 22 | 65.5 | −1.6 | −1.8 |
| 18 | 14 | 64.1 | 5 | 65.6 | −1.5 | 17 | 66.3 | 25 | 66.6 | −0.3 | −1.1 |
| 19 | 20 | 66.2 | 5 | 66.4 | −0.2 | 21 | 66.7 | 50 | 66.4 | +0.3 | 0.0 |
| 20 | 15 | 65.9 | 7 | 63.8 | +2.1 | 51 | 67.1 | 95 | 66.2 | +0.9 | +0.8 |

| | Females | | | | | | | | | | | |
|---|---|---|---|---|---|---|---|---|---|---|---|---|
| | Born in Baltimore | | | | | Born Outside Baltimore | | | | | | |
| | Born Free | | Born Slave | | | Born Free | | Born Slave | | | | |
| | (9) | (10) | (11) | (12) | | (13) | (14) | (15) | (16) | | | Percent |
| Age | N | Height | N | Height | (10) − (12) | N | Height | N | Height | (14) − (16) | ΔH | Urban[a] |
| 5 | 3 | 40.7 | 2 | 39.6 | | | | 3 | 42.3 | | −0.6 | 55% |
| 6 | 5 | 39.7 | | | | | | | | | | |
| 7 | 7 | 46.0 | 1 | 42.0 | | 3 | 47.8 | 3 | 49.0 | | −0.7 | 44 |
| 8 | 4 | 46.0 | | | | 2 | 47.8 | 2 | 45.0 | | −0.4 | 33 |
| 9 | 4 | 48.9 | 2 | 47.0 | | 1 | 49.8 | 4 | 51.8 | | −1.1 | 43 |
| 10 | 7 | 54.9 | | | | 2 | 53.5 | 2 | 53.3 | | +1.3 | 67 |
| 11 | 6 | 54.8 | | | | 1 | 51.8 | 3 | 54.1 | | +0.3 | 53 |
| 12 | 13 | 56.6 | | | | 3 | 56.3 | 7 | 57.5 | | −0.9 | 52 |
| 13 | 12 | 59.4 | 2 | 60.9 | | 5 | 62.1 | 11 | 59.7 | | +0.3 | 42 |
| 14 | 24 | 58.4 | 1 | 57.0 | | 10 | 60.1 | 19 | 60.6 | 0.5 | −1.5 | 42 |
| 15 | 27 | 60.6 | 5 | 60.1 | +0.5 | 33 | 61.3 | 24 | 61.6 | −0.3 | −0.4 | 35 |
| 16 | 48 | 61.6 | 11 | 62.2 | −0.6 | 84 | 61.6 | 102 | 61.3 | −0.3 | +0.2 | 23 |
| 17 | 71 | 61.4 | 13 | 62.4 | −1.0 | 112 | 61.6 | 156 | 62.4 | −0.8 | −0.9 | 24 |
| 18 | 84 | 62.3 | 31 | 62.2 | +0.1 | 153 | 62.0 | 171 | 62.6 | −0.6 | −0.5 | 27 |
| 19 | 65 | 62.1 | 21 | 61.3 | +0.8 | 153 | 62.3 | 136 | 62.8 | −0.5 | −0.4 | 24 |
| 20 | 83 | 61.9 | 21 | 62.4 | −0.5 | 128 | 61.9 | 180 | 62.7 | −0.8 | −0.8 | 22 |

*Notes:* N = number of observations. ΔH = height of all those born free minus the height of all those born enslaved.

*Source:* See Table 10.1.

[a]Percentage of children (male and female) born in Baltimore City and County.

**Table 10.5** **Regressions on the Height of Free Black Youth**

| | Females (1) | Males (2) |
|---|---|---|
| Constant | 61.92* | 65.79* |
| Age | | |
| 19 | 0.17 | 0.00 |
| 18 | 0.15 | −0.31 |
| 17 | −0.34** | −1.68 |
| 16 | −0.80* | −2.76* |
| Birth decade | | |
| 1790s | 0.92* | 0.40 |
| 1800s | 0.80* | 0.65 |
| 1810s | 1.06* | 0.91** |
| 1820s | 0.40 | 1.32 |
| Birthplace: Baltimore | −0.60* | −1.46* |
| Born free | −0.50* | 0.32 |
| N | 1,603 | 346 |
| $R^2$ | .06 | .09 |
| $F$ | 9.6* | .46* |

*Note:* The constants refer to a twenty-year-old born as a slave outside Baltimore in the 1830s.
*Source:* See Table 10.1.
*Significant at the 5 percent level.
**Significant at the 10 percent level.

regional trend is similar to the male pattern as the decline in height (between 1800 and 1830) was greater among those born in Baltimore than in the rest of the state, 0.77 versus 0.50 inches (Table 10.10).

An important difference between the trend in the height of males and that of females is that the decline in height among the latter group began earlier. The pattern does not change much if Anne Arundel County is included in the urban group. Baltimore women began to experience a decline in height in the 1810s, while those born in the rest of the state experienced the decline in the 1820s, that is, about a decade before men. As among men, the deterioration in nutritional status of women began earlier in urban areas. In the 1830s, rural women were between 0.5 and 0.8 inches shorter than birth cohorts of the 1800s, while rural men were about equally tall. In other words, in rural areas women appear to have fared worse in nutritional terms than did men, but in the city perhaps about the same.

Regressions on the height of female youth confirm the pattern obtained for adult women (Table 10.5). The fall in nutritional status among females between the ages of sixteen and twenty also started earlier than it did among males and was greater. By the 1830s the decline in the height of female youth appears to have been close to an inch. This is not unreasonable if one notes that the height of youth is more sensitive to nutritional stress than is terminal height because growth can cease at later ages.

Table 10.6 Indexes of the Trend in the Height of the Free Black Population

| | Girls[a] | Youths[b] | Adults[c]: Born in Baltimore | Adults[c]: Born Oustide Baltimore |
|---|---|---|---|---|
| | | *Females (1810 = 100.0)* | | |
| 1760 | | | | 99.7 |
| 1770 | | | | 100.4 |
| 1780 | | | 100.5 | 99.8 |
| 1790 | 98.1 | 99.9 | 99.9 | 100.2 |
| 1800 | 99.2 | 99.7 | 100.5 | 100.0 |
| 1810 | 100.0 | 100.0 | 100.0 | 100.0 |
| 1820 | 99.7 | 99.1 | 99.4 | 99.6 |
| 1830 | 98.5 | 98.4 | 99.2 | 99.2 |
| 1840 | 98.1 | | | 98.6 |
| | | *Males (1820 = 100.0)* | | |
| 1760 | | | | 99.0 |
| 1770 | | | 99.8 | 99.2 |
| 1780 | | | 99.0 | 99.8 |
| 1790 | | 98.6 | 99.8 | 99.4 |
| 1800 | | 99.0 | 99.9 | 99.8 |
| 1810 | | 99.4 | 99.5 | 99.7 |
| 1820 | | 100.0 | 100.0 | 100.0 |
| 1830 | | 98.0 | 98.7 | 99.7 |
| 1840 | | | | 99.3 |

*Sources:* Tables 10.5, 10.8, and 10.10
[a]Between the ages of 5 and 15.
[b]Between the ages of 16 and 20.
[c]For females, between the ages of 18 and 50; for males, between the ages of 21 and 50.

In sum, the downward tendency in nutritional status is evident among both males and females, and among all age brackets for which evidence exists. It is stronger and began earlier among females and in areas experiencing urbanization. These results are consistent with the trend obtained from the Civil War sample. Between the birth cohorts of the 1820s and the 1840s, the estimated diminution in the height of adult black recruits into the Union army is 0.4 inches.[23] This is practically identical to the decline found among West Point cadets, among white Union soldiers, and also among the free blacks of Maryland (Table 10.8).[24] Another similarity between the Union Army and Maryland samples is that the men born in the second half of the 1830s were

23. Margo and Steckel, "The Heights of American Slaves," pp. 526–27; Fogel, *Without Consent or Contract*, p. 361; John Komlos, "The Height and Weight of West Point Cadets: Dietary Change in Antebellum America, 1820–1880," *Journal of Economic History*, 47 (Dec. 1987), pp. 897–927.
24. No evidence has been collected on the height of white females in the antebellum period.

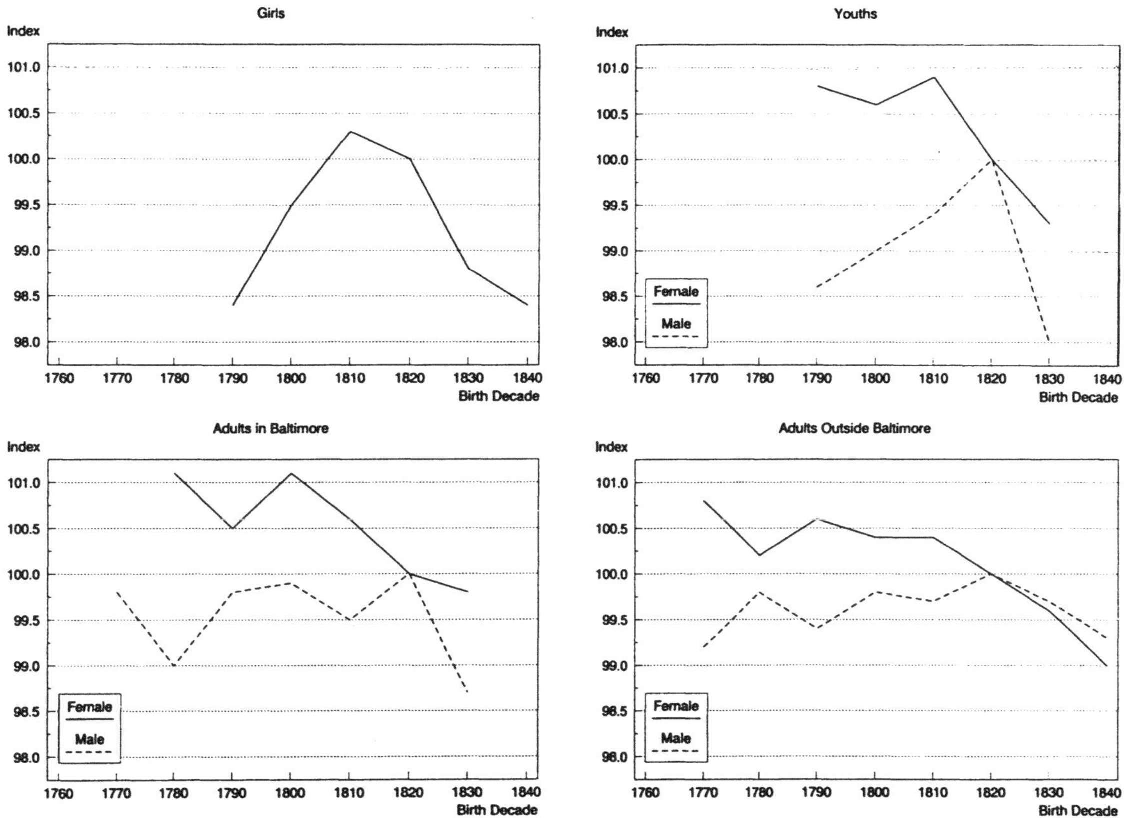


**Fig. 10.3 Secular Trends in the Heights of African-Americans by Birth Decade, 1760–1840**
*Source:* See Table 10.1.

**Table 10.7** **Regressions on the Height of Free Adult Black Males**

| | Entire Sample | | Born Enslaved | Born Free | Entire Sample |
|---|---|---|---|---|---|
| | (1) | (2) | (3) | (4) | (5) |
| Constant | 66.99* | 67.02* | 67.02* | 66.77* | 66.95* |
| Age, 20–22 | −0.21* | −0.20* | −2.6* | −0.12* | 0.22* |
| Birth decade | | | | | |
| 1750 | 0.14 | 0.13 | | | |
| 1760 | −0.35 | −0.37 | −0.32 | | −0.31 |
| 1770 | −0.11 | −0.14 | −0.11 | | −0.04 |
| 1780 | 0.10 | 0.08 | 0.16 | −0.36 | −0.12 |
| 1790 | −0.07 | −0.08 | −0.08 | 0.10 | −0.06 |
| 1800 | 0.16 | 0.16 | 0.19 | 0.22 | 0.17 |
| 1810 | 0.10 | 0.10 | 0.16 | −0.08 | 0.11 |
| 1820 | 0.30* | 0.29* | 0.35* | 0.22 | 0.31* |
| 1840 | −0.16 | −0.19 | −0.50 | 0.02 | −0.18 |
| Region of birth[a] | | | | | |
| Eastern Shore | −0.08 | −0.07 | −0.18 | 0.41* | |
| Baltimore City | −0.83* | −0.84* | 0.72* | } −0.65* | −0.62* |
| Baltimore County | −0.52* | −0.51* | −0.50* | } | −0.55* |
| North[b] | −0.18 | −0.20 | −0.22 | | −0.17 |
| South | −0.17* | 0.14 | −0.18 | 0.08 | −0.05 |
| Color | | | | | |
| Brown | −0.09 | | 0.01 | −0.12 | −0.11 |
| Dark | −0.06 | | −0.07 | −0.03 | −0.03 |
| Light | 0.09 | | 0.00 | 0.23 | 0.15 |
| Yellow | 0.07 | | −0.01 | 0.47 | 0.05 |
| Copper | 0.60* | | 0.59* | 0.85 | 0.56* |
| Chestnut | 0.07 | | 0.08 | 0.40 | 0.03 |
| Mulatto | 0.05 | | 0.04 | 0.12 | 0.10 |
| Bright | 0.54** | | 0.58** | 0.16 | 0.49** |
| Born free | 0.19* | 0.16** | | | 0.17* |
| N | 6,251 | 6,251 | 4,780 | 1,461 | 6,242 |
| $R^2$ | .04 | .04 | .04 | .04 | .04 |
| $F$ | 2.8* | 3.5* | 2.3* | 1.8* | 2.9* |

*Notes:* Dependent variable is the height of adult black males. All values are in inches. Constants refer to a male between the ages of 23 and 50, born in the 1830s; in equations (1) and (3) they refer, in addition, to a man whose place of birth was unknown and who was born a slave, with black skin color; in equation (2), to a man whose place of birth was unknown and who was born a slave; equation (4) is the same as (3), but the constant refers to a man born free instead of as a slave; equation (5) is the same as (1), but region of birth is replaced by region of residence.

*Source:* See Table 10.1

[a]In equation (5), region of birth is replaced by region of residence.

[b]Does not include Baltimore City or County.

*Significant at the 5 percent level.

**Significant at the 10 percent level.

unusually short in both. The terminal height of the Maryland men was identical to the one estimated from the manifest sample, almost the same as that of the Civil War black soldiers, but somewhat shorter than that of the freed slaves of Virginia (Table 10.11).[25]

25. The preliminary (Coopersmith-Steckel) sample of manifests appears to be more homogeneous than its enlarged counterpart, as the standard deviations of its height estimates are within

**Table 10.8** **Height of Adult Black Men by Region of Birth**

| Decade | Eastern Shore | | Baltimore | | North | | South | | Total | | Outside Baltimore | | Urban | | Rural | |
|---|---|---|---|---|---|---|---|---|---|---|---|---|---|---|---|---|
| of Birth | Height | N | Height | N | Height | N | Height | N | Height | N | Height | N | Height | N | Height | N |
| 1760 | 66.80 | 70 | 66.15 | 13 | 65.96 | 13 | 66.32 | 31 | 66.53 | 127 | 66.57 | 114 | | | 66.7 | 101 |
| 1770 | 66.93 | 218 | 66.76 | 61 | 66.58 | 53 | 66.38 | 72 | 66.73 | 404 | 66.76 | 343 | 66.7 | 102 | 66.8 | 300 |
| 1780 | 67.04 | 410 | 66.19 | 117 | 67.13 | 86 | 67.27 | 159 | 66.97 | 772 | 67.11 | 655 | 66.2 | 145 | 67.1 | 556 |
| 1790 | 66.87 | 555 | 66.73 | 75 | 66.80 | 119 | 66.93 | 197 | 66.86 | 946 | 66.87 | 871 | 66.9 | 82 | 67.0 | 631 |
| 1800 | 67.18 | 527 | 66.79 | 96 | 67.08 | 125 | 67.02 | 202 | 67.09 | 950 | 67.12 | 854 | 66.7 | 120 | 67.1 | 595 |
| 1810 | 67.10 | 628 | 66.52 | 60 | 66.64 | 111 | 67.34 | 171 | 67.05 | 970 | 67.09 | 910 | 67.3 | 94 | 67.0 | 579 |
| 1820 | 67.38 | 544 | 66.88 | 93 | 67.20 | 123 | 67.08 | 264 | 67.24 | 1,024 | 67.27 | 931 | 66.9 | 234 | 67.4 | 541 |
| 1830 | 67.09 | 468 | 66.01 | 93 | 66.78 | 125 | 67.08 | 303 | 66.95 | 989 | 67.05 | 896 | 66.7 | 181 | 67.2 | 352 |
| 1840 | 66.91 | 24 | 66.30 | 5 | 66.90 | 5 | 66.72 | 26 | 66.77 | 60 | 66.82 | 55 | | | | |

*Notes:* Adult includes those between the ages of 21 and 50. Baltimore includes Baltimore City and County. Urban includes Baltimore City and County, and Anne Arundel County. N = number of observations.

*Source:* See Table 101.1

**Table 10.9** **Regressions on the Height of Adult Black Females**

| | Entire Sample (1) | Born Enslaved (2) | Born Free (3) |
|---|---|---|---|
| Constant | 62.00* | 61.92* | 61.73* |
| Age, 20–22 | 0.04 | 0.23* | −0.21 |
| Birth Decade | | | |
| 1760 | 0.31 | 0.52 | |
| 1770 | 0.76* | 0.92* | } |
| 1780 | 0.46* | 0.62* | } 0.56* |
| 1790 | 0.57* | 0.70* | } |
| 1800 | 0.57* | 0.62* | 0.78* |
| 1810 | 0.43* | 0.57* | 0.53* |
| 1820 | 0.22* | 0.37* | 0.20 |
| 1840 | −0.46 | | |
| Region of birth | | | |
| Eastern Shore | −0.16** | 0.08 | 0.46** |
| Baltimore City | −0.28* | −0.31* | −0.10 |
| Baltimore County | −0.26 | −0.18 | −0.18 |
| North | 0.06 | 0.00 | 2.17 |
| South | −0.17 | 0.21** | −0.05 |
| Color | | | |
| Brown | −0.28* | −0.35* | −0.07 |
| Dark | 0.01* | −0.01 | 0.03 |
| Light | 0.05 | −0.02* | 0.12 |
| Yellow | 0.15** | 0.01 | 0.65 |
| Copper | 0.25 | 0.22 | 0.81 |
| Chestnut | 0.03 | −0.05 | 0.13 |
| Mulatto | 0.20 | 0.22 | 0.18 |
| Bright | 0.38 | 0.33 | 0.46 |
| Born free | 0.02 | | |
| *N* | 6,061 | 4,247 | 1,812 |
| $R^2$ | .01 | .01 | .01 |
| *F* | 3.8* | 2.5* | 2.7* |

*Notes:* Dependent variable is the height of adult black females. All values are in inches. For definition of constants see Table 10.7.

*Source:* See Table 10.1

*Significant at the 5 percent level.

**Significant at the 10 percent level.

---

the acceptable range (Steckel, "Slave Height Profiles," pp. 364, 368). The trend in heights from the initial sample agrees with the trend outlined above, and also with that of the Civil War sample in many respects: a) male stature increased in the 1820s and decreased in the 1830s; b) the decline in female adult stature began earlier than among males and continued into the antebellum period; and c) the decline among female youth was larger than among adults in the 1840s (Steckel, "Slave Height Profiles," p. 377). The only noteworthy difference between the first manifest sample and the Maryland results is that, according to the former, the height of female youth rose until the 1830s instead of declining monotonically as in the Maryland sample. Apparently, something important happened in the expansion of the manifest sample that influenced the estimated trends.

**Table 10.10** **Height of Adult Black Women By Region of Birth**

| Decade of Birth | Eastern Shore | | Baltimore | | North | | South | | Total | | Outside Baltimore | | Urban | | Rural | |
|---|---|---|---|---|---|---|---|---|---|---|---|---|---|---|---|---|
| | Height | N | Height | N | Height | N | Height | N | Height | N | Height | N | Height | N | Height | N |
| 1760 | 62.65 | 35 | 61.77 | 9 | 61.75 | 3 | 62.20 | 25 | 62.35 | 72 | 62.42 | 63 | | | 61.7 | 45 |
| 1770 | 62.93 | 139 | 67.81 | 28 | 62.70 | 19 | 62.53 | 77 | 62.7[illegible] | 263 | 62.78 | 235 | 62.7 | 76 | 62.8 | 187 |
| 1780 | 62.56 | 326 | 62.78 | 91 | 62.80 | 45 | 62.19 | 221 | 62.49 | 683 | 62.44 | 592 | 62.4 | 232 | 62.5 | 460 |
| 1790 | 62.91 | 434 | 62.15 | 156 | 62.91 | 76 | 62.32 | 342 | 62.59 | 1,008 | 62.67 | 852 | 62.3 | 409 | 62.8 | 664 |
| 1800 | 62.68 | 515 | 62.48 | 386 | 62.53 | 72 | 62.45 | 384 | 62.55 | 1,357 | 62.58 | 971 | 62.4 | 755 | 62.8 | 715 |
| 1810 | 62.73 | 429 | 62.20 | 374 | 62.11 | 81 | 62.42 | 267 | 62.44 | 1,151 | 62.56 | 777 | 62.3 | 791 | 62.7 | 593 |
| 1820 | 62.35 | 342 | 61.81 | 160 | 62.33 | 98 | 62.32 | 248 | 62.24 | 848 | 62.33 | 648 | 61.9 | 457 | 62.4 | 519 |
| 1830 | 62.07 | 257 | 61.71 | 77 | 61.92 | 98 | 62.17 | 208 | 62.04 | 640 | 62.08 | 563 | 61.9 | 310 | 62.0 | 479 |
| 1840 | 61.37 | 15 | 61.26 | 5 | 62.38 | 4 | 61.79 | 26 | 61.66 | 50 | 61.70 | 45 | 61.2 | 71 | 62.2 | 96 |

*Notes:* Adult includes women between the ages of 18 and 50. Baltimore includes Baltimore City and County. Urban includes Baltimore City and County, and Ann Arundel County. N = number of observations.

*Source:* See Table 10.1.

Table 10.11 **Various Estimates of Heights of Adult African-Americans**

| | | | Male | | Female | | |
|---|---|---|---|---|---|---|---|
| Status | Birthplace | Period[a] | N | Height | N | Height | Source |
| 1) Free[b] | Virginia | 1782–1861 | 433 | 67.9 | 400 | 63.0 | |
| 2) Soldiers | Atlantic[c] | 1820s | 3,651 | 67.1 | | | Army |
| 3) Slaves | Baltimore[d] | 1820s[e] | 9,726 | 67.3 | 6,745 | 62.8 | Manifests |
| 5) Free | Maryland | 1820s | 6,251 | 67.3 | 6,048 | 62.2 | Certificates[f] |
| 6) Slaves[g] | Upper South | 1780–1835 | 743 | 67.2 | 256 | 63.1 | Manifests |
| 7) Slaves[h] | Upper South | 1780–1835 | 359 | 67.2 | 260 | 62.4 | Manifests |
| 8) Slaves[i] | Upper South | 1780–1815 | 313 | 67.1 | 87 | 63.0 | Certificates[j] |

*Sources:* See Tables 10.6 and 10.8; Robert Margo and Richard H. Steckel, "Heights of American Slaves: New Evidence on Slave Nutrition and Health," *Social Science History*, 6 (Fall 1982), pp. 519, 520, 526, 533; Jonathan Pritchett and Herman Freudenberger, "A Peculiar Sample: The Selection of Slaves for the New Orleans Market" (manuscript, Tulane University, 1990).

[a]Periods, except row 1, refer to years of measurement.

[b]Numbers of males and females are approximate. Total number of observations is 833.

[c]Upper South Atlantic.

[d]Transported on ships departing from Baltimore.

[e]For women, 1780–1840.

[f]Maryland Certificates of Freedom.

[g]Sold in New Orleans.

[h]Disembarked but not sold in New Orleans.

[i]Transported to Louisiana overland and by ship.

[j]Certificates of Good Character.

### 10.2.3 Regional Pattern

The variation in height between urban and rural regions accords with practically all samples studied thus far. Until improvements in agricultural technology and food distribution toward the end of the nineteenth century, when the urban epidemiological environment also became healthier, the nutritional status and physical stature of urban dwellers were consistently lower than for the rural population. The relative price of nutrients was also higher in towns, and the urban population was less likely to produce even part of its food requirement.

Thus, the free adult male blacks of Baltimore were as much as 0.5 to 0.8 inches shorter than those born in rural areas (Tables 10.7 and 10.8). Among females the difference was somewhat smaller, but it was significant (Tables 10.3, 10.5, 10.9 and 10.10). The rural advantage increased over time. Before 1810, Baltimore men in the sample were about 0.36 inches shorter than rural men, while those born thereafter were 0.67 inches shorter. The pattern is similar among women: the rural-urban difference increased from 0.17 to 0.33 inches during the same period.

Nineteenth-century Maryland had three distinct economic regions. Industrial activity was located in the northern counties of the state. According to the census of 1850, manufacturing output per capita in Northern Maryland, where 90 percent of the industrial workers were located, was about seven times as large as in the rest of the state. Baltimore held a dominant position in manufacturing. In the other parts of the northern region, truck farming, animal husbandry, and grain-growing were important agricultural pursuits. Just two of the counties produced one-third of the state's wheat crop. The value of an improved acre of farmland was almost twice that of the other counties. Slavery was tangential to this economic system; blacks were not more than one-sixth of the population.[26]

In contrast, the counties of Southern Maryland remained economically less developed. This was a relatively backward agricultural region, with the exception of areas around Washington, D.C., and Annapolis, the state capital with a population of 3,000 in 1850. Tobacco remained the primary crop, but wheat production did make some inroads. Here, the number of whites declined during the antebellum decades, and blacks constituted more than half of the population by the end of the period under consideration.

The third main region, the Eastern Shore, occupied an intermediate economic position. Less industrialized and urbanized than northern Maryland and not as dependent on tobacco as the southern counties, farmers of the Eastern Shore cultivated cereal. Blacks represented two-fifths of the population by 1850. In the rural economy, free blacks generally earned a living as hired agricultural laborers; in Baltimore they were concentrated in menial occupations.

Although heights did not vary much among the three main regions of Maryland, they did so at the county level (Tables 10.7 to 10.10). The rank order of the mean height by counties was ascertained in two ways. First, simple averages of adult heights in the entire sample were considered (Table 10.12). As these were quite similar for both sexes, more precise estimates were made using regression analysis, holding the decade of birth constant. Rank orders are similar to those based on the raw means. The final rank was calculated by averaging the rank orders of both sexes as obtained through the regressions.

Baltimore was excluded from further spatial analysis because, as noted above, it is obvious why it had the shortest population in the state. The remainder of the available sample was divided into the three main regions of Maryland, the industrially and economically developed North, the more backward tobacco-producing South, and the grain-producing Eastern Shore. Within each region the county with the tallest population (G1) was compared with the other counties (G3). The counties of the Eastern Shore were divided into three groups, instead of two, as there were enough counties represented in the sample to create a middle category, G2. Heights in the G1 counties all ranked in the top half of the fourteen counties represented in the sample. Av-

26. Blacks constituted a somewhat higher proportion of the labor force; see Barbara J. Fields, *Slavery and Freedom on the Middle Ground: Maryland during the Nineteenth Century* (New Haven, 1985).

Table 10.12 **Heights of Maryland Free Blacks by County**

| | | | Raw Means[a] | | | | | | Regression Results[b] | | | | | |
|---|---|---|---|---|---|---|---|---|---|---|---|---|---|---|
| | | | Male | | | Female | | | Male | | Female[c] | | Final Rank[d] | |
| | | County | N | Rank | Height | N | Rank | Height | Rank | Height | Rank | Height | A | B |
| No | G1 | Harford | 164 | 1 | 68.09 | 148 | 1 | 63.41 | 1 | 1.74* | 1 | 1.22* | 1 | 1 |
| ES | G1 | Dorchester | 890 | 4 | 67.26 | 598 | 2 | 62.81 | 5 | .76* | 3 | 0.72* | 2 | 2 |
| So | G1 | Prince George | 346 | 5 | 67.16 | 244 | 5 | 62.68 | 4 | .81* | 4 | 0.52* | 3 | 4 |
| No | G1 | Washington | 118 | 3 | 67.42 | 77 | 6 | 62.63 | 3 | .83* | 6 | 0.45 | 4 | 5 |
| So | G1 | St. Mary's | 497 | 6 | 67.08 | 268 | 4 | 62.76 | 7 | .58* | 2 | 0.82* | 5 | 3 |
| ES | G1 | Talbot | 618 | 6 | 67.10 | 513 | 3 | 62.78 | 6 | .59* | 5 | 0.47* | 6 | 6 |
| ES | G1 | Kent | 250 | 2 | 67.52 | 153 | 12 | 62.16 | 2 | .88* | 10 | 0.17 | 7 | 7 |
| ES | G2 | Queen Ann | 264 | 10 | 66.93 | 195 | 9 | 62.35 | 10 | .37* | 8 | 0.19 | 8 | 8 |
| ES | G2 | Caroline | 601 | 9 | 66.98 | 360 | 11 | 62.24 | 8 | .38* | 11 | 0.12 | 9 | 9 |
| No | G3 | Frederick | 641 | 11 | 66.81 | 408 | 10 | 62.31 | 11 | .31* | 9 | 0.22 | 10 | 10 |
| ES | G3 | Somerset | 561 | 12 | 66.56 | 344 | 7 | 62.38 | 13 | −.00 | 7 | 0.36* | 11 | 12 |
| So | G3 | Anne Arundel | 486 | 8 | 67.02 | 1,146 | 13 | 62.15 | 9 | .37* | 13 | 0.02 | 12 | 11 |
| No | | Baltimore County | 525 | 14 | 66.52 | 1,347 | 8 | 62.37 | 14 | −.02 | 12 | 0.12 | 13 | 13 |
| No | | Baltimore City | 260 | 13 | 66.59 | 247 | 14 | 61.96 | 12 | 66.47 | 14 | 61.68 | 14 | 14 |

*Notes:* N = number of observations. No = North; So = South; ES = Eastern Shore. G1 = counties that had the tallest populations in their respective regions; G2 = counties whose population had intermediate physical stature within the region; G3 = counties with the shortest population.

[a]Adults only.

[b]Other coefficients of the regression are not reported. The variables included were: decade of birth, born free, and age 20–22. Results for Baltimore City are given for the birth cohorts of the 1830s. The results for the other counties are the average for the whole period by which heights exceeded that of Baltimore.

[c]Between the ages of 16 and 50; hence, in addition to the variables included in the male regression, dummy variables were included for age below 20.

[d]Final rank: A) Average rank of the two sexes. B) The rank according to the mean of the two regression coefficients.

*Significant at the 5 percent level.

eraged across the two sexes, heights in the G1 group were more than half an inch greater than those of the free black population of Baltimore City. In contrast, heights in the other counties were at most about a quarter inch above those of Baltimore. In other words, in nutritional terms there was a noticeable gap between G1 and the other counties.[27]

The black populations of Anne Arundel, Somerset, and Frederick counties (referred to as the G3 counties) were the shortest (Table 10.12).[28] These counties also tended to experience faster population growth (or slower population decline) than the G1 counties (Table 10.13). This is also true if the population is divided into three components, whites, free blacks, and slaves. Thus, the biological standard of living was higher in counties where demographic expansion did not excessively strain the resource base and did not unduly increase competition in the labor market.[29] An exception is found among the white population of Northern Maryland, which grew faster in G1 counties than in G3 counties. Yet, despite the more rapid population growth in Washington and Harford counties (G1), the per capita output of nutrients remained close to that of Frederick County. For Southern Maryland and the Eastern Shore, without exception per capita nutrient production was considerably higher in G1 and G2 than in G3 counties for both 1840 and 1850 (Table 10.13). This confirms the pattern found in other data sets, namely, that during the early stages of industrialization propinquity to nutrients improved nutritional status.

Increased money income from manufacturing employment was, by itself, not inimical to nutritional status, that is, as long as it did not displace agricultural production. Thus, in the G1 counties of Northern Maryland, per capita industrial output was higher than in the G3 county, without having a deleterious effect on nutritional status.[30] Yet manufacturing income did not compensate for, or was not as good as, agricultural income for the maintenance of nutritional status. The slightly higher per capita industrial output of Somerset County did not compensate for its lower per capita output of nutrients. The

27. In other studies, skin color has been found to be a significant correlate of slave heights, inasmuch as light-skinned slaves were born in the New World. Thus, in the Civil War sample, light-skinned slaves were 0.17 inches (and significantly) taller than average (Margo and Steckel, "The Heights of American Slaves," p. 520). This result is not found in the Maryland data, perhaps because in this sample most blacks were born in America.

28. The eighteenth-century residents of Somerset County were among the poorest of the region. Tobacco was a staple, but its land was less fertile than other areas of the Chesapeake. See Russell R. Menard and Lorena S. Walsh, "The Demography of Somerset County, Maryland: A Progress Report," *The Newberry Papers in Family and Community History* (July 1981). The discussion of food consumption based on census records should include the caveat that there is no available evidence on some items, such as game and seafood production.

29. On competition between white and free black labor, see Wright, *The Free Negro in Maryland*, pp. 159, 172. Free blacks also had to compete with slaves who were hired out and accepted smaller wages than whites; see John H. Russell, *The Free Negro in Virginia, 1619–1865* (New York, 1913), pp. 146–47.

30. The regional pattern in northern Maryland must be considered tentative because sample sizes are very small and are mostly for the early part of the century, before nutritional status declined. In addition, evidence is not available for several counties of the region.

**Table 10.13 Demographic and Economic Correlates of Regional Variation in Height within Maryland**

Demographic Indicators

| Region | % Change in Population, 1790–1850 | | | |
|---|---|---|---|---|
| | White | Black | | Total |
| | | Free | Slave | |
| Eastern Shore | | | | |
| G1 | −2% | 324% | −29% | 5% |
| G2 | 14 | 484 | −42 | −3 |
| G3 | 62 | 1,200 | −21 | 44 |
| North | | | | |
| G1 | 64 | 448 | −10 | 63 |
| G3 | 24 | 1,665 | 8 | 33 |
| Southern Shore | | | | |
| G1 | −17 | 446 | −4 | −4 |
| G3 | 42 | 472 | 11 | 43 |

Economic Indicators, Per Capita

| | Output | | | | | | | |
|---|---|---|---|---|---|---|---|---|
| | Wheat | Corn | Tobacco | Cattle | Swine | Improved Acres | Animals Slaughtered | Industrial Output |
| **1840** | | | | | | | | |
| Eastern Shore | | | | | | | | |
| G1 | 10.6 | 35.9 | 0 | 0.77 | 1.3 | | | |
| G2 | 6.8 | 37.9 | 0 | 0.72 | 1.1 | | | |
| G3 | 3.5 | 13.1 | 0 | 0.52 | 0.9 | | | |
| North | | | | | | | | |
| G1 | 17.8 | 21.3 | 0 | 0.62 | 1.3 | | | |
| G3 | 20.2 | 19.4 | 9 | 0.58 | 1.4 | | | |
| South | | | | | | | | |
| G1 | 4.5 | 23.3 | 370 | 0.63 | 1.3 | | | |
| G3 | 7.0 | 19.1 | 136 | 0.47 | 1.1 | | | |
| **1850** | | | | | | | | |
| Eastern Shore | | | | | | | | |
| G1 | 13.7 | 40.3 | 0 | 0.68 | 1.0 | 8.0 | 5.5 | 5.8 |
| G2 | 10.2 | 49.7 | 0 | 0.71 | 1.0 | 13.0 | 5.5 | 5.8 |
| G3 | 7.0 | 16.9 | 0 | 0.44 | 0.7 | 5.7 | 5.0 | 6.9 |
| North | | | | | | | | |
| G1 | 9.8 | 17.6 | 0 | 0.47 | 0.8 | 5.6 | 4.8 | 47.9 |
| G3 | 17.9 | 19.9 | 4 | 0.45 | 0.9 | 6.2 | 5.9 | 39.1 |
| South | | | | | | | | |
| G1 | 11.0 | 30.4 | 288 | 0.60 | 1.0 | 8.3 | 5.4 | 13.7 |
| G3 | 11.1 | 28.6 | 140 | 0.43 | 0.9 | 6.9 | 1.9 | 28.9 |

*Notes:* For region and group designations see Table 10.12. Grain output is measured in bushels, animal stock in heads, tobacco in pounds, industrial output in dollars.

*Sources:* U.S. Census Office, Seventh Census, 1850, *Statistical View of the United States . . . Being a Compendium of the Seventh Census, etc.*, J. D. B. DeBow, ed. (Washington, D.C., 1854), p. 178.

pattern is particularly revealing in Southern Maryland. There, the per capita output of industrial products was twice as high in Anne Arundel (the G3 county) as in the G1 counties (St. Mary's and Prince George). If one were to add the value of tobacco output to the value of manufactured products, the G1 and G3 counties would have about the same amount of non-nutrient output by value. Nonetheless, the population in the G1 counties was taller, even though their per capita output of nutrients was not much greater (with the exception of livestock slaughtered) than that of Anne Arundel County (Table 10.13).[31] This suggests, once again, that being close to the source of nutrients confers nutritional advantages. While tobacco and industrial goods were both inedible, tobacco was complementary to nutrient production while industrial goods were not to the same extent. Farmers who produced tobacco and raised foodstuffs did not have to pay the cost of transporting food and could supplement their earnings by gardening, which might not be reflected in the census figures. In addition, manufacturing was connected with commerce, and consequently with the movement of people, thereby increasing the exposure to childhood diseases which may have increased nutritional stress among the young.

## 10.3 The Maryland and Manifest Samples

As far as the adult parts of the two samples are concerned, the results are practically identical (Table 10.11), but the children's components differ. A careful comparison of the two samples is warranted, because nutritional evidence on black children is quite rare. It is difficult to determine which sample is closer to reproducing the nutritional reality of the average slave child, because both samples have weaknesses and neither was randomly drawn from the African-American population.

The Maryland sample size is small for the young. Yet the consistency with which the height of free children exceeds that of the transported children suggests that it is not the small number of observations that is causing the results to diverge (Figures 10.1 and 10.2). If sample size were the crucial factor, one would expect much more variation in height about the manifest estimates. Yet, it is possible that the free black children were taller than the slave children whose heights were recorded in the manifests on account of their legal status. To be included in the Maryland sample would require a child to have had a free mother or to have been set free by his or her master.[32] In the latter case, manumission probably depended on the religious, ethical, and political con-

31. The propinquity to Baltimore was an additional drain on nutrients for Anne Arundel County.

32. If the mother was a slave, the child became a slave even if the father was free. Manumissions were most often effective either upon the death of the owner or at some specified age of the slave (Wright, *The Free Negro in Maryland*, p. 43). In addition, a free black could purchase the freedom of slaves, and a slave could redeem him or herself as well.

victions of the owner rather than on the physical attributes of the child.[33] If masters with such convictions also treated their slaves better than was typical, then one would expect that manumitted children would have higher nutritional status than those belonging to less benevolent masters.

However, children who were born free would not have benefited from such largess. Their parents were probably common laborers whose work in rural areas complemented that of the slaves. To be sure, there was some social differentiation among free blacks, as among the slaves themselves. Some free blacks occupied skilled positions, a few accumulated property, and fewer still even became well-to-do. While the level of per capita income among free blacks is not known precisely, their average socio-economic position was probably no better than that of slaves: "the environment seemed to foster the preservation of conditions existing before they became free," and in the agricultural sector they were "offered arrangements which were strikingly like those of involuntary servitude."[34]

Slaves had certain advantages over free blacks, since they had a more secure access to medical attention, even if rudimentary, and they were not

33. Kulikoff, *Tobacco and Slaves*, p. 419. The restrictions on manumissions were eased in 1790. The Quakers and Methodists became particularly devoted to the abolitionist cause. See Wright, *The Free Negro in Maryland*, pp. 39, 44–46.

34. Wright, *The Free Negro in Maryland*, pp. 43, 152–53, 157, 160, 171, 239, 243, 247. A bushel of meal and 15–20 pounds of meat allotment for contract laborers was on the order of slave rations, but below that of white men (p. 164). "The allowances for the slaves thus became a sort of standard to which free negroes aimed to attain when providing for themselves. The majority fared apparently about as well as did the slaves excepting sometimes in the winter season" (p. 241). In 1860 about 10 percent of the free black families owned some property, but the average amounted to no more than $13 per capita, or 2.5 percent of white wealth (p. 185). "They seemed in many cases to have believed that their material condition would have been just as good in slavery as in freedom" (p. 258). "The two classes shared . . . the same standards of living . . . it seems certain that the average slave was better provided for than was many a free negro. Had the 83,942 free negroes exchanged places with the 87,189 slave negroes in 1860, but little difference in the material welfare of the majority of either class would probably have resulted" (p. 259). Also see Russell, *The Free Negro in Virginia, 1619–1865*, pp. 130, 145. "The free negro was not infrequently a better 'slave' than his kinsman in bondage" (p. 148). "The occupations of persons of this class [free blacks] are nearly the same as those of slaves" (George Tucker, *Progress of the United States in Population and Wealth in Fifty Years, as exhibited by the Decennial Census from 1790 to 1840* [New York, 1855], p. 139, as cited in Russell, p. 150). Barbers were supposedly the most prosperous occupational group among free blacks, see Russell, p. 151. See also Ira Berlin, *Slaves without Masters: The Free Negro in the Antebellum South* (New York, 1974), p. 218; and Leonard P. Curry, *The Free Black in Urban America, 1800–1850: The Shadow of the Dream* (Chicago, 1981). The poverty of free blacks exposed them to a higher incidence of cholera: "In Philadelphia, the case rate among Negroes was almost twice as great as that among whites—probably a reliable, if informal, index to the poverty in which the North's free Negroes lived" (Charles E. Rosenberg, *The Cholera Years: The United States in 1832, 1849, and 1866*[(2d edn., Chicago, 1987], pp. 59–60). Moreover, children who were born free could have had a slave father, blurring further the material significance of the legal distinction. In addition, some free children were neglected sufficiently to become wards of the county, and some were apprenticed out by county authorities. For a contemporary debate over the mental health of free blacks compared to that of slaves see William Stanton, *The Leopard's Spots: Scientific Attitudes Toward Race in America, 1815–1859* (Chicago, 1959), p. 58.

threatened by cyclical unemployment to the same extent as free blacks.[35] Hence, free-born black children probably did not fare better than their slave counterparts. Consequently their heights should approximate the overall slave average.[36] If, however, manumitted children came from benevolent owners, their nutritional status could have been higher than that of children who were born free. Perhaps this is why children in the Maryland sample who were born free tended to be shorter than those who were born enslaved (Tables 10.3 and 10.4). The pattern is most vivid among female youth, whose height difference is close to half an inch in favor of those born into slavery (Table 10.5). The pattern does not hold among adults; with more time to reach terminal height, adults' nutritional status affected mainly the tempo of growth. The nutritional advantage of having been born a slave is clearer and more consistent among rural children than among urban ones (Tables 10.3 and 10.4). Higher urban food prices may have induced urban slave owners to be more stringent than their rural counterparts. Alternatively, urban slaves, who were hired out more frequently than rural slaves, may have been fed more frequently by employers.

The comparison of the two results is confounded by the fact that the slaves in the manifests were more diverse in regional origin than in the free sample, which is limited to blacks born in Maryland. A shortcoming of the manifests is that they do not contain information on the provenance of the slaves, only on the port of embarkation.[37] This is a problem because there was considerable regional variation in stature for which we cannot control. With such a heterogeneous composition of the manifest sample of slave children it would, indeed, be possible to obtain both a small average height for a particular age and a large variance in heights. If all slave children were malnourished, one would not expect the standard deviations to be as high as the records in the manifests indicate.[38] Because height varied considerably by place of birth, it is possible that the Maryland children were taller than those in the manifests on account of their geographic distribution.

35. Wright, *The Free Negro in Maryland*, pp. 131, 133, 246; Russell, *The Free Negro in Virginia*, p. 155. Another factor to consider is that, after their emancipation, the labor force participation rates of blacks declined considerably, and it is conceivable that a similar pattern obtained among free blacks in the antebellum period.

36. From this it also follows that children imported into Louisiana would have been about as well nourished as the average, but children whose heights appear on the manifests would be shorter than the average slave child by as much as 3.4 percent (Table 10.3, rows 6, 7 and 13). This inference requires further substantiation.

37. Margo and Steckel, "The Heights of American Slaves," p. 521. Hence African-born slaves could be in the early part of the sample. See Bancroft, *Slave Trading in the Old South*, p. 22.

38. Habsburg children enrolled in schools run by the military at the turn of the nineteenth century were about two inches shorter than slave children reported in the manifest sample. Yet the standard deviation of Habsburg children's height is well within the normal range. In other words, the distribution of height is generally not affected by nutritional deprivation, as long as that deprivation is evenly distributed; see Komlos, *Nutrition and Economic Development*, p. 91.

An unpublished analysis of the manifest sample by Richard Steckel reveals a North-South gradient in children's stature along the Atlantic seaboard, with children in Virginia and Maryland being taller than those in the Carolinas. Consequently, the difference between the height of the free black children of Maryland and that of the slave children entering interregional trade from the Chesapeake region is likely to be less than between the entire samples.[39] Part of the disparity shown in Figures 10.1 and 10.2 might, therefore, be accounted for by the regional divergence between the Maryland and manifest samples. Another reason the results may differ is that neither height profile has been standardized for year of birth.[40] Because stature declined over the course of the antebellum decades, a greater share of children born in the latter part of the period could also make the manifest children appear shorter than those in Maryland.[41]

Although one might argue that freedom conferred nutritional advantages on children, this hypothesis is called into question by the fact that children in the Maryland sample who were born free were shorter on average than those born into slavery. The pattern holds for both sexes and also among those born in urban and in rural areas (Tables 10.3 and 10.4).[42] Admittedly, we do not know how long children born as slaves spent in bondage before they were freed, but it was clearly more time than for children of the same age who were born free. Thus, if freedom guaranteed a better diet, one would expect those children who gained their freedom earlier to have been taller. Since this is not the case, it is improbable that the free black children of the Maryland sample were taller on account of their legal status than those who ended up in the manifest sample.[43]

Another aspect of the manifest sample to consider is that the very high

39. Personal communication from Richard Steckel.

40. Although both samples cover the same period, the distribution within the period may vary somewhat. The number of children in the Maryland sample is too small to do a statistically adequate job of standardization. The distribution of the children under the age of seventeen by birth cohorts was: 1790s, 8.8 percent; 1800s, 15.2 percent; 1810s, 31.0 percent; 1820s, 21.0 percent; 1830s, 13.9 percent; 1840s, 9.9 percent.

41. Height tended to be greater inland, that is, in newly settled regions. But children in the Maryland sample and children in the manifests originated largely from the Atlantic seaboard. (Those who lived farther inland were more likely to have been taken South overland.) Consequently, from the perspective of the East-West height gradient, the Maryland-born children were not at a nutritional advantage compared with the slaves represented on the manifests. It might also be mentioned that urbanites tended to be shorter than those born in rural areas, but it is improbable that urban slaves appeared on the manifests more frequently than in the freedom certificates. In fact, even urban free black children were taller than the children of the manifests (Table 10.3, rows 9 and 13). It is, therefore, improbable that the urban-rural mix in the two samples accounts for the differences in stature.

42. The effect is more pronounced among the rural than among the urban born (Table 10.3). While the difference among the former is between 0.5 and 1.0 percent, among the latter it is 0.1–0.2 percent.

43. Having been born free did not necessarily confer privileges since the father could still have been a slave. See Kulikoff, *Tobacco and Slaves*, p. 375; Russell, *The Free Negro in Virginia*, p. 131.

standard deviations of children's height throughout childhood and adolescence indicate a high level of diversity in the composition of the sample over time and space and perhaps nutritional experience. Yet, the standard deviation of heights in a population is generally in the neighborhood of 2.7 inches.[44] While deviations from this standard of a few tenths of an inch are not extraordinary, the ones found in the manifests are often twice their expected value (Table 10.2). Such a large variance about the mean has never been documented in homogeneous populations, and consequently is an indication that the manifest records contain observations from several distinct segments of the slave child population. This possibility is underscored by the fact that in the preliminary version of the manifest sample, confined to just two ports of disembarkation, standard deviations were close to the normal range and the height of the children was consistently above that in the enlarged sample.[45] Moreover, a recent decomposition of the height of the slave children who disembarked at New Orleans by those who were sold there and those who were not, had variances well within the normal range, as did the heights found in the recently discovered Louisiana Certificates of Good Character.[46] In contrast, it is reassuring that standard deviations in the Maryland sample lie within the normal range.[47] Moreover, the differences between the standard deviations of height at a particular age in the two samples and the differences in heights have an almost perfectly linear relationship.[48] This means that as the standard deviations of the manifest sample approach the normal range, the average heights of the transported children approach those found in the certificates of freedom.

44. The variance in height does depend to some extent on the distribution of income, but in the case of the manifest sample all observations refer to the same class of subjects. One might expect to obtain large variances if the height observations of lower-class London paupers and upper-class gentry boys of the same age were mixed into the same sample. Note that although Trinidad slave children were even shorter than American slaves, the standard deviations of their height are much smaller than those of the heights in the manifest sample, even though they were often still somewhat above the normal range. See Barry W. Higman, "Growth in Afro-Caribbean Slave Populations," *American Journal of Physical Anthropology*, 50 (Mar. 1979), pp. 373–85. The standard deviation increases with age until adolescence and then levels off. At the time of the adolescent growth spurt the standard deviation can reach as high as 3.3 inches. See J. M. Tanner, R. H. Whitehouse, and M. Takaishi, "Standards from Birth to Maturity for Height, Weight, Height Velocity, and Weight Velocity: British Children, 1965, Part II," *Archives of Disease in Childhood*, 41 (1966), pp. 613–35.

45. Steckel, "Slave Height Profiles," pp. 364, 368. The Coopersmith-Steckel sample of slave manifests appears to be more homogeneous than its enlarged version, perhaps because it was limited to two ports of disembarkation, Mobile and New Orleans. The height of the children in that sample was also consistently greater than in its enlarged version.

46. Jonathan B. Pritchett and Herman Freudenberger, "A Peculiar Sample: The Selection of Slaves for the New Orleans Market" (manuscript, Tulane University, 1990).

47. The extraordinarily high variances found in the manifest sample need not necessarily bias the estimated mean heights, but they have that potential. The variances indicate that the sample consists of observations from different segments of the slave children population, and it is impossible to verify whether the number of observations collected from each stratum is appropriate relative to the size of the population in that stratum.

48. That is, a regression of the differences in standard deviations on the differences in height is linear.

## 10.4 Conclusion

The physical stature of free blacks in Maryland declined in the antebellum decades in close agreement with the trend found for other segments of the American population, both black and white. This suggests that the decline in nutritional status was widespread.[49] The cumulative decline in the height of free black males between the 1820s and the 1840s of about half an inch is similar to that experienced by whites.

The widespread decline in height among whites and blacks appears to have been caused by an acceleration in the growth of the urban population and the industrial sector. The relatively slow expansion of the agricultural labor force and the absence of technological breakthroughs in food production, coupled with the increased demand for food, meant that the price of nutrients increased both absolutely and relative to industrial products in the antebellum decades. Although incomes rose, they did not rise fast enough for many workers to compensate fully for the rise in food prices. Hence, per capita calorie and protein consumption (particularly animal protein) declined. Yet, because of out-migration, the population of Maryland was growing at a slower rate than the average for the entire United States prior to 1840. Compared with the norm of about 30 percent per decade, Maryland's growth was only about 8 percent. In the subsequent decade, however, its demographic expansion was very close to the national average. Perhaps this is why the height of men declined slowly in rural Maryland. In contrast, Baltimore's population expanded from 13,500 in 1790 to 169,000 in 1850, and that rapid growth could very well account for the deterioration in nutritional status among both its free black men and women.[50]

Relative to modern standards, free black girls were taller than boys in both the manifest and the Maryland samples, as were adult women prior to the deterioration in nutritional status. Females born in the first decade of the new century reached the 29th centile, while males reached the 27th centile of modern standards. By the 1830s, however, the rank order was reversed, 21st for women and 25th for men. The early female nutritional advantage might be attributed to two factors. Female youth may have been more likely to be engaged in domestic service, thereby gaining access to nutrients within the white household.[51] Furthermore, the sex ratio favored women. Among both

49. The relationship among the rapid economic growth of the antebellum decades, the increase in the relative price of food, and the concomitant decline in the consumption of nutrients is outlined in Komlos, "The Height and Weight of West Point Cadets," p. 919. Among free blacks, the highest per capita property holdings were in Caroline, Kent, and Queen Anne's counties, in that order. These counties were in the middle of the rank order by height, occupying the seventh, eighth, and ninth place (Table 10.12). See Wright, *The Free Negro in Maryland*, p. 185.

50. The growth of the nearby Philadelphia market would also have served as a source of demand for foodstuffs; see Diane Lindstrom, *Economic Development in the Philadelphia Region, 1810–1850* (New York, 1978), p. 62.

51. For the nutritional advantages of domestic service in Vienna see W. Peter Ward, "Weight at Birth in Vienna, 1865–1930," *Annals of Human Biology*, 14 (Nov.–Dec. 1987), pp. 495–506. Servants owned by wealthy planters had a higher social status in the black community than poor free blacks; see Russell, *The Free Negro in Virginia*, p. 133.

races, females were scarce in the eighteenth century, as there were 12 percent more men than women in the population (Table 10.14).[52] The advantage diminished over time, however, and by 1840 it had vanished altogether. The decline in the sex ratio was greatest among free blacks. Compared with the 8 percentage point decline among slaves and a 10 point decline among whites, the diminution in the sex ratio among the free blacks was 33 points. The demand for the services of black women, therefore, including those who were free, would not have been growing as fast as the supply, possibly explaining why their nutritional status began to decline earlier than it did for black men.

To the extent that the evidence above corroborates earlier findings, the following aspects of the anthropometric history of African-Americans now appear beyond debate. 1) All samples consistently indicate that heights of adult slaves, whether manumitted, born free, transported i ' interregional trade, or recruited into the Union Army, fell within a narrow range: about 67 inches for males and a little over 62 inches for females. By both African and European standards of the time, American blacks were quite tall, as were whites. The attainment of such physical stature is indicative of the nutritional advantages of the New World. But by modern standards, African-Americans appear to have been less well nourished. The average adult slave reached between the 20th and 25th centile of the height distribution of today's industrialized nations. Even compared with their white American contemporaries, blacks fared less well, reaching, on average, about the 35th centile of the antebellum white height distribution. 2) The nutritional status of all groups studied in the American population declined in the antebellum period. 3) Black children were shorter for their age than were black adults. Undernutrition meant that the tempo of slave growth was slower than that of whites.[53] 4) The lower and upper bounds of the black children's growth profile have now been identified as lying between about the first and tenth centile of modern standards. The height of children transported from the Upper to the Lower South on water might be considered the lower bound of all slave children. In contrast, the height of the manumitted slave children could very well constitute the upper bound. 5) Steckel's suggestion of an acceleration of growth in the late teenage years is confirmed by the Maryland sample, but seems to have been somewhat smaller than previously thought. This means that malnutrition among blacks was perhaps less severe and less widespread than the manifest sample indicates. This finding is corroborated by a number of samples recently analyzed.[54] 6) Proximity to the source of nutrients had biological advantages in

52. Earlier in the eighteenth century the sex imbalance was much greater among both races, reaching as high as 40 percent (Russell R. Menard, *Economy and Society in Early Colonial Maryland* [New York, 1985], p. 265).

53. On anthropological evidence on slave undernutrition see Ted A. Rathbun, "Health and Disease at a South Carolina Plantation: 1840–1870," *American Journal of Physical Anthropology*, 74 (Oct. 1987), pp. 239–53.

54. Pritchett and Freudenberger, "A Peculiar Sample." An intense catch-up growth during late adolescence has now been found among runaway slaves as well as among runaway white apprentices in eighteenth-century America. This pattern suggests that a late adolescent growth spurt was

Table 10.14 Sex Ratios in Maryland, 1755–1840

| | | Black | | |
|---|---|---|---|---|
| | White | Slave | Free | Total |
| 1755 | 1.10 | 1.14 | 1.22 | 1.12 |
| 1790 | 1.06 | | | |
| 1820 | 1.03 | 1.10 | 0.89 | 1.03 |
| 1840 | 1.00 | 1.06 | 0.89 | 0.99 |

*Note:* The sex ratio is the ratio of males to females in the population.
*Sources:* Various federal population censuses: James M. Wright, *The Free Negro in Maryland, 1634–1860* (New York, 1921), p. 85.

the early industrial period. Rural populations were taller than urban ones. Higher per capita output of food was more important in determining nutritional status than was money income.

Some findings need further confirmation on the basis of other data sets to broaden their applicability beyond the Maryland free black population. Among these are: 1) In the antebellum United States the nutritional status of black females began to decline earlier than it did among males, possibly because of a decline in the relative market value of their labor. 2) The nutritional status of females relative to males depended on the population's sex ratio. When the sex ratio favored women their nutritional status, too, was greater. 3) The decline in nutritional status affected urban black populations before it affected rural populations. 4) Owners who set their slaves free may have been especially benevolent, and consequently may have provided a better-than-average diet for their young slaves prior to manumission. If so, this part of the sample should not be considered representative of all slave children. Because black children who were born free would not have benefited from such largess, their height is probably a better proxy for the slave average than that of either the manumitted children or the children in the manifests. 5) The exact beginning of the decline in nutritional status is difficult to determine because it was region, age, and gender specific. In this regard even place of residence begins to play a role. But once one divides the sample in so many ways, the number of observations becomes insufficient to determine accurately the turning point at the county level. It appears, however, that in rural areas of Maryland the nutritional status of black men improved for those born in the 1820s, and then fell among those born in the latter half of the 1830s. Urban men were less likely to benefit from the improvements of the 1820s; on the contrary, in some cases a decline was already evident, as was the case among men living

an integral part of the human experience of the New World, probably until the end of the nineteenth century; see John Komlos, "A Malthusian Episode Revisited: The Height of Indentured Servants in Colonial America," and "The Height of Runaway Slaves in Colonial America" (manuscripts, University of Pittsburgh, 1991).

in Baltimore and in the bordering Anne Arundel County, home of the state capital, Annapolis. 6) Although the catch-up growth of the late teenage years is now beyond question, the degree of malnourishment among children prior to puberty needs further exploration.

Female height did not increase among the birth cohorts of the 1820s. The deterioration in female nutritional status began at least a decade earlier than it did for men. Like men, urban women were more likely to suffer from nutritional stress earlier than their rural counterparts. Because changes in the epidemiological environment probably would have affected men and women equally, it is unlikely to have caused the earlier onset of the decline in stature among women. The pattern, therefore, supports the notion that the decline in nutritional status was related more to a fall in nutritional intake than to an increase in the claims on the nutrients.

# XVI

# The Height of Runaway Slaves in Colonial America, 1720–1770

State of North Carolina, Graven County, by Richard Ellis and Will Tisdale, esquires, two of the justices for the said county. "[W]e do hereby, by virtue of an act of assembly of this state concerning servants and slaves, intimate and declare, if the said slave doth not surrender himself and return home immediately after the publication of these presents, that then any person may kill or destroy the said slave, by such means as he or they may think fit, without accusation or impeachment of any crime or offence for so doing, or without incurring any penalty or forfeiture thereby."

*The North Carolina Gazette*, New Bern, June 27, 1777

"Pewter, 44 years, runaw (sic) Charles City: The said negro is outlawed and I will give 10 pounds to any person or persons that will kill him and bring me his head separate from his body."

*Virginia Gazette*, 1769

"As the said fellow ran away without receiving any abuse, the taker up is desired to give him ten lashes every ten miles."

*Virginia Gazette*, 1769

Recent efforts to understand the nutritional experience of various populations during the course of the last two centuries have focused on analyzing both temporal changes and cross-sectional differences in human stature. The examination of the stature of African Americans has not only been a part of this research program but it actually gave the impetus for its inception.[1] These investigations have increased our knowledge of the material conditions of the bound labor force prior to the abolition of property rights in human beings.[2]

I am indebted to Peter Coclanis for bringing the availability of these data to my attention; to Richard Steckel and Kenneth Sokoloff for their helpful comments on an earlier version of this chapter; to Paul Mageli for helping to collect the data; to Philip Sidel of the University of Pittsburgh's Social Science Computer Research Institute for assistance in processing the data; to Tim Cuff for help in the analysis; and to the University of Pittsburgh's Office of Research and Development for its financial support.

1. Steckel, "Slave Height Profiles from Coastwise Manifests."

2. Margo and Steckel, "Heights of American Slaves"; Engerman, "Height of U.S. Slaves"; Friedman, "Heights of Slaves in Trinidad"; Eltis, "Nutritional Trends in Africa and the Americas."

The University of Chicago Press, Chicago 60637
The University of Chicago Press, Ltd., London

The mean height of adult male slaves turns out to have been about 67.2 inches in the late eighteenth and early nineteenth centuries, perhaps an inch above that of the European peasantry.[3] Blacks fared less well, however, relative to the white population of the United States. Male slaves and urban white laborers were about the same height, but slaves were approximately one inch shorter than whites on average and even shorter relative to southern whites.[4] Because data are also available on slaves of African birth, it has been shown that in the nineteenth century, African Americans were taller than the populations from which they originated. American-born slaves were, on average, 1.6 inches taller than even the Yoruba, the tallest ethnic group in Africa on which data are available.[5]

The conclusion to be drawn from these studies is that the nutritional status of adult slaves was adequate even relative to the work effort demanded of them, although by no means at the level of their masters.[6] This cannot, however, be said with the same degree of certainty about children. Until the teenage years, their diet could well have been deficient; it was probably below that of adults.[7] The average adult would have been in the twenty-fifth percentile of today's population, while children would hardly have reached the fifth centile.[8]

Thus, while several studies have explored the physical stature of American slaves, none have reached very far back into the colonial period. Our knowledge of these issues can be extended into an earlier time period on the basis of runaway slave advertisements found in newspapers.[9] These records

3. Margo and Steckel, "Heights of American Slaves," 521; Komlos, *Nutrition and Economic Development*.

4. Margo and Steckel, "Heights of Native Born Northern Whites"; Sokoloff and Villaflor, "Early Achievement of Modern Stature in America."

5. Eltis, "Welfare Trends among the Yoruba."

6. Needless to say, this evidence by no means indicates that the slave diet was palatable, but only that given the disease environment of the New World, it provided a higher level of caloric and protein intake than was available to the inhabitants of Africa. On the nutritional adequacy of the slave diet just prior to emancipation, see Fogel and Engerman, *Time on the Cross*, 2:95.

7. Steckel, "A Peculiar Population." The analysis of human skeletal remains supports the notion of an unhealthy slave childhood. According to one such study, "slave children . . . were vulnerable to recurrent metabolic stress between the ages of 2 and 4, probably from increased exposure to parasites and infections." Moreover, there is evidence that "acute development arrest [was] followed by a marked recovery period." Rathbun, "Health and Disease at a South Carolina Plantation," 245.

8. Steckel, "Growth Depression and Recovery."

9. Windley, *Runaway Slave Advertisements*, a compilation, was also used in this study. Note that Windley excluded advertisements placed by jailkeepers.

are the oldest extant data on slave stature and are likely to remain the only ones available for the early part of the century. In fact, only a handful of height observations on American blacks has been found for the colonial period.[10]

Hitherto three major data sets on the height of African Americans have been analyzed: slave-shipping manifests, Maryland manumission records, and muster rolls of soldiers who served in the Union Army.[11] All of these pertain essentially to the nineteenth century, although the birth cohorts of the late eighteenth century are also found in the first two sets of documents. The data have additional limitations. The manifests were shipping documents which masters of ships filled out when slave imports were prohibited after 1808. They do not contain information on the slaves' place of birth, and consequently Africans cannot be compared with African Americans using this source. In addition, the manifest records as well as the Maryland Certificates of Freedom pertain to only a part of slave society, and therefore might not be representative of the whole population. The Civil War records are no doubt better in this regard, but begin with the birth cohorts of the 1820s and are therefore irrelevant as far as both the eighteenth century and African-born slaves are concerned.

It is important to learn more about the cross-sectional pattern of and secular changes in the material condition of slaves during an epoch that witnessed the rapid expansion of slavery. Because of the extreme paucity of eighteenth-century data, the advertisements provide a valuable source for extending our knowledge of this population backward through time.

### Quality of the Data

The only known major source on the height of eighteenth-century African Americans is provided by the newspaper advertisements for runaways.[12] These were placed by slave owners to try to apprehend slaves who had fled in the hope of gaining their freedom. The notices began to appear with the very first newspapers in the colonies, but seldom contained numeric information on

10. These height observations have provided the evidential basis for the assertion that African-born slaves living in the colonies were shorter than their brethren born in the New World as early as the eighteenth century.

11. Margo and Steckel, "Height of American Slaves"; Steckel, "Slave Height Profiles"; Sokoloff and Villaflor, "Early Achievement of Modern Stature"; Komlos, "Toward an Anthropometric History of African-Americans." However, new data sets are being found; see Freudenberger and Pritchett, "The Domestic United States Slave Trade." There is other evidence for Caribbean slaves; see Higman, "Growth in Afro-Caribbean Slave Populations."

12. *Early American Newspapers*. The list of newspapers used for the sample is available from the author on request.

stature until midcentury, although descriptive references such as "tall, medium, or short" were made even before then.

To be sure, the evidence gleaned from the advertisements has its own limitations. After all, runaways might not be representative of all slaves, and the advertisements might not be representative of all slaves who fled. For instance, one can by no means rule out *a priori* that the stronger, hence taller slaves might have been more likely to take the risks associated with fleeing.[13] Yet, such a bias would have a relatively small effect on the analysis: it would tend to influence the intercept of the regressions, but not the other coefficients, as long as it was uncorrelated with any of the independent variables.[14] This point is crucial, because the main purpose of this study is to estimate both the trend and the cross-sectional differences in height. Luckily, the period under consideration overlaps with the manifest and manumission records, and thus the extent of the bias can be estimated.

Another problem is that the reference to height in the advertisements was based on the memory of the masters and not on actual measurements. To be sure, owners did have an incentive to represent the height of runaways as accurately as possible, since it could affect the likelihood of the slaves' return. Some of the runaways were caught in flight and then jailed. Their misfortune turns out to be useful for the analysis, because the jailkeepers advertised for the owner of the runaway, and the height mentioned in such notices would not have been based on recollection but on actual observation, perhaps even measurement. Consequently, one can ascertain the extent to which the jailers' height advertisements deviated systematically from those placed by the owners themselves. Similar problems exist with regard to the age of the runaways. Most of the analysis is based on adult stature, and after the terminal height has been reached, any discrepancy in reporting age is immaterial.

## Results

More than ten thousand observations on males and one thousand records on females were collected (table 6.1).[15] Even though the advertised heights

13. Moreover, the physical descriptions might have been mentioned in the advertisements for those slaves whose height was a distinguishing feature, that is, those who were very tall or very short.

14. That is, as long as the degree of bias, if any, remained unchanged over time or was not a function of the place of birth of the slaves.

15. This is an indication that males were much more likely to take the risks associated with running away than were females. This probably had to do with the need to take care of children as well as with the greater risks women faced in the wilderness. *Note:* tables and figures appear at the end of this chapter, pp. 104–16.

were not based on actual measurements, the observations are internally consistent with *a priori* expectations. Adults were taller on average than youth, for example (table 6.2). Slaves born in America were taller than those born in the Caribbean, who in turn were taller than those born in Africa. Yet the extremely unusual height distributions indicate that the evidence is not fully reliable inasmuch as there is an unexpectedly large number of observations in the six-feet-tall category, irrespective of birthplace (figures 6.1, 6.2). In addition, some heaping on round number of inches and even on the nearest foot is apparent.[16] Such height distributions are not characteristic of human populations.[17] Among younger slaves the five-feet-tall cell has an excessively large number of observations (table 6.3). However, the predominance of the number of five-feet-tall observations among the 18-year-old youth vanishes by age 21 in favor of the six-feet-tall cell (figure 6.3).[18] Because such rounding to the nearest foot was frequent, the advertisements could overestimate adult heights and perhaps underestimate the height of young adults.

However, the distributions did improve over time, and the number of observations in the six-feet-tall category diminished during the course of the century. Perhaps the owners were becoming more knowledgeable in judging the physical stature of their slaves (figure 6.1). Jailers' advertisements also improved as a data source during the century, as evinced by the fact that the height distributions became less skewed. (figure 6.4). These notices included fewer slaves described as being six feet tall than did the masters' advertisements (4 percent vs. 9 percent). Consequently, slaves in the custody of a jailer were said to be shorter than the runaways at large. The discrepancy is quite large among slaves born in America: it is more than one inch and is statistically significant (tables 6.4, 6.5). Among African-born slaves, however, the difference was smaller, about 0.3 inches, and was insignificant (table 6.6). Because African-born slaves were shorter than slaves born in America, fewer were close to six feet tall, and thus their masters were probably less likely to round upward their height estimates to the six-feet-tall category.

The average height of adult male runaway slaves born in America who

16. Heaping on even numbers is evident even in modern measurements. Fogel, "Physical Growth."

17. For example, these distributions do not match that found among slaves intercepted by British vessels enforcing the embargo on shipments of Africans to the New World. See Eltis, "Nutritional Trends in Africa and the Americas," 456.

18. Among American-born adults, 14 percent were described as being six feet tall (table 6.2). To be sure, this percentage includes slaves who were designated to be "about," "near," or "over" six feet tall.

were in the custody of a jailer varied between 67.3 and 66.5 inches, declining during the course of the century, while for those born in Africa it was between 67.1 and 66.0 inches (tables 6.5, 6.6).[19] American-born slaves were taller than African-born ones by between 0.2 and 0.5 inches.[20] This difference in height is found not only among adults but among younger runaways as well, and is independent of time (table 6.4).[21]

The stature of slaves born both in America and in Africa appears to have declined somewhat during the second half of the century (tables 6.7, 6.8).[22] This tendency is apparent both among slaves who were in jail at the time of the advertisement and those who were not (table 6.4). The inference is thus warranted that the trend is not caused by an improvement in the owners' ability to recall the height of their slaves with increased accuracy. The diminution among American-born slaves in the custody of a jailkeeper is as much as one inch. Regression analysis confirms this pattern even if the six-feet-tall observations are excluded from the sample (tables 6.5, 6.6), suggesting that the negative trend is not a mere artifact of the rounding-to-the-nearest-foot effect.[23] Among the African-born men, the decline in height is first evident among the birth cohorts of the 1760s, while among those born in America the pattern begins to appear a decade later.[24]

19. A number of advertisements were found for native-born American male slaves. The height of twenty-five adult males was given as 66.9 inches. Since this average is unadjusted for bias, it is comparable to the height of African-born slaves.

20. The American height advantage remains after the exclusion of the six-feet-tall observations (tables 6.5 and 6.6, regression [2]). The height of those whose provenance was not given in the advertisement is approximately the average of those of African and of American birth (table 6.2).

21. The divergence in height between American- and African-born slaves is greater in the sample as a whole than among the runaways who were jailed at the time of the advertisement, inasmuch as rounding to the nearest foot was more prevalent among the former group. To the extent that the height of American-born slaves was more likely to be nearer to six feet than that of those of African birth, one would expect that the discrepancy would, indeed, be greater in the entire sample than in the subsample of those jailed.

22. This was not the case among those whose birthplace is unknown. However, that result could easily be influenced by changes in the geographic composition of the sample.

23. The result also holds when the residence or the ethnic composition of the slaves are held constant (tables 6.5, 6.6, regression [3]). It is true that the birthplaces of many of the Africans are unknown, and it cannot be ruled out that the decline in height among the slaves might have been due in part to a changing composition of their provenance. In the nineteenth century, for example, "The slave trade declined on that part of the coast where Africans were taller—in Senegambia." (Eltis, "Nutritional Trends in Africa and the Americas," 455.)

24. That the negative trend does not appear among those of unknown provenance might be an indication that the share of those born in America in this part of the sample was increasing.

An exploration of regional diversity shows that among those of African birth, the Angolans alone appear to have been shorter than average by more than two inches (table 6.6). This finding is consistent with other evidence that men of the Congo region were relatively short.[25] Among slaves born in the colonies, those residing in the Upper South were the tallest.[26] That this was roughly the case among whites as well is indicative of the effect of the environment on both races regardless of social status.[27]

The analysis so far has been confined to those runaways who were not described as "mulattos." These are analyzed separately in order to standardize the genetic composition of the population studied to the extent possible. Their adult height was about equal to that of other American-born slaves, although among the 19–22 year olds they were somewhat taller (table 6.2). Their height distribution is also skewed, although the trend in their height is not negative, as is the case in the other part of the sample (table 6.7). Mulattos who were in jail at the time of the advertisement were 0.6 inch shorter than those who were not (table 6.4).

The average height of adult females who were in the custody of a jailer was 62.7 inches. This is about the same as the height of slave women in other samples. The cross-sectional pattern among females is the same as the one found for male slaves: American-born women were taller than those of African birth. The difference appears to have been greater for the women, but this might be the artifact of errors introduced by rounding to the nearest foot. A decline in height is evident among the females, beginning in the 1780s (table 6.9).

The few observations available on youth in this sample are especially valuable, because such evidence is relatively scarce and the growth pattern among slaves has been a controversial issue. Richard Steckel has found that slave children were shorter than expected based on their final attained height.[28] Slave youth experienced a short period of unusually rapid growth in their late teens. This remarkable phase of "catch-up growth" has subsequently been observed among runaway white apprentices as well, thereby increasing the plausibility of such a pattern.[29] The height-by-age profile of runaway slaves older

25. Eltis, "Nutritional Trends in Africa and the Americas," 459.

26. Slaves in New England were the shortest, but note that this is based on a very small number of observations (table 6.1).

27. Komlos, "Height and Weight of West Point Cadets."

28. Steckel, "Growth Depression and Recovery."

29. The author's calculation of runaway white apprentices is also based on data from newspaper advertisements.

than 16 is quite similar to the one posited by Steckel, and thus supports the notion of an acceleration in the tempo of growth in the late teens among both boys and girls (table 6.10).[30] At age 16 the runaway boys were about the same height as were free blacks of Maryland, but taller than those who were shipped to the Lower South in interregional trade. For children between the ages of 17 and 20, all samples so far examined show a remarkably consistent growth profile, and therefore confirm the hypothesis of a period of intense catch-up growth.

The rounding of heights upward to six feet begins to affect the height estimates in the 21-year-old category.[31] The growth profile of the runaway girls over age 16 is remarkably consistent with the profiles found for other African-American populations of the nineteenth century.[32] Among the female runaways, too, catch-up growth in the late teens appears to have been quite noticeable (table 6.10).

## Conclusion

Because masters were relying on either their own memory or on that of their overseers in recalling the physical stature of their runaway slaves, the present sample must be used with caution.[33] The height distributions of runaway white indentured servants were by no means as skewed as those of the slaves. This difference is noteworthy in itself insofar as it points to the greater social distance separating white masters from their black slaves than from their white indentured servants.

Although the data are of inferior quality to the anthropometric evidence so far reported in the literature, they are nonetheless very important insofar as evidence on the physical stature of African and African-American populations

30. For runaway slaves below age 16, the limited number of observations combined with the rounding problem renders this sample of limited value.

31. The 15-year-old boys, however, are shorter than the slaves whose height was recorded on the manifests. This should be considered an anomaly, because if the rounding to the nearest foot had operated in this age category as well, then one would expect that some of the heights would have been rounded up to 60 inches, and therefore these heights should be upwardly biased. The 14-year-old boys are taller than the boys on the manifests, as these considerations would lead one to think. Those below the age of 14 are shorter than the manifest boys, which could be an artifact of rounding to 48 inches.

32. The younger runaways are shorter than the girls shipped in interregional trade. Because of the small number of observations at ages 14 and 15, no great significance should be attached to this result.

33. The large number of observations in the six-feet-tall category is probably due more to rounding to the nearest foot than to any propensity to advertise for very tall slaves. If the latter were the case, one would expect that the height distribution among the African-born slaves would be the same as, or at least closer to, that of American-born slaves.

prior to 1800 is extremely scarce. It is particularly noteworthy in this regard that in spite of the fact that the observations were not based on actual measurements, the cross-sectional, longitudinal, and age-cohort patterns are all remarkably consistent. This increases the confidence with which one can use these data to infer trends in the actual physical stature of the slave population. Information on the runaways who were caught in flight is particularly valuable. Their heights were advertised by the jailer in search of the slaves' owners, and therefore have the advantage of having been based on personal observation, rather than on memory.

The evidence indicates that blacks born in America were taller than those of African birth practically from the beginning of the system of human bondage in the New World. The 0.4-inch height advantage implies that even the slave population benefited to some extent from the superior nutritional and epidemiological environment of the New World. The pattern is consistent with nineteenth-century evidence on height and with investigations of the basic adequacy of the slave diet which indicate that adult slaves probably consumed more meat per annum than contemporary East-European serfs did throughout their youth.

There is some evidence that nutritional status might have diminished during the second half of the century among both African- and American-born runaways. The diminution in height would be consistent with the contemporaneous decline in the height of Europeans caused by Malthusian population pressures and might be an indication that population pressure was not confined to the European continent. The inference is strengthened by the fact that in the early nineteenth century the height of African men (in Africa) was between 65.0 and 65.6 inches, that is, at the lower end of the range found among the sample of American jailed runaway slaves.[34] This pattern supports the notion that the decline in nutritional status in Africa persisted beyond the period under consideration.[35]

A possible decline in the nutritional status of American slaves in the late colonial and early federal periods would be consistent with a considerable body of evidence that points to the "economic difficulty and perhaps depressed living standards for the new republic"[36] from which slaves, at the bottom of the socioeconomic hierarchy, would not have been isolated. The growing popula-

34. Eltis, "Nutritional Trends in Africa and the Americas," 459, and "Welfare Trends among the Yoruba," 526.

35. An alternative but less likely interpretation might be that at the beginning of the eighteenth century, taller Africans were more likely to have been sold into slavery than shorter ones.

36. Shepherd and Walton, "Economic Change after the American Revolution," 420.

tion pressure in New England and the decline in agricultural productivity in Pennsylvania in the late colonial period would not have affected a significant share of the slave population.[37] However, the decline in per capita export earnings was more likely to have done so, insofar as the slave economy was integrated into the world market to a considerable degree.[38]

Furthermore, the war was particularly disruptive of commerce and inflicted significant damage to capital such as the dykes and rice fields of South Carolina.[39] Consequently, rice exports plummeted. In fact, exports from the whole of the South stagnated during the 1770s and 1780s.[40] Thus, it would not be surprising if entrepreneurs caught in this depression reacted by decreasing the living standards of their bound labor force.[41] In fact, slave imports during the 1770s were minimal.[42]

Another pertinent issue to the well-being of slaves is that the price of foodstuffs was increasing rapidly even in America during the second half of the eighteenth century.[43] This could have induced slave owners to replace some of the more nutritious foods such as grains and animal products with corn in the diet of slaves. Moreover, food prices were rising faster than those of the commodities produced by slaves.[44] During the War of Independence there were at times severe food shortages even in urban areas. On George Washington's

37. Lockridge, "Land, Population, and the Evolution of New England Society"; Ball and Walton, "Agricultural Productivity in Eighteenth Century Pennsylvania," 100.

38. Shepherd, "British America and the Atlantic Economy."

39. Henretta, "The War of Independence and American Economic Development." Many of the tidewater plantations in the Georgia and South Carolina low country were damaged, and slaves were taken or were freed by the British. See Quarles, *The Negro in the American Revolution,* 158–81; Klein, *Rise of the Planter Class,* 114–15.

40. Shepherd and Walton, "Economic Change after the American Revolution," 413.

41. Carr and Walsh, "Economic Diversification and Labor Organization in the Chesapeake." There is evidence of intensification of the labor requirements in the Chesapeake.

42. Fogel and Engerman, *Time on the Cross,* 1:25. Also worthy of mention is the fact that per capita wealth was declining during and after the Revolutionary War and wealth distribution was becoming more skewed. Engerman and Gallman suggest that it is quite possible "that the war (and perhaps the 1780s, as well) was truly disastrous, from an economic point of view. More than 20 years after the end of the war per capita wealth remained below the level achieved in 1774." The inference is drawn by comparing Jones' wealth estimates for 1774 to early nineteenth century ones. See Jones, *Wealth of a Nation to Be,* as cited in Engerman and Gallman, "U.S. Economic Growth, 1783–1860," 19.

43. Farm prices in Massachusetts increased by 89 percent between 1750–54 and 1796–1800. See Rothenberg, "A Price Index for Rural Massachusetts"; Bezanson, Gray, and Hussey, *Prices in Colonial Pennsylvania,* 422, and *Wholesale Prices in Philadelphia,* 61, 248.

44. Between 1784–88 and 1796–1800, the price of corn increased by 50 percent, while the price of tobacco increased by only 25 percent; even the price of cotton declined. See Bezanson, Gray, and Hussy, *Wholesale Prices in Philadelphia,* 61, 67, 242, 248.

plantation, slaves allegedly consumed 2,800 calories per day, a low intake by subsequent standards reached in America.[45]

While economic processes in late colonial and early federal America could have had an adverse effect on the nutritional status of American slaves, the hypothesis needs further corroboration. The main reason for this caveat is that the diminution in height is not evident among mulattos.[46] Until the inconsistencies between mulattos and blacks can be explained independently, the hypothesis of a diminution in physical stature during the second half of the eighteenth century on American soil ought to be viewed with reservation.

It is noteworthy that the height of jailed runaways born in the second half of the eighteenth century, at about 66.8 inches, differed only slightly from that of slaves shipped in interregional trade (67.0) or that of the free blacks of Maryland (66.9). This result supports the notion that runaway slaves were not taller than average.

Slaves residing in the Lower South were shorter than those in the Upper South. This finding is in keeping with the notion that work intensity was greater and the epidemiological environment was worse (and hence more inimical to the nutritional status of slaves) in the rice culture of South Carolina and Georgia than it was in the tobacco culture of Virginia and Maryland.[47] Furthermore, there is a high correspondence between the local production of foodstuffs and the physical stature of the slaves. For instance, New England was the only region that had to import a substantial portion of its food requirements; it also had the shortest slaves. The Lower South produced less food per capita than did the Upper South; slaves in the Lower South were also shorter than those from the Upper South.[48]

The stature of these runaways corroborates the hitherto accepted notion that even the slaves benefited from the propitious nutritional environment of

45. U.S. Department of Commerce, *Historical Statistics of the United States*, 2:1175.

46. Diminution of height is also minute among adults of unknown provenance. To be sure, this result might be due to changes in the regional composition of the runaway sample, but the mulattos were nearly all born in the New World.

47. Even if the task system (by providing quotas) allowed the slaves some flexibility in work routine and did not make full use of their labor power, slaves used their "free" time to work on their own behalf. Thus, the combined workload must have led to long workdays with high calorie requirements, especially since rice cultivation is quite strenuous, requiring work in swamps. The concomitant exposure to mosquito-borne diseases undoubtedly resulted in onerous labor conditions. See Coclanis, *The Shadow of a Dream;* Morgan, "Work and Culture."

48. Lindstrom, "Southern Dependence upon Interregional Grain Supplies," and "Domestic Trade and Regional Specialization"; Klingaman, "Food Surpluses and Deficits in the American Colonies."

North America.[49] This proposition can now be extended back to the early part of the eighteenth century. Whereas in Europe class differences translated into substantial height differences, so that the aristocracy was considerably taller than the peasantry, in America abundant food supplies meant that even substantial differences in legal status did not have similarly strong nutritional implications.

49. The runaway slave sample provides further evidence on the growth profile of slave youth. It shows that a few years of intense catch-up growth in the late teens, found among both blacks and whites, was biologically feasible in North America. Possibly some interaction of dietary intake with the disease encounters of childhood and adolescence gave rise to this premodern growth profile. That such a growth pattern was also found among white apprentices indicates that the unusual growth profile was not a unique feature of the slave experience.

TABLE 6.1 Characteristics of the Male Runaway Slave Sample

| | *N* | Percentage |
|---|---|---|
| *Race* | | |
| Black | 9,501 | 88.4 |
| Mulatto | 1,187 | 11.1 |
| Indian | 54 | 0.5 |
| Total | 10,742 | 100.0 |
| *Decade of birth*[a] | | |
| 1600s | 2 | |
| 1700s | 9 | 0.1 |
| 1710s | 68 | 0.6 |
| 1720s | 300 | 2.8 |
| 1730s | 896 | 8.5 |
| 1740s | 1,561 | 14.8 |
| 1750s | 1,446 | 13.7 |
| 1760s | 1,418 | 13.4 |
| 1770s | 1,643 | 15.5 |
| 1780s | 1,306 | 12.3 |
| 1790s | 1,058 | 10.0 |
| 1800s | 599 | 5.7 |
| 1810s | 271 | 2.6 |
| | 10,577 | 100.0 |

| | *N* | Percentage |
|---|---|---|
| *Place of birth* | | |
| Africa | | |
| Angola | 59 | |
| Congo | 24 | |
| Gambia | 15 | |
| Guinea | 149 | |
| Ibo | 26 | |
| Other | 383 | |
| Subtotal | 656 | 6.1 |
| Mainland colonies | | |
| New England | 17 | |
| Georgia | 11 | |
| Maryland | 71 | |
| New Jersey | 7 | |
| New York | 41 | |
| North Carolina | 31 | |
| Pennsylvania | 13 | |
| South Carolina | 25 | |
| Virginia | 231 | |
| Other | 8 | |
| Unspecified | 676 | |
| Subtotal | 1,131 | 10.5 |
| Europe | 69 | 0.6 |
| West Indies | 150 | 1.4 |
| Unknown | 8,736 | 81.3 |
| Total | 10,742 | 100.0 |

SOURCE: Advertisements in colonial American newspapers.
NOTE: $N$ = number of observations.
[a]Assumes that the runaways of unknown age were 27 years old.

TABLE 6.2 Height of Male Slaves Born between 1690 and 1810 by Place of Birth (inches)

| | Over 22 Years Old | | | | 19–22 Years Old | | | | Unknown Age | | | |
|---|---|---|---|---|---|---|---|---|---|---|---|---|
| Place of Birth | *N* | Height | S.D. | *A*[a] | *N* | Height | S.D. | *A*[a] | *N* | Height | S.D. | *A*[a] |
| Africa | 314 | 67.0 | 3.2 | 10 | 66 | 66.0 | 3.2 | 0 | 250 | 67.3 | 3.2 | 14 |
| Caribbean | 66 | 67.6 | 2.9 | — | 26 | 66.1 | 3.4 | — | 52 | 67.1 | 3.4 | — |
| Mainland colonies | 583 | 68.1 | 2.8 | 14 | 228 | 66.9 | 3.2 | 7 | 240 | 68.0 | 3.3 | 15 |
| Europe | 36 | 68.6 | 2.5 | — | 7 | 66.8 | 4.4 | — | 18 | 67.6 | 4.2 | — |
| Unknown | 4,589 | 67.8 | 2.8 | 9 | 1,711 | 67.2 | 2.9 | 5 | 1,777 | 67.6 | 3.3 | 11 |
| Total | 5,588 | 67.8 | 2.8 | 10 | 2,038 | 67.2 | 2.9 | 5 | 2,337 | 67.6 | 3.3 | 11 |
| In jail | 435 | 66.8 | 2.8 | 4 | 150 | 66.1 | 2.8 | 2 | 62 | 67.6 | 3.2 | 5 |
| Mulatto | 552 | 68.2 | 2.6 | 9 | 269 | 67.3 | 2.9 | 6 | 228 | 67.4 | 3.3 | 6 |
| Indian | 18 | 66.6 | 4.4 | — | 7 | 67.5 | 2.1 | — | 18 | 64.6 | 3.9 | — |

NOTE: S.D. = standard deviation.
[a]Percentage of observations recorded as six feet tall.

TABLE 6.3 Runaway Male Slaves Either Five or Six Feet Tall (%)[a]

| Age | 60 inches | 72 inches |
|---|---|---|
| 17 | 13.0 | 0.5 |
| 18 | 6.3 | 1.7 |
| 19 | 4.1 | 3.4 |
| 20 | 3.3 | 6.1 |
| 21 | 1.4 | 6.4 |
| 22 | 2.4 | 6.1 |
| 23 | 1.8 | 8.4 |
| 24 | 2.0 | 7.5 |
| 25 | 1.6 | 8.4 |
| 26 | 1.1 | 11.4 |
| 27 | 1.0 | 12.6 |

[a]African-born slaves and mulattos are not included.

TABLE 6.4 Height of Runaway Slaves by Place of Birth, Custody Status, and Age

| Place of Birth and Custody Status | Year of Birth | | | | | |
|---|---|---|---|---|---|---|
| | 1700–1759 | | 1760–1810 | | 1700–1810 | |
| | *N* | Height | *N* | Height | *N* | Height |
| *Adult males* | | | | | | |
| Africa | | | | | | |
| In jail | 15 | 67.7 | 32 | 65.7 | 47 | 66.3 |
| At large | 199 | 67.4 | 68 | 66.4 | 267 | 67.1 |
| Mainland | | | | | | |
| In jail | 25 | 67.3 | 46 | 66.3 | 71 | 66.7 |
| At large | 410 | 68.4 | 98 | 67.6 | 508 | 68.2 |
| Mulatto | | | | | | |
| In jail | | | | | 27 | 67.6 |
| At large | | | | | 525 | 68.2 |
| Unknown | | | | | | |
| In jail | 25 | 67.2 | 260 | 66.9 | 285 | 66.9 |
| At large | 1,677 | 67.7 | 2,572 | 67.9 | 4,249 | 67.8 |
| *19- to 22-year-old males* | | | | | | |
| Mainland | | | | | | |
| In jail | | | 33 | 65.4 | | |
| At large | | | 66 | 67.4 | | |
| *Adult females* | | | | | | |
| Unknown | | | | | | |
| In jail | | | | | 74 | 62.7 |
| At large | | | | | 400 | 63.3 |

TABLE 6.5 Determinants of Height of Male Runaway Slaves Born in the Mainland Colonies[a]

| | (1) | (2) | (3) |
|---|---|---|---|
| Constant | 68.3* | 67.6* | 68.2* |
| Decade of birth | | | |
| 1720s | −0.0 | −0.5 | −0.2 |
| 1730s | 0.1 | 0.0 | 0.0 |
| 1740s | 0.3 | 0.2 | 0.2 |
| 1750s | 0.2 | 0.3 | 0.2 |
| 1770s | −0.5 | −0.2 | −0.5* |
| Age | | | |
| 19–22 | −1.0* | −0.8* | −1.1* |
| Unknown | −0.3 | −0.4 | −0.2 |
| In jail | −1.3* | −1.0* | −1.2 |
| Residence[b] | | | |
| New England | | | −1.3 |
| Upper South | | | 0.4 |
| Lower South | | | −0.2 |
| Other | | | 1.7 |
| $F$ | 8.0* | 4.8* | 6.6* |
| $R^2$ | .06 | .04 | .07 |
| $N$ | 1,051 | 913 | 1,051 |

NOTE: Slaves under 19 years old were excluded from the regressions. Intercept represents the height of an adult slave, born in North America during the 1760s, who was not in jail at the time the advertisement appeared. Regression (2) does not include those whose height was given as 72 inches. In (3) the constant also refers to one whose residence was the Mid-Atlantic region.
*Significant at the .05 level.
[a]Mulattos excluded. [b]State in which advertisement appeared.

TABLE 6.6 Determinants of the Heights of Runaway Slaves Born in Africa

| | (1) | (2) | (3) |
|---|---|---|---|
| Constant | 66.3* | 65.9* | 66.7 |
| Date of birth | | | |
| Before 1750 | 1.1* | 0.8* | 1.1* |
| Age | | | |
| 19–22 | −0.9* | −0.9* | −0.9 |
| In jail | −0.3 | −0.2 | 0.2 |
| Region of origin | | | |
| Nigeria[a] | | | 0.0 |
| Gambia[b] | | | 0.3 |
| Congo[c] | | | −2.3* |
| Other | | | −0.3 |
| $F$ | 5.6* | 4.3* | 7.5* |
| $R^2$ | 0.04 | 0.04 | 0.09 |
| $N$ | 380 | 344 | 630 |

NOTE: Slaves under 19 years old were excluded from the regressions. Intercept represents an adult slave (over 22 years old) born after 1760 who was not in jail at the time of the advertisement. Regression (2) does not include heights above 72 inches. In (3) the constant also refers to a slave originally from Guinea.
*Significant at the .05 level. [b]Includes Senegambia.
[a]Includes Bight of Benin, Biafra, and Ibo. [c]Includes Angola.

TABLE 6.7 Height of Adult Male Slaves[a] by Decade and Place of Birth

| Blacks: Decade of Birth | Africa | | | | Mainland Colonies | | | | Unknown | | | |
|---|---|---|---|---|---|---|---|---|---|---|---|---|
| | *N* | Height | S.D. | A[b] | *N* | Height | S.D. | A[b] | *N* | Height | S.D. | A[b] |
| 1710s | | n.a. | | | 9 | 68.2 | 3.5 | 33 | 34 | 67.7 | 2.6 | 15 |
| 1720s | 25 | 67.6 | 2.6 | 8 | 37 | 68.4 | 2.7 | 24 | 146 | 67.8 | 3.2 | 14 |
| 1730s | 48 | 68.1 | 2.9 | 17 | 103 | 68.1 | 2.8 | 15 | 360 | 67.9 | 3.1 | 12 |
| 1740s | 83 | 67.1 | 3.0 | 7 | 119 | 68.7 | 2.7 | 19 | 560 | 67.5 | 2.9 | 9 |
| 1750s | 54 | 67.2 | 3.3 | 15 | 94 | 68.4 | 2.5 | 16 | 602 | 67.8 | 2.8 | 8 |
| 1760s | 39 | 65.0 | 3.3 | 3 | 73 | 68.1 | 2.7 | 15 | 699 | 68.0 | 2.6 | 8 |
| 1770s | 27 | 66.9 | 4.6 | 15 | 63 | 67.2 | 3.0 | 6 | 731 | 67.9 | 2.9 | 11 |
| 1780s | 21 | 67.2 | 2.2 | 10 | 43 | 67.3 | 2.7 | 2 | 626 | 67.8 | 2.7 | 9 |
| 1790s | 11 | 66.0 | 1.0 | 0 | 30 | 66.7 | 2.7 | 0 | 467 | 67.7 | 2.7 | 6 |
| 1800s | | n.a. | | | | n.a. | | | 232 | 67.7 | 2.7 | 10 |
| 1810s | | n.a. | | | | n.a. | | | 77 | 68.1 | 2.7 | 10 |

| Mulattos: Decade of Birth | *N* | Height | S.D. |
|---|---|---|---|
| 1720s | 26 | 68.1 | 2.8 |
| 1730s | 66 | 68.2 | 2.6 |
| 1740s | 49 | 67.7 | 3.4 |
| 1750s | 32 | 68.7 | 2.1 |
| 1760s | 59 | 67.8 | 2.7 |
| 1770s | 96 | 68.5 | 2.5 |
| 1780s | 84 | 68.5 | 2.5 |
| 1790s | 74 | 68.4 | 2.1 |
| 1800s | 43 | 68.2 | 2.7 |
| 1810s | 15 | 69.0 | 1.6 |

[a]Over 22 years old.
[b]Percentage of sample 72 inches tall.

TABLE 6.8 Height of Male Slaves 19 to 22 Years Old by Decade and Place of Birth[a]

| Decade of Birth | Place of Birth: Africa | | | | Mainland Colonies | | | | Unknown | | | |
|---|---|---|---|---|---|---|---|---|---|---|---|---|
| | *N* | Height | S.D. | *A*[b] | *N* | Height | S.D. | *A*[b] | *N* | Height | S.D. | *A*[b] |
| 1720s | | | | | | | | | 8 | 68.1 | 3.8 | 38 |
| 1730s | 6 | 66.3 | 3.1 | 0 | 9 | 66.8 | 2.8 | 11 | 32 | 68.0 | 3.2 | 19 |
| 1740s | 13 | 66.8 | 2.9 | 8 | 34 | 66.8 | 4.5 | 15 | 122 | 67.0 | 2.8 | 5 |
| 1750s | 19 | 66.4 | 4.0 | 21 | 44 | 67.6 | 2.9 | 9 | 179 | 66.9 | 3.1 | 5 |
| 1760s | 8 | 64.8 | 2.1 | 0 | 38 | 67.4 | 2.4 | 5 | 199 | 67.4 | 3.0 | 8 |
| 1770s | 5 | 64.5 | 2.3 | 0 | 46 | 66.8 | 2.7 | 2 | 339 | 67.6 | 2.5 | 5 |
| 1780s | 9 | 66.2 | 3.3 | 0 | 13 | 65.2 | 3.4 | 0 | 266 | 67.3 | 2.9 | 7 |
| 1790s | 5 | 65.9 | 1.6 | 0 | 26 | 65.9 | 2.4 | 0 | 265 | 67.2 | 2.9 | 4 |
| 1800s | | n.a. | | | 12 | 68.7 | 2.0 | 8 | 172 | 66.8 | 3.0 | 1 |
| 1810s | | n.a. | | | | n.a. | | | 96 | 67.2 | 2.7 | 5 |

[a]Mulattos not included.
[b]Percentage of sample 72 inches tall.

TABLE 6.9 Characteristics of the Female Runaway Slave Sample[a]

| *Place of birth* | *N* | Percentage |
|---|---|---|
| Africa | 58 | 6.1 |
| Mainland | 62 | 6.5 |
| West Indies | 10 | 1.0 |
| Unknown | 828 | 86.4 |
| Total | 958 | 100.0 |

| *Decade of birth*[b] | *N* | Percentage |
|---|---|---|
| 1730s | 45 | 4.8 |
| 1740s | 84 | 9.0 |
| 1750s | 116 | 12.4 |
| 1760s | 114 | 12.2 |
| 1770s | 142 | 15.2 |
| 1780s | 141 | 15.1 |
| 1790s | 126 | 13.5 |
| 1800s | 112 | 12.0 |
| 1810s | 54 | 5.8 |

| *Height by age* | Blacks | | Mulattos | |
|---|---|---|---|---|
| | *N* | Height | *N* | Height |
| 19–22 | 168 | 62.6 | 20 | 63.2 |
| 23–50 | 464 | 63.3 | 51 | 63.3 |
| Unknown | 180 | 63.7 | 19 | 63.1 |

| *Height by place of birth*[c] | *N* | Height |
|---|---|---|
| Africa | 25 | 61.1 |
| Mainland | | |
| Black | 33 | 64.5 |
| Mulatto | 51 | 63.3 |
| All | 84 | 63.8 |
| Unknown | 400 | 63.3 |
| In jail | 74 | 62.7 |

| *Height by date of birth*[c] | *N* | Height |
|---|---|---|
| 1730–79 | 282 | 63.5 |
| 1780–1810 | 178 | 63.0 |

[a]Unless otherwise mentioned, mulattos excluded.
[b]Assumes that the women whose age was not given in the advertisement were 26.1 years old, the average age of those whose age was given.
[c]Adults only.

TABLE 6.10 Height Profile of Runaway Slaves by Birthplace

| | Africa | | | Mainland Colonies | | | Unknown | | | All | | |
|---|---|---|---|---|---|---|---|---|---|---|---|---|
| Age | *N* | Height | S.D. | *N* | Height | S.D. | *N* | Height | S.D. | *N* | Height | S.D. |
| *Males* | | | | | | | | | | | | |
| 10 | | n.a. | | | n.a. | | 2 | 48.5 | 0.7 | 2 | 48.5 | |
| 11 | | n.a. | | | n.a. | | 6 | 50.8 | 2.5 | 6 | 50.8 | |
| 12 | | n.a. | | | n.a. | | 8 | 51.8 | 3.4 | 8 | 51.8 | |
| 13 | 1 | 56.0 | 0.0 | | n.a. | | 11 | 53.9 | 4.3 | 12 | 54.1 | |
| 14 | 2 | 55.0 | 0.0 | 2 | 48.0 | 0.0 | 26 | 58.4 | 3.2 | 30 | 57.5 | |
| 15 | | | | 4 | 59.5 | 2.9 | 40 | 58.8 | 4.2 | 44 | 58.9 | |
| 16 | 5 | 62.6 | 2.7 | 23 | 61.4 | 3.4 | 85 | 63.6 | 3.5 | 113 | 63.6 | |
| 17 | 3 | 64.3 | 1.5 | 20 | 65.1 | 1.8 | 163 | 64.4 | 3.4 | 186 | 64.5 | |
| 18 | 15 | 63.1 | 3.8 | 31 | 66.5 | 2.8 | 318 | 66.2 | 2.9 | 364 | 66.1 | |
| 19 | 12 | 64.9 | 2.1 | 27 | 66.6 | 2.8 | 263 | 66.6 | 2.9 | 302 | 66.5 | |
| 20 | 22 | 66.1 | 3.7 | 76 | 66.3 | 4.0 | 544 | 67.0 | 3.0 | 642 | 66.9 | |
| 21 | 8 | 64.5 | 2.4 | 55 | 66.8 | 2.8 | 383 | 67.7 | 2.7 | 446 | 67.5 | |
| 22 | 24 | 66.9 | 3.1 | 70 | 67.8 | 2.6 | 521 | 67.5 | 2.8 | 615 | 67.5 | |
| *Females* | | | | | | | | | | | | |
| 14 | | | | | | | | | | 11 | 57.9 | 3.5 |
| 15 | | | | | | | | | | 23 | 59.0 | 3.1 |
| 16 | | | | | | | | | | 32 | 61.2 | 3.8 |
| 17 | | | | | | | | | | 40 | 62.6 | 2.7 |
| 18 | | | | | | | | | | 58 | 61.5 | 4.5 |
| 19 | | | | | | | | | | 42 | 62.8 | 2.8 |
| 20 | | | | | | | | | | 69 | 62.7 | 3.7 |
| 21 | | | | | | | | | | 23 | 62.7 | 3.5 |
| 22 | | | | | | | | | | 54 | 62.4 | 3.3 |
| 23 | | | | | | | | | | 41 | 63.2 | 3.8 |

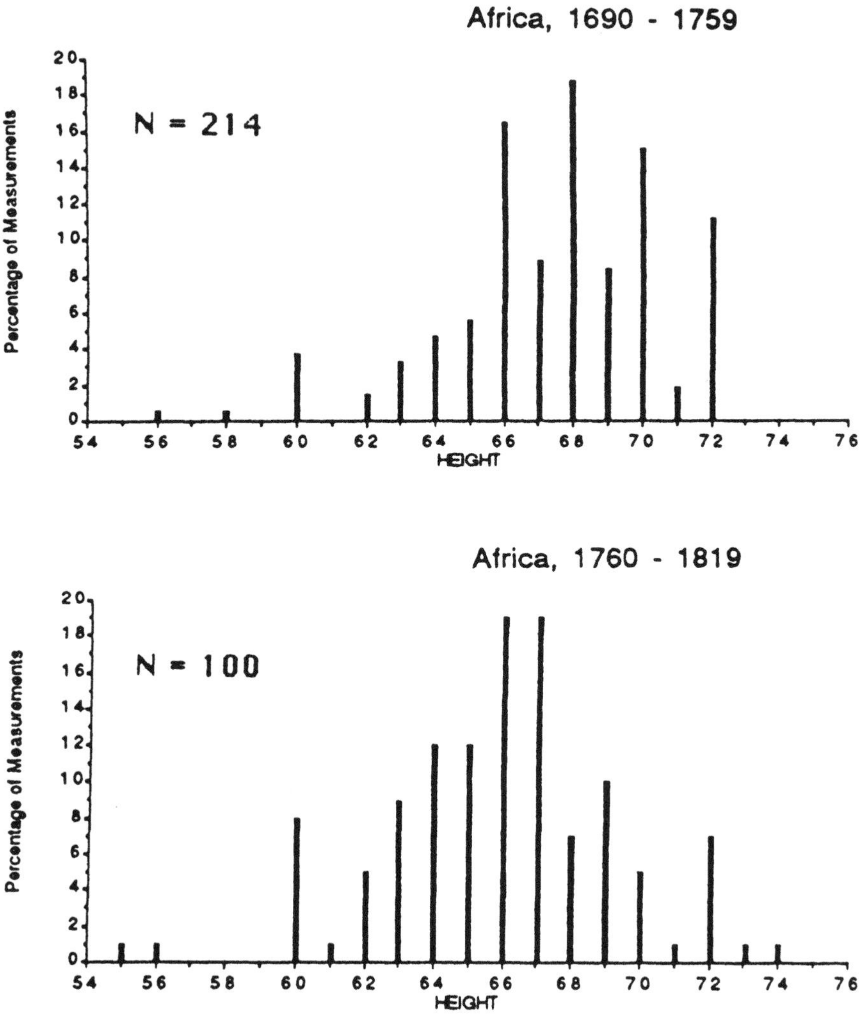


FIG. 6.1 Height distributions of adult male runaway slaves by place and date of birth

America, 1740 - 1759

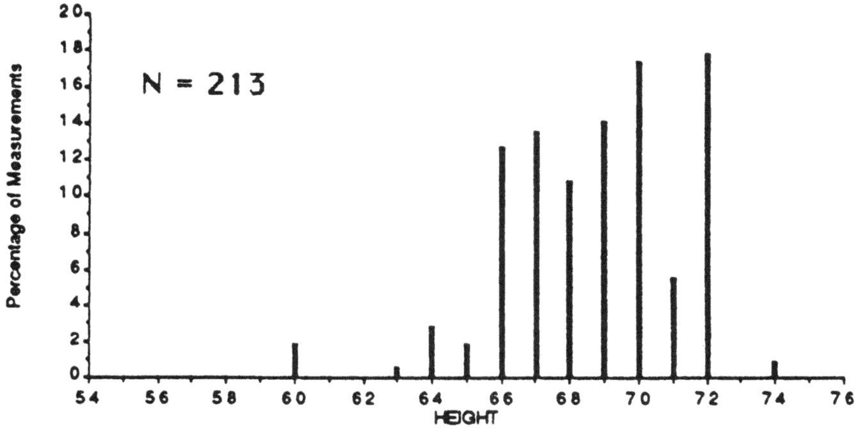


America, 1760 - 1819

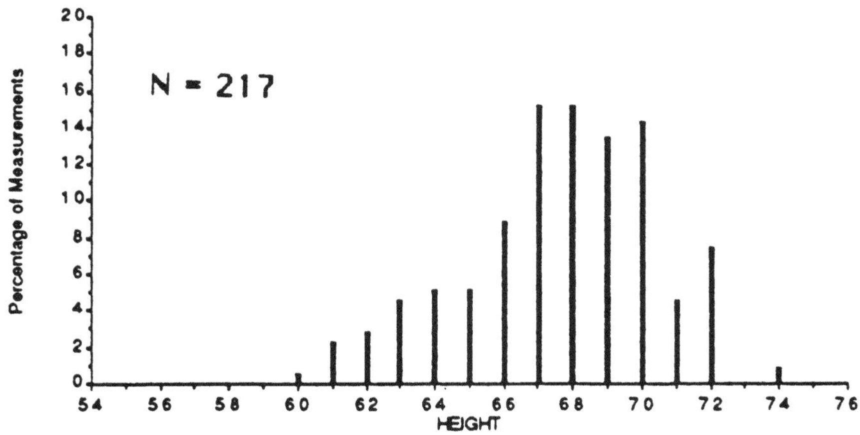

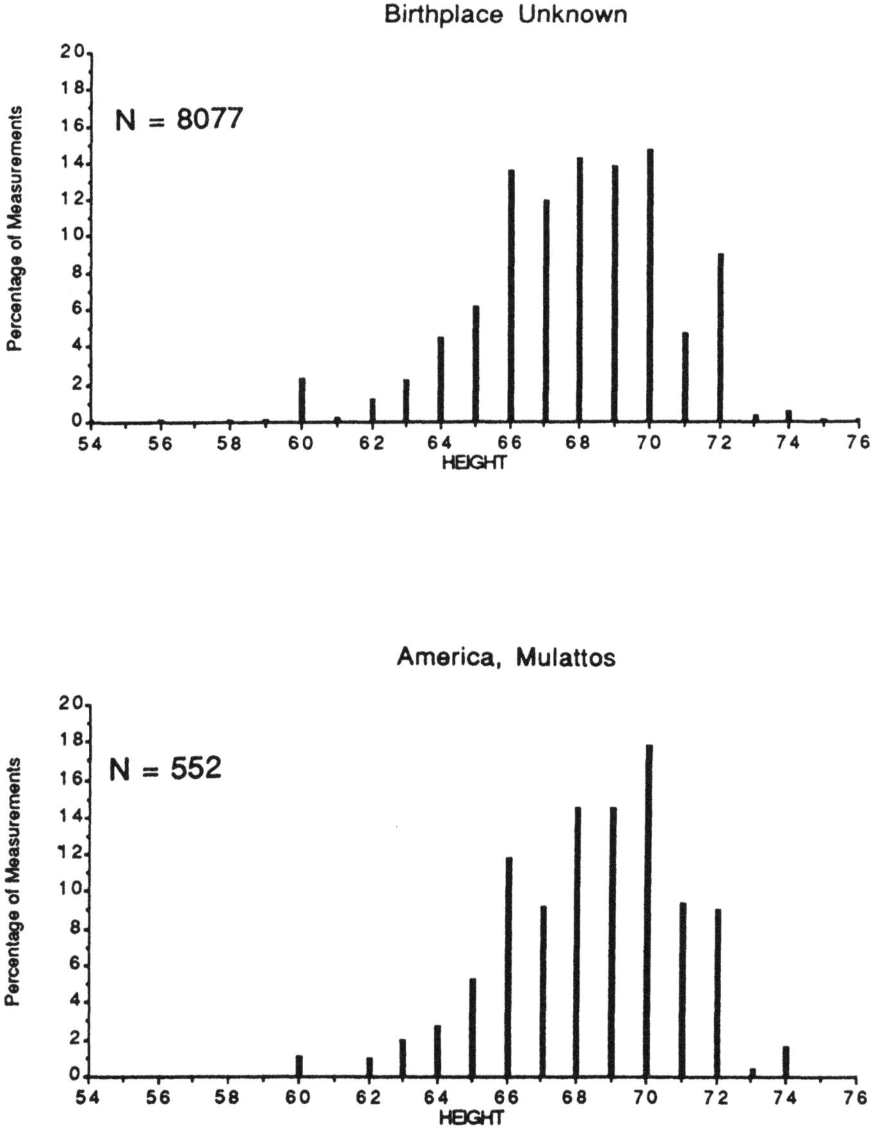


FIG. 6.2 Height distributions of adult male runaway slaves by place of birth (all years)

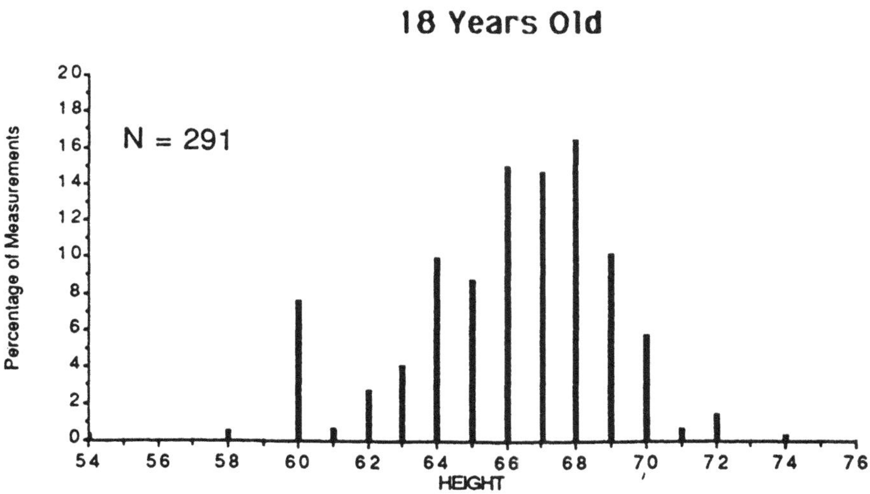


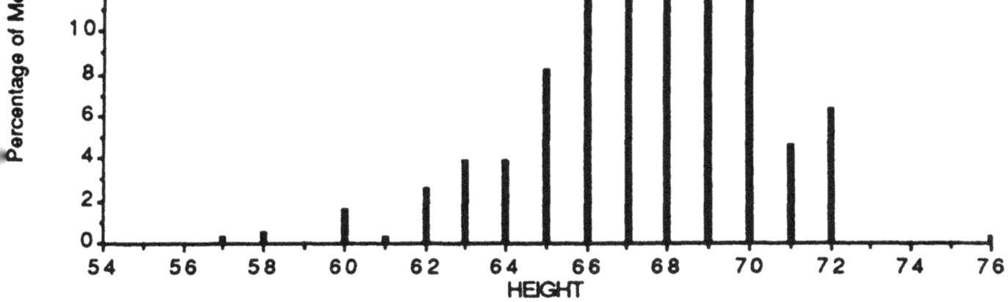


FIG. 6.3 Height distributions of 18- and 21-year-old male runaway slaves (all years, all birthplaces)

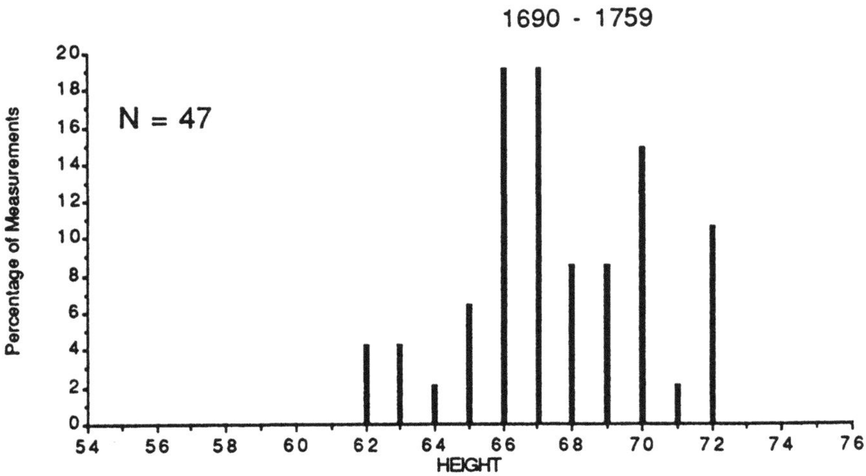


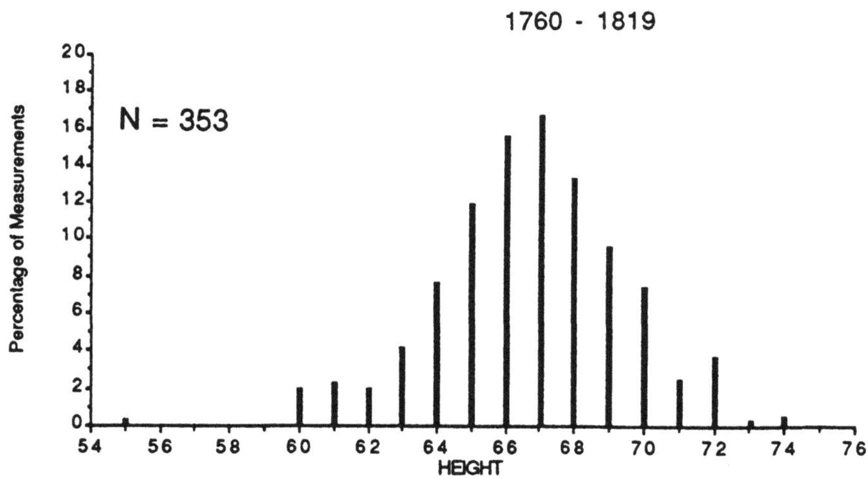


FIG. 6.4 Height distributions of adult male runaway slaves in custody of a jailer by date of birth (all birthplaces)

# INDEX

For Product Safety Concerns and Information please contact our EU
representative GPSR@taylorandfrancis.com
Taylor & Francis Verlag GmbH, Kaufingerstraße 24, 80331 München, Germany

www.ingramcontent.com/pod-product-compliance
Ingram Content Group UK Ltd.
Pitfield, Milton Keynes, MK11 3LW, UK
UKHW022205190926
475702UK00003B/213

* 9 7 8 0 8 6 0 7 8 4 5 7 9 *